Principles of Genetics and Molecular Epidemiology

Juan Carlos Gomez-Verjan
Nadia Alejandra Rivero-Segura
Editors

Principles of Genetics and Molecular Epidemiology

Springer

Editors
Juan Carlos Gomez-Verjan
Dirección de Investigación
Instituto Nacional de Geriatría (INGER)
Ciudad de México, Mexico

Nadia Alejandra Rivero-Segura
Dirección de Investigación
Instituto Nacional de Geriatría (INGER)
Instituto Nacional de Geriatría
Ciudad de México, Mexico

ISBN 978-3-030-89600-3 ISBN 978-3-030-89601-0 (eBook)
https://doi.org/10.1007/978-3-030-89601-0

This Springer imprint is published by the registered company Springer Nature Switzerland AG
The registered company address is: Gewerbestrasse 11, 6330 Cham, Switzerland

Foreword

Do not go gentle into that good night,
Old age should burn and rave at close of day;
Rage, rage against the dying of the light.
Though wise men at their end know dark is right,
Because their words had forked no lightning they
Do not go gentle into that good night.
Dylan Thomas (1914–1953)

At the National Institute for Geriatric Medicine, we have aimed to develop a forum for transdisciplinary research with a complex system approach in the field of human aging. During the past 10 years, a diverse group of seasoned researchers (public health specialists, geriatricians, biologists, and social scientists) have been interacting with brilliant young biologists and data scientists, such as the editors of this book, giving rise to the development of a burgeoning research network. This book is an outstanding product of the exchanges and interaction promoted within this network.

The modern approach to human health becomes increasingly complex. Scientific knowledge tends to evolve from a reductionistic approach to a complex system view. The application of molecular biology to the answering of epidemiological questions in this context is commonplace today. Molecular epidemiology allows for the understanding of most of the molecular consequences and implications of diet, lifestyle, and environmental factors, and how these decisions give rise to genetic mutations and how these mutations are distributed among certain populations using biomarkers and genetic information. Molecular epidemiology studies also provide substantial information about previously identified risk factors and disease mechanisms. Specific applications include the molecular surveillance of risk factors, the measurement of their geographical and temporal distribution, and the characterization of pathogens and their evolution, as we see every day today through the COVID-19 pandemic.

While many molecular epidemiological studies still use the conventional disease designation system, evidence grows about the fact that disease progression represents a heterogeneous process that differs from person to person. So, everyone faces eventually a unique disease process that is different from that of any other individual (single disease principle) and becomes even more distinct as we age. The environmental factors and their influence on the process of molecular pathology in everyone contribute significantly to the uniqueness of disease expression. Knowledge about these factors contributes to the understanding of how is it that the economic and social determinants of health get "under the skin" [1]. Beyond that, the single disease principle still causes conflicts with the premise that individuals with the same disease have similar etiologies and processes.

Studies analyzing the relationship between the environment and the pathological footprint of the disease (particularly cancer) have become more common since the last 20 years. However, the use of molecular pathology in epidemiology still faces unique challenges including the lack of standardized methodologies and guidelines as well as the scarcity of multidisciplinary experts and training programs. This book contributes to face this challenge, providing useful, up-to-date information for the research community and for public health and clinical practitioners.

The first chapter relates to the new field of life course epidemiology and its intersection with molecular epidemiology and translational science; it reviews the challenge of real-world application of this knowledge.

In the second chapter, the thoughtful review of both the nucleic acid-based molecular tools (PCR, Karyotyping, and microarray) and the protein-based molecular tools (Immunoassays), commonly used in everyday clinical practice, provides the clinician with a useful update about this diagnostic tool.

The authors of the third chapter focus on the high throughput genomic tools that have been recently incorporated in epidemiological studies for the identification of rare genetic variants, genetic and environmental risk factors, and accurate biomarkers for the diagnosis and treatment of several diseases, focusing in the COVID-19 pandemic.

Epigenetic mechanisms are complex biological mechanisms that modulate the cells' interaction with the environment. In recent years, the study of the epigenetic process in human disorders has grown exponentially, but clinical applications are still an area of opportunity, and the fourth chapter focusses in this particular challenge.

Transcriptome-wide association studies (TWAS) integrate genome-wide association studies (GWAS) and gene expression datasets to identify gene–trait associations. This novel technique gives us the power to analyze what genes are on or off among specific tissues or samples; it is quite powerful and has already several clinical applications, for instance in pharmacology. A thoughtful review of the potential revealed by this technique is given in Chap. 5.

Chapter 6 gives the audience a complete overview of the scope of proteomics, since this has proven to be highly sensitive and specific in a wide range of samples and several diseases, suggesting their potential for being considered in daily clinical practices. A field of growing relevance and with significant advances is geroscience; the identification of plasma proteins that systematically change with age and, independent of chronological age, predict accelerated decline of health is an expanding area of research. Circulating proteins are ideal translational "omics" since they are final effectors of physiological pathways and because physicians are accustomed to use information of plasma proteins as biomarkers for diagnosis, prognosis, and tracking the effectiveness of treatments. Recent technological advancements and these characteristics allow the authors to predict that this field will soon contribute to clinical sciences.

Metabolomics employs non-invasive human biological samples such as serum, breath, and urine to screen and identify novel biomarkers, together with proteomics contributes to a better understanding of underlying pathogenetic processes. Changes in concentrations and fluxes of specific groups of metabolites reflect systemic responses to environmental, lifestyle, and therapeutic challenges. Thus, the study of metabolites is a powerful tool for the characterization of complex disorders and will contribute to the development of the precision medicine concept and the demonstration of the unique disease principle, as it is described in Chap. 7.

As the "omics" grow and differentiate, we need new approaches that allow us to encompass the integrity of the physiological changes and their phenotypic expression. Physiomics employs bioinformatics to construct networks of physiological features that are associated with genes, proteins, and their networks, and phenomics is the systematic measurement and analysis of qualitative and quantitative traits, including clinical, biochemical, and imaging methodologies, for the refinement and characterization of a phenotype. Both are needed in order to constantly advance the field in an orderly manner. Digital health, an emerging field of study at the intersection of healthcare and digital technologies, is deeply interrelated with phenomics. Chapters 9 and 10 allow us to understand its processes and potential.

The progressive refinement of imaging technologies, both to look into the cell and into the whole body and their digitalization, allows a refinement of their use, together with new AI algorithms that allow automation but also for a better understanding of the processes on which they are focused. A remarkable example of this potential is the Alzheimer's Disease Neuroimaging Initiative (ADNI) [2] that has significantly contributed to the advancement of the field. Both tools are discussed in different medical fields and how these have translated into clinical applications in Chaps. 8 and 11.

Complexity in science is hard to approach. Social scientists like Edgar Morin [3] have been prescient about this truth and have contributed to lead us to where we are today. But only today we begin to have the tools we need to approach this complexity in a purposeful manner, the next part of the book encompasses systems biology, bioinformatics, and spatial statistics, which are the tools we need to manage and understand the relations among large datasets at different levels and their interactions. And finally, the last three chapters deal with a holistic approach where human health is integrated from different perspectives, from biomedical to environmental and social, in the fields of pharmacogenomics, epidemiology, and population health surveillance.

I am certain that this volume will contribute significantly to the much-needed dissemination of this novel approach to the research and practice in epidemiology and public health. And this knowledge will also allow us to grow wiser and contribute to a better healthspan even knowing that at the end "....dark is right."

Luis Miguel Gutiérrez Robledo
Dirección General, Instituto Nacional de Geriatría
Ciudad de México, Mexico

References

[1] Vineis P, Avendano-Pabon M, Barros H, Bartley M, Carmeli C, Carra L, Chadeau-Hyam M, Costa G, Delpierre C, D'Errico A, Fraga S, Giles G, Goldberg M, Kelly-Irving M, Kivimaki M, Lepage B, Lang T, Layte R, MacGuire F, Mackenbach JP, ... Zins M. Special Report: The Biology of Inequalities in Health: The Lifepath Consortium. Frontiers in public health. 2020;8:118. https://doi.org/10.3389/fpubh.2020.00118.

[2] Petersen RC, Aisen PS, Beckett LA, Donohue MC, Gamst AC, Harvey DJ, Jack CR Jr, Jagust WJ, Shaw LM, Toga AW, Trojanowski JQ, Weiner MW. Alzheimer's Disease Neuroimaging Initiative (ADNI): clinical characterization. Neurology. 2010;74(3):201–9. https://doi.org/10.1212/WNL.0b013e3181cb3e25.

[3] Morin E. Leçons d'un siècle de vie. Denoël, Paris; 2021.

Preface

According to several authors, the field of molecular epidemiology is a subdivision of medical science and epidemiology that involves potential environmental and genetic risk factors, recognized at the molecular level, which altogether improve our knowledge about the pathogenesis of disease and lead to the design of more efficient therapeutic strategies against them. However, with the constant evolution and improvement of novel technologies such as the omics sciences, microscopy, and others, it is quite important to update health professionals in this field and build a bridge between physicians and researchers to face the upcoming challenges that represent the pathogenesis of the disease. In this sense, the current book, entitled *Principles of Genetics and Molecular Epidemiology*, urges the need to update all health professionals (biomedical, clinical, and social) on the ongoing advances of molecular biology and genetics in the field of epidemiology. As seen during the COVID-19 pandemic, the need to increase scientific knowledge of health professionals in the field of genetics (genomics, transcriptomics, epigenetics, DNA and RNA research, proteomics, metabolomics, systems biology, and others) becomes evident, as well as the way these advances can be applied to the different branches in epidemiology, such as epidemiological surveillance, public health, populations, infectious diseases drug development, vaccines, digital health, and imaging, among others. It also urges the need to sustain scientific knowledge over the huge amount of crossed or erroneous information generated in the networks or the media without any scientific support of the so-called infodemics. Through its 17 chapters, we resume the most outstanding advances performed to date on genetics applied in the epidemiological field. For instance, we cover in several chapters the state of the art in systems biology, biomedicine, medicine, and epidemiology. We recover the most important principles on public health and then we follow by describing the principles of genomics, transcriptomics, proteomics, metabolomics, and imaging; therefore, any health professional interested in applying these technologies on clinics could start by reading these chapters. Additionally, it is important to mention that microscopy has benefited from advances in molecular biology; thus, we dedicate an entire chapter to the basics of microscopy and the enormous potential that such technology has on epidemiology and therefore on public health. The book delves into the state-of-the-art of themes such as phenomics and digital health where several advances in public health have become more evident with the advances in computation and data science, and where knowledge is being generated day by day, and are hot topics on epidemiology. In the final sections of the book, we cover the most recent advances in molecular pharmacology, which needs to be considered by epidemiologists and public health professionals for the development of drugs against the most important diseases that afflict the population and that depend so much on the characteristics and health problems of each population knowledge generated at the epidemiological and clinical level and that some times are seen as two separated fields. Finally, we conclude by giving an introduction to the genomic surveillance so important right now not only for the control of the SARS-CoV-2 lineages and how they spread around the world but also for the control of other infectious diseases such as tuberculosis, HIV, and Ebola; as well as for the surveillance of chronic diseases on populations such as diabetes or cancer. We hope that our book will be useful to any health professional (doesn´t matter if they are starting their career or they are advanced in their professional life). This book allows to the understanding of topics that seem quite complicated and sophisticated at first glance, making them easily digested, and arouse interest to delve and eventually apply them on field. Paraphrasing a giant in the world of genomic research, Dr. Eric Lander, *we are experiencing a historical revolution in molecular biology*; therefore, we can simply stay and watch or be part of it.

Mexico City, Mexico — Juan Carlos Gomez-Verjan
Mexico City, Mexico — Nadia Alejandra Rivero-Segura

Acknowledgements

Dra. Nadia Rivero-Segura

To my parents, my sister, my grandma, who gave me the courage and the strength to face any challenge in my life.

To Pergd, my partner in crime, my couple, my soulmate, and my biggest support through this path called research. This wouldn't have been possible without you.

To my nieces for sharing with me their curiosity and amazement about their surroundings. Please never stop exploring.

To all those who contributed to the production of this volume in the middle of the COVID-19 pandemic.

To the Instituto Nacional de Geriatría for supporting our projects.

"*A life in research can be a most enjoyable life with many frontiers to explore.*" – Earl Sutherland

Dr. Juan Carlos Gomez-Verjan

To my parents for all the support and unrestricted love they have given me during all these years. Without their support, nothing would be possible.

To my little apple for being my ideal accomplice for all our live projects and for her love and understanding.

To all the authors who supported us in the writing of this book despite the fact that we were in the contingency caused by the COVID-19 pandemic.

To the National Institute of Geriatrics for supporting us as researchers.

Contents

Contributors

Neftali Eduardo Antonio-Villa Faculty of Chemistry, National Autonomous University of Mexico, Mexico City, Mexico

J. G. Avila-Acevedo Laboratorio de Fitoquímica, Unidad de Biología y Prototipos (UBIPRO), FES-Iztacala, Universidad Nacional Autónoma de México, Estado de México, Mexico

José Alberto Avila-Funes Department of Geriatrics, Instituto Nacional de Ciencias Médicas y Nutrición Salvador Zubirán, Mexico City, Mexico

Lizeth Avila-Gutierrez Department of Biomedical Engineering and Gerontology, National Institute of Geriatrics, Mexico City, Mexico

Oscar Salvador Barrera-Vázquez Departamento de Farmacología, Facultad de Medicina, Universidad Nacional Autónoma de México (UNAM), Mexico City, Mexico

Omar Yaxmehen Bello-Chavolla Research Division, Instituto Nacional de Geriatría, Mexico City, Mexico

María Isabel Coronado-Mares Hospital Regional Tlalnepantla, ISSEMyM, Mexico City, Mexico

Esteban Cruz-Arenas Hospital Epidemiological Surveillance Unit, Instituto Nacional de Rehabilitación "Luis Guillermo Ibarra Ibarra", Mexico City, Mexico

Marco Antonio Delaye-Martinez (PECEM) Program, Faculty of Medicine, National Autonomous University of Mexico, Mexico City, Mexico

A. M. Espinosa-González Laboratorio de Fitoquímica, Unidad de Biología y Prototipos (UBIPRO), FES-Iztacala, Universidad Nacional Autónoma de México, Estado de México, Mexico

E. A. Estrella-Parra Laboratorio de Fitoquímica, Unidad de Biología y Prototipos (UBIPRO), FES-Iztacala, Universidad Nacional Autónoma de México, Estado de México, Mexico

Carlos Alberto Fermín-Martínez (PECEM) Program, Faculty of Medicine, National Autonomous University of Mexico, Mexico City, Mexico

Luisa Fernández-Chirino Faculty of Chemistry, National Autonomous University of Mexico, Mexico City, Mexico

Edgar Flores-Soto Departamento de Farmacología, Facultad de Medicina, Universidad Nacional Autónoma de México (UNAM), Mexico City, Mexico

A. M. García-Bores Laboratorio de Fitoquímica, Unidad de Biología y Prototipos (UBIPRO), FES-Iztacala, Universidad Nacional Autónoma de México, Estado de México, Mexico

Carmen García-Peña Health Research Department, National Institute of Geriatrics, Mexico City, Mexico

Alfredo García-Venzor Department of Life Sciences, Ben-Gurion University of the Negev, Beer Sheva, Israel

Alma Delia Genis-Mendoza Laboratorio de Genómica de Enfermedades Psiquiátricas y Neurodegenerativas, Instituto Nacional de Medicina Genómica, CDMX, Mexico

Juan Carlos Gomez-Verjan Dirección de Investigación, Instituto Nacional de Geriatría (INGER), Ciudad de México, Mexico

Enrique C. Guerra (PECEM) Program, Faculty of Medicine, National Autonomous University of Mexico, Mexico City, Mexico

Otto Hahn-Herrera Dirección de Investigación, Instituto Nacional de Geriatría (INGER), Mexico City, Mexico

Karla Daniela Rodríguez-Hernández Laboratorio de Estudios sobre Tripanosomiasis, Departamento de Inmunología, Instituto de Investigaciones Biomédicas, UNAM, Mexico City, Mexico

C. T. Hernández-Delgado Laboratorio de Bioactividad de Productos Naturales, Unidad de Biología y Prototipos (UBIPRO), FES-Iztacala, Universidad Nacional Autónoma de México, Estado de México, Mexico

Virgilio Alejandro Hernández-Ruiz Univ. Bordeaux, Inserm, Bordeaux Population Health Research Center, UMR 1219, Bordeaux, France

Victor Takeshi Landero-Yoshioka Experimental Surgery Department. Centro Médico Nacional "20 de Noviembre" ISSSTE, Mexico City, Mexico

Edna Ayerim Mandujano-Tinoco Laboratory of Connective Tissue, Centro Nacional de Investigación y Atención de Quemados, Instituto Nacional de Rehabilitación "Luis Guillermo Ibarra Ibarra", Mexico City, Mexico

Alejandro Márquez-Salinas (PECEM) Program, Faculty of Medicine, National Autonomous University of Mexico, Mexico City, Mexico

Mónica Itzel Martínez-Gutiérrez (PECEM) Program, Faculty of Medicine, National Autonomous University of Mexico, Mexico City, Mexico

José Jaime Martínez-Magaña Laboratorio de Genómica de Enfermedades Psiquiátricas y Neurodegenerativas, Instituto Nacional de Medicina Genómica, CDMX, Mexico

Sandra Lizbeth Morales-Rosales Posgrado en Biología Experimental, Universidad Autónoma Metropolitana, Unidad Iztapalapa, Mexico City, Mexico

Karla Moreno-Tamayo Research Unit in Epidemiology and Health Services, Area of Aging, National Medical Center Siglo XXI, Mexican Social Security Institute, Mexico City, Mexico

Humberto Nicolini Laboratorio de Genómica de Enfermedades Psiquiátricas y Neurodegenerativas, Instituto Nacional de Medicina Genómica, CDMX, Mexico

E. Nolasco-Ontiveros Laboratorio de Fitoquímica, Unidad de Biología y Prototipos (UBIPRO), FES-Iztacala, Universidad Nacional Autónoma de México, Estado de México, Mexico

Gibrán Pedraza-Vázquez Posgrado en Biología Experimental, Departamento de Ciencias de la Salud, Universidad Autónoma Metropolitana Unidad Iztapalapa, Mexico City, Mexico

I. Peñalosa-Castro Laboratorio de Fitoquímica, Unidad de Biología y Prototipos (UBIPRO), FES-Iztacala, Universidad Nacional Autónoma de México, Estado de México, Mexico

Ricardo Quiroz-Baez Dirección de Investigación, Instituto Nacional de Geriatría (INGER), Mexico City, Mexico

Ricardo Ramírez-Aldana Dirección de Investigación, Instituto Nacional de Geriatría (INGER), Mexico City, Mexico

Ixchel Ramírez-Camacho Dirección de Investigación, Instituto Nacional de Geriatría, Ciudad de México, Mexico

Eliseo Ramírez-García Research Unit in Epidemiology and Health Services, Area of Aging, National Medical Center Siglo XXI, Mexican Social Security Institute, Mexico City, Mexico

Ruth Rincón-Heredia Microscopy Core Unit, Instituto de Fisiología Celular, Universidad Nacional Autónoma de México, Mexico City, Mexico

J. C. Rivera-Cabrera Laboratorio de Cromatografía de Líquidos, Departamento de Farmacología, Escuela Médico Militar, Mexico City, Mexico

Nadia Alejandra Rivero-Segura Dirección de Investigación, Instituto Nacional de Geriatría (INGER), Instituto Nacional de Geriatría, Ciudad de México, Mexico

Sergio Sánchez-García Research Unit in Epidemiology and Health Services, Area of Aging, National Medical Center Siglo XXI, Mexican Social Security Institute, Mexico City, Mexico

Alejandro Sicilia-Andrade (PECEM) Program, Faculty of Medicine, National Autonomous University of Mexico, Mexico City, Mexico

Elizabeth Sulvaran-Guel Dirección de Investigación, Instituto Nacional de Geriatría (INGER), Mexico City, Mexico

Licenciatura en Ciencias Genómicas, Universidad Nacional Autonoma de Mexico, Mexico City, Mexico

Licenciatura en Ciencias Genómicas UNAM, Mexico City, Mexico

Arsenio Vargas-Vázquez (PECEM) Program, Faculty of Medicine, National Autonomous University of Mexico, Mexico City, Mexico

Pamela Tella-Vega Department of Clinical Epidemiology Research, National Institute of Geriatrics, Mexico City, Mexico

Juan Carlos Yustis-Rubio Departamento de Ecología Funcional, Instituto de Ecología, UNAM, Mexico City, Mexico

Dirección de Investigación, Instituto Nacional de Geriatría (INGER), Mexico City, Mexico

Principles of Modern Epidemiology and Public Health

1

Carmen García-Peña, Lizeth Avila-Gutierrez, Karla Moreno-Tamayo, Eliseo Ramírez-García, Sergio Sánchez-García, and Pamela Tella-Vega

Introduction

In recent decades, increased life expectancy and falling fertility rates have led to a growing world population, which creates more complex challenges in the public health area. These challenges include global environmental changes, profile transformation of the main causes of death, and changes in the infectious disease patterns. In this scenario, epidemiological research plays a relevant role through the increasingly detailed recognition and identification of factors at different levels: at the population, individual (lifestyles), and genetic level.

Over time, epidemiology has benefited from scientific and technological contributions of statistics and informatics, as well as from its interrelation with social and political sciences, economics, anthropology, and other sciences in the field of communication and education. In the context of new challenges faced by decision-makers and thus researchers, this should not be seen as an isolated science, despite being the basis for the study of population health conditions causes. The integration of a holistic approach in this research area is necessary to achieve a balance between variables at each of the aforementioned levels. It is also necessary to include other variables derived from the macro-environment to develop conceptual frameworks and analytical studies that contribute to the improvement of individual health conditions [1, 2].

In the population perspective, the identification of causal factors associated with health conditions through epidemiological prediction models is insufficient if the aim is to move toward risk prediction at the individual level [3]. The incorporation of epidemiological studies from a life-course approach can provide a perspective where the individual per se, and through data collection of variables since birth and throughout his life, allows a better understanding of individual trajectories and their health outcomes [4].

The addition of genetic data to this area offers an encouraging perspective to solve the problem of identifying small causal associations at new levels of study. In the era of the human genome plan, the last two decades have been decisive to incorporate genetic knowledge in epidemiological research and in the context of public health. This paves the way for its translation toward its different uses and benefits in the clinical area [5]. The development and implementation of new knowledge from a molecular biology, genetics, and bioinformatics perspective, the incorporation of new collection methods, information storage, and innovations in communication mechanisms will allow the generation of great benefits through constant evolution. In this context of changes and challenges, both conventional and new, it is evident that the practice of public health and epidemiology is more complex. Therefore, the need to continue providing scientific evidence from the approach of different study fields to create and implement health care and treatment models is fundamental.

C. García-Peña (✉)
Health Research Department, National Institute of Geriatrics, Mexico City, Mexico
e-mail: mcgarcia@inger.gob.mx

L. Avila-Gutierrez
Department of Biomedical Engineering and Gerontology, National Institute of Geriatrics, Mexico City, Mexico
e-mail: lavilag@inger.gob.mx

K. Moreno-Tamayo · E. Ramírez-García · S. Sánchez-García
Research Unit in Epidemiology and Health Services, Area of Aging, National Medical Center Siglo XXI, Mexican Social Security Institute, Mexico City, Mexico
e-mail: sergio.sanchezga@imss.gob.mx

P. Tella-Vega
Department of Clinical Epidemiology Research, National Institute of Geriatrics, Mexico City, Mexico
e-mail: ptella@inger.gob.mx

J. C. Gomez-Verjan, N. A. Rivero-Segura (eds.), *Principles of Genetics and Molecular Epidemiology*,
https://doi.org/10.1007/978-3-030-89601-0_1

Life-Course Epidemiology

In the framework of epidemiology, the life course is "the study of long-term biological, behavioral, and psychosocial processes linking health and disease risk in adults with physical or social exposures, which occur during pregnancy, childhood, adolescence, early adult life or across generations" [6].

The life-course approach in epidemiological studies has a relatively recent life. In the late 1990s, Kuh and Ben-Shlomo [7] sought to explain the complex phenomena of the health-disease development in humans. Consequently, they initiated a disciplinary field for epidemiology, inspired by Barker's research [8] on the fetal origins of diseases in adulthood and based on Elder's contributions [9] made from social sciences. So far, this disciplinary field allows documenting the diverse and complex relations through which the genotype, socioeconomic, demographic, psychosocial, and environmental factors shape the health-disease process of human populations over time and by generation [6].

The incorporation of a life perspective in the field of epidemiology contributes to etiological research, genetic networks, and biological and systemic patterns in the study of processes and factors influencing the development of diseases throughout the human life [6]. This comprehensive approach helps to identify causal pathways and describes how social settings and behavioral effects could promote disease development [10]. Therefore, there has been a growing interest in linking epidemiological studies with a life-course approach and genomic studies in areas such as environmental epigenetics and social genomics [11].

Principles of Life-Course Research

Different conceptual bases have been proposed on the health-disease relationship over time. According to Elder et al. [12], the life-course approach supports its application in five principles; these have been adapted according to the progress and use given to them from the interaction of different disciplines [13].

Development over Time

It is indisputable that what we call life course implies a cumulative process from birth to death, so the only way to capture and understand it is by having a long-term vision [12]. In classical epidemiological studies, this principle has been pivotal and is identified in longitudinal designs; however, there is a lack of information on contextual changes over time. From the point of view in which the life-course approach is included, it is known that various environmental and individual factors, as well as the context to which people are exposed during the prenatal stage and childhood, contribute to the health conditions of adulthood life; such is the case of coronary heart disease, type 2 diabetes, and high blood pressure [6, 8].

Time and Place

Since the actions of individuals are influenced by the historical context and place, this principle is fundamental in the interpretation of results, in which it is important to locate the social, political, and economic context of certain historical periods belonging to the population groups under study [12]. The life trajectories can vary according to the geographic space and the period time. These may cause common experiences to all members of a population but in turn can have differential effects for certain subgroups of the same.

Timing

The importance of the timing of an event lies in the different ways it can impact individuals [14]. The particular roles and behaviors of individuals are related to different biological and psychosocial aspects associated with age [15]. In the context of epidemiology, the conceptual model of critical or sensitive period suggested by Ben-Shlomo et al. [6] is characterized by emphasizing this principle. Under this conceptual framework, the time at which an event is observed can be decisive for the trajectory development and how it modifies the spheres of an individual's life.

Linked Lives

The interdependence of domains and experiences in the life of an individual belonging to a family line may have implications in the life course of other members [12]. Family relationships related to this principle may even span several generations. In this regard, research with a life perspective is abundant [16]. Within the framework of epidemiological studies, there is a rise of multigenerational studies analyzing the transmission of mental health diseases between two or more generations. In a comprehensive review on the intergenerational transmission of mental health, Warner and Weissman [17] described these studies collecting information from generations of biologically related relatives. Among the mechanisms that elucidate these links, the effect of the social environment in which a certain generation grows up is proposed. In turn, the expression of genes regulating behavioral and endocrine responses may also have an impact.

Free Will or Agency

In the life course of individuals, the decisions and actions they take shape their trajectories [12]. Decision-making is not made independently or apart from the events in which people are immersed, since it involves the interaction of the social categories to which they belong (gender, socioeconomic status, or ethnicity) and the social time in which they live. This network of factors facilitates or imposes restrictions that people will deal with throughout their lives [12, 14, 18].

Challenges for Life-Course Epidemiology

Although the life course requires observing individuals from different stages, it has not always been feasible to create studies that provide follow-up of the same participants since birth. Studies starting in later stages lack the close follow-up that could have occurred if they had started earlier. However, the implementation of biographical questionnaires to collect information on domains of life such as work, family care, etc., has been used to rebuild trajectories and analyze them in relation to the physical health of older adults, for example [19, 20].

Currently, the systematization of large databases and even the registration of biological samples of diverse origin favor the design of longitudinal studies with a life-course approach. However, this involves both economic feasibility and technical-methodological difficulties.

The longitudinal information obtained from the life course requires proper data handling and processing. Statistical analysis is essential, and researchers must be trained in up-to-date knowledge, in order to use the most appropriate statistical techniques.

Translational Epidemiology

In recent years, there has been a greater focus on evidence on etiology and the mechanisms involved in the history of the disease. Their application in the development and improvement of related health outcomes has received less attention. The emergence of translational research focused on the "bench to bedside" term (about the translation from basic research toward clinical applications) needs epidemiology and other population sciences to integrate an overview of novel scientific discoveries, visualized through the evidence of population health [21].

Translational epidemiology is considered a "fundamental" bidirectional discipline, which encompasses different observational, experimental, and theoretical epidemiological methods to link clinical and laboratory aspects with population research [22]. In order to understand the natural history of a disease and identify risks or factors, it is necessary to understand the epidemiological principles in clinical and public health practice. Therefore, the objective of translational epidemiology is the effective transfer of new scientific discoveries with evidence-based approaches for disease treatment, prevention, and control with novel interventions at the individual level and in the planning of health programs and public policies [21, 22].

The dynamic process of translation focuses on the links between epidemiology, fundamental sciences, and involved parties (decision-makers) through different themes and priority areas. According to Windle et al. [23], the main domains that contribute to the efficiency of knowledge translation include the following:

- Adequate inclusion of the research question into the appropriate methods.
- Efficient communication between researchers and decision-makers.
- Education through different abilities and skills.
- Implementation of interventions and measurements associated with research monitoring and evaluation.
- Efficient understanding of public health needs.

The Khoury's proposal [24], in which the translational research process consists of four phases, is shown in Fig. 1.1, taking the development context of a new vaccine and its interconnection with the clinical trial phases as an example [25]:

1. P1 "From basic research into clinical application": In this preclinical stage, translation of laboratory results into the first testing in humans is performed. Clinical trials are small and last only a few months. Phase I and II clinical trials.
2. P2 "From clinical trial results into practice and decision-making": Larger clinical trials are conducted with hundreds of participants. Duration can be up to 2 years. Results obtained in this phase can be used to determine the composition, dose, and profile of adverse reactions. Phase III clinical trials.
3. P3 "Translation of recommendations into clinical practice": This stage can last several years, as clinical trials conducted during this stage compare a larger number of study groups. It begins after the vaccine is authorized and is recommended for its use. Phase IV clinical trials.
4. P4 "Outcomes in the population and analysis at the public health level": This stage includes monitoring the vaccine benefits and risks. Components such as costs, quality, accessibility, organization, and financing are examined, allowing to understand structure, processes, and results for their distribution and impact on the population health.

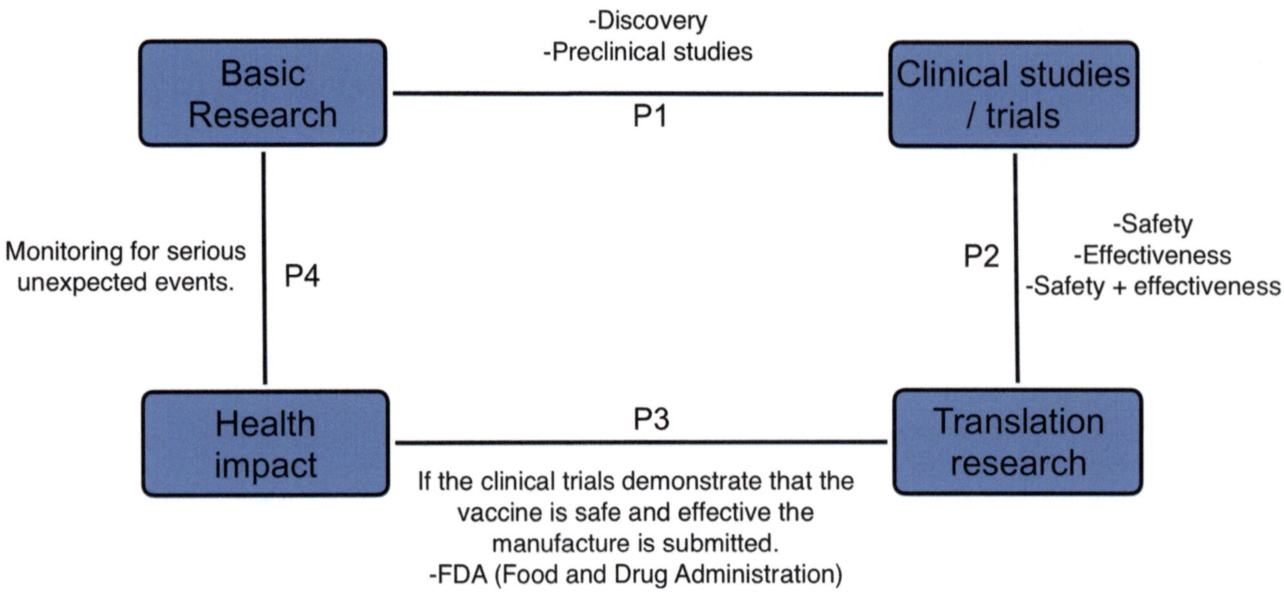

Fig. 1.1 Example of the vaccine development process applied to the Translational Research phases. (Adapted from Khoury et al. [24] and Agurs-Collins et al. [27])

Postmodern Epidemiology

Because the evolution of knowledge on epidemiology is a constant, the current paradigms on which this discipline is ruled will not be definitive. Epidemiology went from a "conventional" to a "modern epidemiology" approach. The first paradigm is focused on the prevention of diseases and population health needs, while the second focuses on risk factors. It is assumed that this transition of approaches had the analysis level from the population to the individual as the main change. Hence, this generated a greater interest in risk factors and individual lifestyles and disinterest in population factors as causes of disease, such as social, economic, cultural, historical, and political factors depending on or determining the social structure of each population [26].

It is possible that the displacement of population aspects in modern epidemiology is due in part to the need of seeking explanations or "actual" theories about the causes of diseases focused on individuals, relying on innovative knowledge generated by basic sciences such as biology, genetics, immunology, and pharmacology. It is assumed that the topics or findings indicated by these disciplines in different fields of research concentrate their applicable results as interventions or treatments focused on specific factors or causes. These disciplines have a predominantly clinical research trend focused on addressing the causes indicated by the basic sciences. Nevertheless, the aspects that are not located at the individual level are omitted. Given these perspective changes in epidemiology throughout history, it has been suggested that the approach of "postmodern epidemiology" should contemplate the restoration of a population perspective and its reintegration into public health while using recent methodological developments [26, 28].

Need for a Refocus on Postmodern Epidemiology

One of the main goals of numerous scientific investigations is to point out the usefulness of their findings, which has led to an approach focused on clinical aspects that undermine the understanding of population patterns on the occurrence of diseases [26]. Such is the case of controlled clinical trials aimed to improve the therapeutic stratification of patients. They are usually considered the maximum reference for research in health sciences, due to their close approach to experimental research and thus to what is verifiable or "true." In an overview where there is rigorous control of variables and even the suppression of others, it is possible to approach "actual" events under a particular research context. This is useful when looking for a treatment or intervention to solve problems of specific populations or with particular characteristics. However, the study design exhibits great limitations by excluding a great variety of conditions; it does not take into consideration that each individual is exposed to different environmental, sociodemographic, political, and historical factors and/or contexts. The study of individual risk factors or mostly clinical aspects can lead to decontextualized findings of low usefulness or low applicability in a reality where social determinants gain relevance [26].

For these reasons, postmodern epidemiology is intended not only to become just a set of generic methods to study the prevalence or incidence of diseases and their solutions [29] but also to include a distinctive theory that allows the understanding of the population patterns of diseases in the same way [26]. Consequently, epidemiologists must consider the nature of causality and work with whatever concept seems most useful and not just seek a lifestyle approach to social politics, where the main responsible for health is often the individual [30].

Recent epidemic chronic diseases – such as diabetes mellitus 2 – or infectious diseases – such as COVID-19 – which have not been addressed quickly and effectively, are a clear example of the great challenge that epidemiology must solve. This should be achieved through comprehensive strategies, designed according to population characteristics, socioeconomic, human, technological, political, and historical resources, to name a few. Therefore, a multidisciplinary approach is needed with greater involvement of social sciences and appropriate study designs that fit the public health issue to be addressed [26, 31]. In other words, an appropriate methodology should be used instead of making the problem fit the method [32].

Analysis Levels to be Considered in Postmodern Epidemiology

The causes of disease can be studied at different levels such as environmental and socioeconomic factors, lifestyle, exposure to biological or chemical agents, or the genetics of individuals, just like diseases in a population [33]. Although certain specific risk factors seem to act directly at the individual level, exposure and susceptibility may be due to a wide range of political, economic, and social factors. Hence, any analysis of the causes of disease in populations must integrate the individual-biological and population analysis levels without overlapping one into the other or denying that any of them exist [26].

Two main approaches have been identified to address the analysis of these different levels: the first "bottom-up" focuses on understanding the individual components of a process from lower to higher levels of an organization. An example of this is the understanding of diseases at a molecular level and the use of this knowledge in public policies (application of screening tests to detect diseases). The main difficulty with this approach lies in the challenge of the complexity and dynamism of the origin of diseases since this is only at the individual level.

On the other hand, the so-called top-down approach uses a structural model of causality, focused on underlying processes and structures as generators of event occurrence. This approach considers that causality results from internal mechanisms of the population under study and not just regular associations between independent objects.

As the epidemiology paradigms have evolved, so has its way of analyzing information. In postmodern epidemiology, the search for relationships between exposures and effects is not prioritized, but rather the analysis of systems within which mechanisms that contribute to the development of a disease in the population can be identified. In order to analyze the behavior of population diseases, a dynamic system has been applied, which considers the interactions between individuals. However, this creates the need to seek other mathematical or statistical analysis tools [33] with a broader view of the disease and its causes located at multiple levels referring to the complexity theory. In this sense, the trend is toward a contextual or multilevel analysis aimed to study the effects observed at an individual level, based on the characteristics of the community or group. The multilevel analysis method is considered appropriate for the new approach in epidemiology since it allows going beyond the study of individual epidemiological factors. This in addition to the need to analyze the role of each component involved in a complex interaction in the health-disease process, where it is important to include the relationship between people and elements of their environment. Multilevel analysis models hierarchical relationships, reduces biases in hypothesis testing, and provides practical estimates of the variability and replicability of regression coefficients across contexts. In addition, it has the potential to emphasize the role of variables from the individual to the macro-level on the configuration of health and disease in populations. Consequently, this type of analysis requires units to be well-identified and structured in hierarchies or levels where the lowest unit of analysis is contained in the next one, generating a higher level of complexity in the data [34].

Postmodern Epidemiology Perspective

Over the past three decades, advances in the field of information technology have provided new opportunities in all areas of science, including the healthcare area, and the substantial expansion of the access to information. This technological advance opened at least three expansion areas for epidemiological analyses: (1) improvement in the speed and reliability of statistical analyses; (2) practical feasibility of multivariate analysis; and (3) development of all inference methods, based on simulation and resampling. It is evident that new technologies play a fundamental role in the future development of epidemiology and epidemiological research [33].

Given the great technological advances in different areas of science, the continuous generation of genetic information and the evolution of computational systems allow for greater data recording and storage. It is necessary for epidemiology to adopt the approach of constantly growing data, both in the number of observations and in the number of variables, as well as in its different health-related levels and disciplines. The new technologies used, for example, in genome-wide studies or those that use imaging techniques applied to population studies, are setting a trend that could redirect the interests of epidemiological analysis. This in order to approach data of different nature (structured, unstructured, and semi-structured data, to name a few), according to the type of format [35, 36].

Electronic data processing, which includes all procedures to acquire, archive, retrieve, and transmit data, has developed at an exponential rate to date. It has had a high-impact trend in epidemiology through the development of data science and increasing access to large and heterogeneous health-related data. However, this large data generation currently places them at a level where they have become unmanageable with currently available technologies. This has led to the creation of the term "big data" to describe data that is large and unmanageable [36] with massive amounts of information. Big data has become a topic of special interest during the last two decades; it offers great potential, the idea being that the more data obtained, the greater the understanding of healthcare processes. Furthermore, they can provide a wealth of information that often remains hidden or unidentified in smaller experimental or observational methods. Although the big data approach could offer new findings, data creation will depend on the different levels at which the events of interest are studied, from the basic to the population level. Using this type of data analysis can be complementary and adapted to the current needs of epidemiology, which faces the challenge of the multiple levels through which health-disease processes occur in populations. The use of big data is changing the observational research, as well as public health surveillance and population health monitoring through an increasingly growing spectrum of applications and new methods [28], which are expected to enrich healthcare strategies [36].

Final Considerations

The integration of new methodological and analytical approaches represents a crucial challenge for current epidemiology applied to public health problems. Scientists, clinicians, and decision-makers will have to adapt and incorporate these new methods and knowledge based on a transdisciplinary vision.

In this context, the acknowledgment of social and global changes over time plays an important role in solving population health problems. The exponential growth of genomic knowledge in different related areas has enabled the inclusion of novel and robust methodologies that are currently applied in the study of complex diseases. Clearly, both the study main objective and methods have been evolving. This trend has allowed the scope and benefits of epidemiology to have an increasingly greater impact on the study of population health.

References

1. Bergonzoli G. Epidemiología y genética: ¿Alianza estratégica en el nuevo milenio? Pan Am J Public Heal. 2005;17:38–45. https://doi.org/10.1590/s1020-49892005000100006.
2. Thacker SB, Buffington J. Applied epidemiology for the 21st century. Int J Epidemiol. 2001;30:320–5. https://doi.org/10.1093/ije/30.2.320.
3. Lau B, Duggal P, Ehrhardt S, Armenian H, Branas CC, Colditz GA, et al. Perspectives on the future of epidemiology: a framework for training. Am J Epidemiol. 2020;189:634–9. https://doi.org/10.1093/aje/kwaa013.
4. Smith GD. Epidemiology, epigenetics and the "gloomy Prospect": embracing randomness in population health research and practice. Int J Epidemiol. 2011;40:537–62. https://doi.org/10.1093/ije/dyr117.
5. Smith GD, Ebrahim S, Lewis S, Hansell AL, Palmer LJ, Burton PR. Genetic epidemiology and public health: Hope, hype, and future prospects. Lancet. 2005;366:1484–98. https://doi.org/10.1016/S0140-6736(05)67601-5.
6. Ben-Shlomo Y, Cooper R, Kuh D. The last two decades of life course epidemiology, and its relevance for research on ageing. Int J Epidemiol. 2016;45:973–88. https://doi.org/10.1093/ije/dyw096.
7. Kuh D, Ben-Shlomo Y. A life course approach to chronic disease epidemiology: tracing the origins of ill-health from early to adult life. 1st editio. Oxford, UK: Oxford University Press; 1997.
8. Barker DJP. The fetal and infant origins of disease. Eur J Clin Investig. 1995;25:457–63. https://doi.org/10.1111/j.1365-2362.1995.tb01730.x.
9. Elder GH. Age differentiation and the life course. Annu Rev Sociol. 1975;1:165–90. https://doi.org/10.1146/annurev.so.01.080175.001121.
10. Koehly LM, Persky S, Philip Shaw, Bonham VL, Marcum CS, Sudre GP, et al. Social and behavioral science at the forefront of genomics: Discovery, translation, and health equity. Soc Sci Med. 2019. https://doi.org/10.1016/j.socscimed.2019.112450.
11. Shanahan MJ. Social genomics and the life course: opportunities and challenges for multilevel population research. In: New directions in the sociology of aging. Washington, D.C.: National Academies Press (US); 2013.
12. Elder GH, Johnson MK, Crosnoe R. The emergence and development of life course theory. Boston, MA: Springer; 2003. https://doi.org/10.1007/978-0-306-48247-2_1.
13. Bernardi L, Huinink J, Settersten RA. The life course cube: a tool for studying lives. Adv Life Course Res. 2019;41:100258. https://doi.org/10.1016/j.alcr.2018.11.004.
14. Blanco M. El enfoque del curso de vida: orígenes y desarrollo. Rev Latinoam Población Asoc Latinoam Población. 2011;5:5–31.
15. Hutchison ED. Dimensions of human behavior: the changing life course. 6th ed. SAGE: Thousands Oaks, California; 2019.
16. Gilligan M, Karraker A, Jasper A. Linked lives and cumulative inequality: a multigenerational family life course framework. J Fam Theory Rev. 2018;10:111–25. https://doi.org/10.1111/jftr.12244.
17. Warner V, Weissman MM. Intergenerational transmission. In: Koenen KC, Rudenstine S, Susser E, Galea S. A life course approach to ment disord. 2nd Oxford, UK: Oxford University Press; 2014, p. 273–290. https://doi.org/10.1093/acprof:oso/9780199657018.001.0001.
18. Alwin DF. Integrating varieties of life course concepts. Journals Gerontol Ser B Psychol Sci Soc Sci. 2012;67B:206–20. https://doi.org/10.1093/geronb/gbr146.

19. Benson R, Glaser K, Corna LM, Platts LG, Di Gessa G, Worts D, et al. Do work and family care histories predict health in older women? Eur J Pub Health. 2017;27:1010–5. https://doi.org/10.1093/eurpub/ckx128.
20. Lu W, Benson R, Glaser K, Platts LG, Corna LM, Worts D, et al. Relationship between employment histories and frailty trajectories in later life: evidence from the English longitudinal study of ageing. J Epidemiol Community Health. 2017;71:439–45. https://doi.org/10.1136/jech-2016-207887.
21. Khoury MJ, Gwinn M, Ioannidis JPA. The emergence of translational epidemiology: from scientific discovery to population health impact. Am J Epidemiol. 2010;172:517–24. https://doi.org/10.1093/aje/kwq211.
22. Marrone M, Schilsky RL, Liu G, Khoury MJ, Freedman AN. Opportunities for translational epidemiology: the important role of observational studies to advance precision oncology. Cancer Epidemiol Biomark Prev. 2015;24:484–9. https://doi.org/10.1158/1055-9965.EPI-14-1086.
23. Windle M, Lee HD, Cherng ST, Lesko CR, Hanrahan C, Jackson JW, et al. From epidemiologic knowledge to improved health: a vision for translational epidemiology. Am J Epidemiol. 2019;188:2049–60. https://doi.org/10.1093/aje/kwz085.
24. Khoury MJ, Gwinn M, Yoon PW, Dowling N, Moore CA, Bradley L. The continuum of translation research in genomic medicine: how can we accelerate the appropriate integration of human genome discoveries into health care and disease prevention? Genet Med. 2007;9:665–74. https://doi.org/10.1097/GIM.0b013e31815699d0.
25. Centers for Disease Control and Prevention. U.S. Vaccine Safety - Overview, History, and How It Works 2020. https://www.cdc.gov/vaccinesafety/ensuringsafety/history/index.html#anchor_1593624461068 (Accessed 8 March 2021).
26. Pearce N. Traditional epidemiology, modern epidemiology, and public health. Am J Public Health. 1996;86:678–83. https://doi.org/10.2105/AJPH.86.5.678.
27. Agurs-Collins T, Khoury MJ, Simon-Morton D, Olster DH, Harris JR, Milner JA. Public health genomics: translating obesity genomics research into population health benefits. Obesity. 2008;16:S85–94. https://doi.org/10.1038/oby.2008.517.
28. Chiolero A. Post-modern epidemiology: Back to the populations. Epidemiologia. 2020;1:2–4. https://doi.org/10.3390/epidemiologia1010002.
29. Krieger N. Epidemiology and the web of causation: has anyone seen the spider? Soc Sci Med. 1994;39:887–903. https://doi.org/10.1016/0277-9536(94)90202-X.
30. Vandenbroucke JP, Broadbent A, Pearce N. Causality and causal inference in epidemiology: the need for a pluralistic approach. Int J Epidemiol. 2016;45:1776–86. https://doi.org/10.1093/ije/dyv341.
31. McMichael AJ. The health of persons, populations, and planets: epidemiology comes full circle. Epidemiology. 1995;6:633–6. https://doi.org/10.1097/00001648-199511000-00013.
32. McKinlay JB. The promotion of health through planned sociopolitical change: challenges for research and policy. Soc Sci Med. 1993;36:109–17. https://doi.org/10.1016/0277-9536(93)90202-F.
33. Saracci R. Informazione per la salute e deformazioni della salute: uno sguardo critico sull'epidemiologia postmoderna. Epidemiol Prev. 2007:239–46.
34. Diez-Roux AV. Bringing context back into epidemiology: variables and fallacies in multilevel analysis. Am J Public Health. 1998;88:216–22. https://doi.org/10.2105/AJPH.88.2.216.
35. Hofman A. Editorial: new studies, technology, and the progress of epidemiology. Eur J Epidemiol. 2010;25:851–4. https://doi.org/10.1007/s10654-010-9531-8.
36. Dash S, Shakyawar SK, Sharma M, Kaushik S. Big data in healthcare: management, analysis and future prospects. J Big Data. 2019;6:54. https://doi.org/10.1186/s40537-019-0217-0.

Molecular Tools for Modern Epidemiology: From the Concepts to Clinical Applications

2

María Isabel Coronado-Mares, Elizabeth Sulvaran-Guel, Karla Daniela Rodríguez-Hernández, and Nadia Alejandra Rivero-Segura

Abbreviations

ALP	Alkaline phosphatase
BiFC	Bimolecular fluorescence complementation
BSA	Bovine serum albumin
BSE	Bovine spongiform encephalopathy
CEIA	Capillary electrophoresis immunoassay
CMA	Chromosomal microarray analysis
aCGH	Comparative genomic hybridization arrays
cDNA	Complementary deoxyribonucleic acid
CNV	Copy number variants
DNA	Deoxyribonucleic acid
dPCR	Digital polymerase chain reaction
ELISA	Enzyme-linked immunosorbent assay
EtBr	Ethidium bromide
FISH	Fluorescence in situ hybridization
FRET	Fluorescence resonance energy transfer
GFP	Green fluorescent protein
HRP	Horseradish peroxidase
HGP	Human Genome Project
HIV	Human immunodeficiency virus
LIF	Laser-induced fluorescence
LAMP	Loop-mediated isothermal amplification
MHC	Major complex of histocompatibility
MSP	Methylation-specific PCR
MWB	Multiplex western blot
NMR	Nuclear magnetic resonance
PBMC	Peripheral blood mononuclear cells
PBS	Phosphate buffer solution
PS	Phosphatidylserine
pNPP	P-Nitrophenyl phosphate
PCR	Polymerase chain reaction
PVDF	Polyvinylidene difluoride
qPCR	Quantitative polymerase chain reaction
RFLP	Restriction fragment length polymorphisms
RT-PCR	Reverse transcription polymerase chain reaction
RNA	Ribonucleic acid
SNP	Single-nucleotide polymorphism
SDS-PAGE	Sodium dodecyl sulfate polyacrylamide gel electrophoresis
CFSE	Succinimidyl-carboxyfluorescein ester
TUNEL	TdT dUTP nick-end labeling
UPR	Unfold protein response
WB	Western blot

M. I. Coronado-Mares
Hospital Regional Tlalnepantla, ISSEMyM, Mexico City, Mexico

E. Sulvaran-Guel
Licenciatura en Ciencias Genómicas UNAM, Mexico City, Mexico

K. D. Rodríguez-Hernández
Laboratorio de Estudios sobre Tripanosomiasis, Departamento de Inmunología, Instituto de Investigaciones Biomédicas, UNAM, Mexico City, Mexico
e-mail: xandy411@comunidad.unam.mx

N. A. Rivero-Segura (✉)
Dirección de Investigación, Instituto Nacional de Geriatría (INGER), Instituto Nacional de Geriatría, Ciudad de México, Mexico
e-mail: nrivero@inger.gob.mx

Introduction

Since the beginning of the century, the advances in laboratory technologies have allowed the acquisition of valuable molecular information regarding human health [1]. In 2003, the Human Genome Project (HGP) concluded with the publication of over 90% of the DNA sequence in the human genome. Before the HGP, very few loci were associated with diseases. However, the sequencing and annotation of the human genome and further analysis in several diseases allowed the establishment of these associations, firstly with diseases following Mendelian heritage rules and, more recently, with complex diseases which manifest as a result of the combination of several factors, including genetics and environmental cues [2].

Nowadays, it is possible to assess a human genome in order to find disruptions that could be causative of a particular disease [1]. These disruptions range from base pair modifications to

J. C. Gomez-Verjan, N. A. Rivero-Segura (eds.), *Principles of Genetics and Molecular Epidemiology*,
https://doi.org/10.1007/978-3-030-89601-0_2

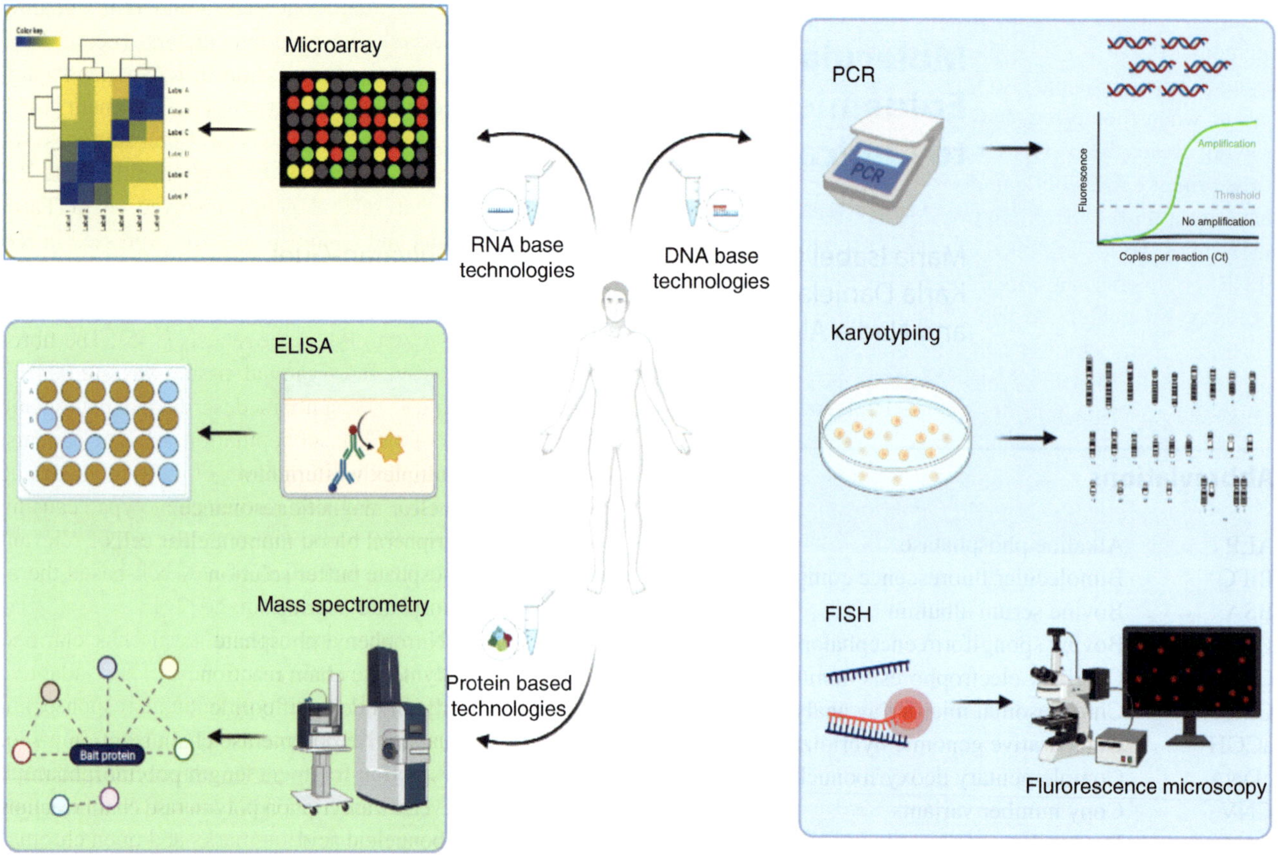

Fig. 2.1 In this chapter we will address the most popular molecular tools that are currently used in clinical practice. For instance, nucleic acids-based technologies such as PCR, microarrays, karyotyping, and FISH; and the protein-based technologies (immunoassays and mass spectrometry)

chromosomal rearrangements and modifications in gene expression [2]. Furthermore, besides chronic and congenital diseases, infectious diseases, such as those caused by bacteria, viruses, or parasites, may be identified rather by changes made in the genome or by the detection of the infectious agent's genome itself. For these reasons, in recent years numerous studies have been done for the discovery of molecular hallmarks in particular diseases for its posterior utilization in clinical contexts.

Molecular tools refer to the whole set of techniques based on properties of nucleic acids, such as DNA or RNA, and proteins for the identification of the causes of a disease. The discovery of molecular hallmarks in a particular disease can permit them to be detected in patients and are useful as a diagnostic tool [1]. Molecular tools have ameliorated the process of disease detection, both by improving efficiency and accuracy of diagnoses, and also its speed, which is of particular interest when the prognosis of the patient depends on the appropriate early diagnosis. Furthermore, molecular tools also allow the collection of useful knowledge for drug development and prevention of diseases, relevant for many of those which have no known cures yet [2].

Although molecular tools have important applications in human disease diagnosis, its utilities go further into the resolution of criminal cases and paternity tests, evolution and population studies of diverse species, microbes identification, and even microbiome studies with the advent of metagenomics. In this chapter, we will review several molecular techniques used in the clinics, with the description of its procedure and medical applications. These encompass PCR, which is commonly used for pathogen identification and deletions or insertions in particular diseases, as well as in forensic medicine and paternal genetic testing; karyotyping, FISH, and microarrays for the detection of chromosomal abnormalities; and finally, immunoassays and proteomics, for the evaluation of aberrant gene expression and protein isoforms or localization in diseases (Fig. 2.1 summarizes the reviewed molecular tools and its applications) [2].

PCR

The polymerase chain reaction, most commonly known as PCR, is a process created in 1983 by Kary Mullis for DNA molecule amplification in vitro [3, 4]. PCR uses polymerases, the enzymes in charge of DNA replication found in all living organisms [4]. Polymerases synthesize DNA by adding the complementary base pair for each position in a whole fragment [3]. To do this, they need a primer, which is a sequence complementary to the DNA fragment but shorter than it: when the correct base pair is incorporated into the growing chain, the 3'-OH molecule exposed in the pentose of the primer carries out a nucleophilic attack to the triphosphate group in the incoming nucleoside triphosphate, forming the phosphodiester bond [3]. Therefore, when performing a PCR, polymerases are required, as well as a pair of primers, deoxyribonucleotide triphosphates (dNTPs), and other substrates, such as buffers and ions, and they are all set in an apparatus called thermal cycler, which raises and lowers the temperature [5]. Since PCR requires heating, a special thermostable polymerase is needed, which is usually the Taq polymerase, the enzyme found in a thermophilic bacteria named *Thermus* aquaticus [5].

The PCR process requires three steps: denaturing, annealing, and amplification. During denaturing, the thermal cycler raises the temperature to 93–95 °C, at which the hydrogen bonds in the double-stranded DNA are broken, resulting in two single-stranded DNA molecules [5]. Next, the temperature lowers to ~60 °C, and the primers designed to flank the extremes of the DNA region to be amplified bind to the single-stranded DNA molecules by base pair complementarity [5]. Finally, in amplification, the temperature again rises to 70–75 °C, and the polymerase incorporates the complementary base pair for every position in the interest sequence, now resulting in two double-stranded DNA molecules [5]. The process is repeated, resulting in 4 double-stranded molecules, 8, 16, and so on, until it is done ~30 times, which yields billions of copies for the interest sequence (~2^{30}) [4].

Finally, in the standard PCR procedure known as endpoint PCR, the amplified fragments are observed in an agarose electrophoresis gel, using ethidium bromide (EtBr) for DNA staining and ultraviolet light. DNA fragments of known length must also be used as markers for the identification of the PCR product with the previous knowledge of its molecular weight. However, there are several PCR variants available.

Types of PCR

Multiplex PCR

Multiplex PCR is a variant that allows the amplification of more than one interest DNA fragment by using more than a pair of primers specific for different fragments [6]. It was first used in 1988 for the detection of deletion variants in the human dystrophin gene, which served as a diagnosis for Duchenne muscular dystrophy. Since several primers are used, it is essential to have extensive knowledge of the sequence in the extremes of the fragment that will bind to them to avoid non-specific amplification [6]. Nowadays, multiplex PCR has several applications, including pathogen identification, genetic diseases diagnoses, and forensic analyses [6].

Real-Time PCR

Real-time PCR or quantitative PCR (qPCR) is a modification that allows the quantification of initial DNA molecules [7]. This type of PCR has the advantage that the agarose gel electrophoresis is not needed; instead, the quantitation of the samples is done, while the reaction is happening [7]. qPCR is done in a thermal cycler able to detect fluorescent signals, and fluorescently labeled probes are included in the reaction [7]. For every cycle, the fluorescence signal emitted is directly correlated with the amount of DNA [4]. Therefore, the more initial concentration of DNA, the faster the fluorescence signal will reach a threshold known as CT, allowing the identification of the cycle in which this is achieved and consequently, the initial DNA volume [4].

Reverse Transcription PCR

Reverse transcription PCR (RT-PCR) allows the amplification of a cDNA molecule originating from an RNA sample. The procedure is essentially the same as in a traditional PCR, with the modification that RNA fragments must be retrotranscribed to cDNA. Usually, the RNA fragments to be amplified come from cellular mRNAs; thus, the primers used for retrotranscription are commonly oligo(dT) molecules that bind to the poly-A tail in the mRNA 3′ end [8]. RT-PCR and qPCR can be combined in a methodology known as qRT-PCR, which allows the quantification of specific RNA molecules [2]. This is particularly useful for the quantification of the viral load for several viral infections caused by RNA genome viruses [2].

Digital PCR

Digital PCR (dPCR) is a PCR variant introduced in 1999 which allows the identification of allelic mutants for a gene or specific locus [9]. dPCR relies on the dilution of DNA to be amplified in wells on a plate, so that there are very few molecules in each well [9]. qPCR is done for every well, and the fluorescent probes used are designed to bind to the different alleles expected, each with a different color [9]. This way, mutant alleles can be identified even in a cell population in which the wild-type allele is predominant — a phenomenon common in certain diseases, such as cancer —thanks to the dilution of the initial sample [9]. The analysis of the fluorescent signal will allow the identification of the proportion of the mutant allele in the cell population, by adding the proportion of mutant and wild-type alleles in every well [9].

Lamp

Loop-mediated isothermal amplification or LAMP is a methodology alternative to PCR, first described in 2000

[10]. In contrast with PCR, LAMP does not require alternating temperatures; instead, all the procedure is carried out at the same temperature, and hence, it does not require a thermal cycler [10]. LAMP is performed at ~60 °C, and it uses several primers, from two to four pairs for each end of the region to be amplified, which highly raises the specificity and lowers the false-positive discovery [10]. Furthermore, LAMP uses a polymerase with strand-displacement activity [10]. The LAMP procedure initiates with a primer targeting the middle region of the 3′ end and amplifying the whole fragment from that starting point [10]. Subsequently, the polymerase begins amplification starting on a primer targeting the outermost region of the 3′ end, and displacing the previously amplified sequence [10]. Since the innermost and the outermost regions of each end are complementary, the newly synthesized strand forms a double stem-loop structure, and this allows new primers targeting the middle region of the 3′ end to anneal to it and amplify again starting from that point [10]. This form of amplification enables multiple copies of an interest fragment to be generated rapidly, as well as with a really high specificity [11]. As well as PCR, LAMP can be done qualitatively and with reverse transcription, which is useful for the detection of viral infections [11] (Table 2.1).

Table 2.1 Applications of different classes of PCR

Type	Application	Principle	Reference
Classical PCR	Forensic medicine	Across the genome, there exist several polymorphisms (over 20,000) that correspond to short repeated sequences in tandem (STRs). The specific amplification of these sequences and its posterior length determination by gel electrophoresis create a "DNA fingerprint" almost unique for every person in the world	[4]
RT-PCR or qRT-PCR	Pathogen detection	Primers are designed for amplification of a pathogen's mRNA, and through electrophoresis, it is possible to determine the presence of the pathogen in the sample. qRT-PCR can also be used to determine the load of the pathogen in the sample	[12]
Allele specific PCR	Point mutations detection	When wanting to distinguish between a point mutation and a wild-type allele in a specific position of the genome, PCR primers are generated so that this position is the last nucleotide of the primer. Since the polymerase needs the 3'-OH, this nucleotide needs to be bound to the fragment to be amplified. If the allele is mutated, there would not be amplification, so it is possible to identify point mutations possibly responsible for a disease	[2]
Classical PCR	ChIP	Chromatin immunoprecipitation (ChIP) is a methodology to identify a region of the genome bound to a specific protein. To do this, the proteins are crossed-linked to the DNA, and the DNA is cut in small fragments. The mixture is divided into two: In one of the mixtures, all the proteins are removed, and PCR is performed for the interest region and for a control region where the protein is known not to bind to. The other mixture is immunoprecipitated with an antibody specific for the interest protein, and only the fragments containing it will be selected. Then, the proteins are removed, and again, PCR is performed for the interest region and for the control region. In the first mixture, amplification is expected for both the control region and the interest region. However, if the control was appropriately selected, in the second mixture there should not be amplification of this region. If the protein is bound to the interest region, amplification will be observed. This methodology works for identification of DNA-protein interactions specifically identified in certain contexts	[3]
PCR-RFLP	SNV detection	Restriction fragment length polymorphisms (RFLPs) are length variants of fragments cut by restriction enzymes (REs) present in the genomes of different people. Some REs recognize a single base pair and cut in it, creating two shorter fragments. If this base pair is mutated, the RE won't cut it, leaving the longer fragment. PCR can be used for amplification of some regions, and the products processed by REs and then run in an agarose gel to identify the different lengths of the fragments. In this way, it is possible to identify the base pair present in a region of the genome in both alleles without sequencing and is a commonly used method for identification of SNPs in mtDNA and in the Y chromosome for ancestry studies	[2]
Common PCR	Insertion and deletion detection	Genetic deletions or insertions can be identified without sequencing by PCR. The methodology consists in using primers flanking the interest region or gene, amplifying it and identifying longer or shorter fragments with a gel electrophoresis. Some diseases, such as cystic fibrosis, are characterized by allele insertions or deletions; thus, this methodology may work as a diagnose tool	[13]
Multiplex PCR	Deletion variants detection	As mentioned previously, the dystrophin gene can present deletions in several exons, leading to the development of Duchenne's muscular dystrophy. 98% of these deletions can be identified by multiplex PCR, working as a diagnose tool	[13]
Methylation-specific PCR (MSP)	Detection of methylated regions of the genome	PCR can be used to detect methylated regions in the genome by first treating the samples with bisulfite, which changes unmethylated cytosine to uracil, and methylated cytosines are left unmodified. Then, primers with guanine are designed to pair with methylated DNA, and primers with adenine are designed to pair with unmethylated DNA. A quantitative PCR can be done for detection of the amount of methylated and unmethylated DNA	[14]
qRT-PCR	Gene expression analysis	A quantitative retrotranscription PCR can be performed to analyze cellular mRNAs with altered expression in certain contexts compared to controls. This is particularly interesting in some diseases that show aberrant gene expression in specific genes, such as cancer	[7]

Principles of Karyotyping

The human genome is arranged in 23 pairs of chromosomes, 22 of them are somatic and 1 of them is the sexual pair [5]. The somatic chromosomes are identical between males and females and are numbered from 1 to 22, the longest and the shortest pairs, respectively [5]. Conversely, the sex chromosomes differ between sexes, females possess two X chromosomes, and males possess an X chromosome and a Y chromosome [5]. The visualization of the arrangement of the chromosomes during mitosis is called a karyotype [12]. A normal karyotype consists of 46 chromosomes and a pair of X chromosomes (46XX) or an X and a Y chromosome (46XY) for females and males, respectively [12]. Karyotypes enable the identification of several chromosomal abnormalities, such as aneuploidies, and also deletions, insertions, duplications, and chromosome rearrangements [2, 15].

Initial karyotyping methodologies, introduced in the 1970s, were based on DNA staining for the identification of abnormalities in the chromosomes [15]. These DNA staining methodologies produced light and dark bands patterns in the chromosomes [2]. One of the most common techniques was the G-banding, in which the chromosomes were treated with trypsin and subsequently stained with a chemical dye named Giemsa [2]. Darker regions observed with G-banding correspond to condensed chromatin, which has low gene density and transcriptional activity. Consequently, G bands have lower CG content [2]. The staining can be observed and photographed under a light microscope, and then the karyotype is constructed by arranging the homologous chromosomes [5].

FISH

Fluorescence in situ hybridization (FISH) is a methodology first described in 1969, which is based on the labeling of a DNA or RNA probe with a fluorescent dye and its hybridization with a sample DNA or RNA [16]. Using a fluorescence microscope, the DNA or RNA hybrids can be observed, and the identification of the position of the probe in the sample is allowed [12]. FISH permitted the location determination of several genes by using entire chromosomes as samples during the Human Genome Project, and it is still useful for organisms whose genome annotation or sequence is not available. For the case of humans, FISH is more commonly used in clinical contexts [17, 18].

FISH can be used for the detection of chromosomal abnormalities, such as deletions, insertions, and rearrangements [18]. Furthermore, in comparison with banding techniques used for karyotyping, FISH has a higher sensitivity and may detect abnormalities more easily than these [18]. For example, FISH can be used to diagnose acute febrile neutrophilic dermatosis, in which the genes BCR and ABL are fused by using two different color dyes in the probes for each gene (e.g., red and green) [16]. The expected FISH results should be two separate spots of each color, but the fusion can be detected if there is a yellow signal [16]. Additionally, a specific type of FISH named Multiplex FISH can be used for the labeling of every human chromosome in metaphase [18]. The probes with different colors are designed for DNA regions in a single chromosome, and after hybridization, only one color should be visible for each chromosome [18]. Translocation events can be detected if chromosomes show color stripes [18].

Microarray

Microarrays can also be used for the detection of chromosomal abnormalities, in a variant known as chromosomal microarray analysis (CMA) [19]. CMAs can detect copy number variants (CNVs) of interest genes or loci, with a much higher resolution than banding methodologies [19]. Furthermore, it can use more probes simultaneously than FISH [20]. There are two existing classes of microarrays useful for chromosomal anomalies detection: comparative genomic hybridization arrays (aCGH) and SNP arrays [20]. aCGH requires DNA isolation from reference and test samples, differential labeling, and their hybridization at interest regions [21]. This approach is useful for the identification of CNVs [21]. If the test sample is labeled in red and the reference sample in green, yellow arrays should be expected for all the interest regions [21]. When the array scanning reveals regions with more or less abundance of test DNA (wells in red and green, respectively), it is possible to spot copy number gains or copy number losses [21]. Conversely, SNP arrays only use test DNA hybridization, and the array results are compared to reference DNA [22]. The probes placed in the array are only around 20 base pairs long, enabling the characterization of small regions of DNA across the whole genome [22].

Immunoassays

Disease diagnosis is crucial for correct patient treatment. The development of proteomic technologies has increased the identification of protein biomarkers involved in the immunogenicity of diseases in body fluids, such as blood, urine, saliva, cerebrospinal fluid, and different tissues (biopsies), to predict the course of the disease, information on cellular signaling pathways, monitoring treatment response, adverse effects and the identification of new diagnostic, therapeutic methods, and new targets[23].

Immunoassay methodologies are the most commonly used tools in protein research, using the properties of antibodies to bind different protein domains and to mark them. ELISA and Western blotting are the oldest methods that changed, adapted, and modernized over time, improving their sensitivity and leading to the appearance of new methods and equipment for biomarker investigation and analysis, with the goal of studying more analytes in a single sample, in a shorter time, and with increased accuracy. The reproducibility and reliability of the results are also a goal pursued by manufacturers [24].

Enzyme-Linked Immunosorbent Assay (ELISA)

The gold standard of immunoassays, ELISA, is a very sensitive diagnostic method used to detect and quantify a large variety of protein biomarkers like antibodies, antigens, proteins, peptides, glycoproteins, and hormones. This technique was developed simultaneously in 1971 by Engvall and Perlmann and Van Weemen and Schuurs, and nowadays it continues to be used as a routine analytic tool. The detection of these products is based on the antigen-antibody interactions, and detection is usually done with the help of an enzyme and a substrate. An antibody is a type of protein produced by an individual's immune system and has a specific region that binds to a protein from a foreign source called "antigen." This binding allows identifying a specific disease biomarker with small amounts of sample [25, 26]

In the ELISAs methodology, the primary and specific antibody only binds to the protein of interest, and the secondary detection antibody is a second enzyme-conjugated antibody that binds the primary antibody and, through the addition of a substrate, generates an observable color that indicates the presence of antigen. The most common substrates available for ELISA are horseradish peroxidase (HRP), whose substrate is hydrogen peroxide and results in a blue color change, and the alkaline phosphatase (ALP) that uses P-nitrophenyl-phosphate (pNPP) producing a yellow color of nitrophenol after room temperature incubation. The new ELISA methodologies have developed fluorogenic, quantitative PCR, nonenzymatic and electrochemiluminescent reporters for signal generation [24].

Currently, four major types of ELISA have played a prominent role in the quantitative and qualitative identification of analytes:

Direct ELISA (Antigen-Coated Plate, Screening Antibody)

The simplest type of ELISA, the primary detection antibody, binds directly to the protein of interest. This method begins with the coating of antigen to the ELISA plates. The first binding step involves adding antigen to the plates and incubate overnight at 4 °C; the next step is to wash the plates of any potential unbound antibody and block any unbound sites on the ELISA plate using agents like BSA, ovalbumin, aprotinin, or other animal proteins to prevent the binding of any nonspecific antibodies and avoid a false-positive result. After adding the buffer, the plate is rewashed to remove any unbound antibody and followed by the addition of a substrate/chromophore (AP or HFP), which results in a color change by the hydrolysis of phosphate groups from the substrate AP or by the oxidation of substrates HRP. The advantages of direct ELISA include eliminating secondary antibody cross-reactivity and quantifying a specific molecule with high sensitivity from a wide variety of samples; it is faster than indirect ELISA, but the signal is less amplified compared to the other types of ELISA, and it has a high cost of reaction [27].

Indirect ELISA (Antigen-Coated Plate; Screening Antigen/Antibody)

Indirect ELISA detection is a two-step ELISA which involves a primary antibody and a labeled secondary antibody. The steps of the indirect ELISA are identical to the direct ELISA, except for an additional wash step and the types of antibody added after the buffer is removed. It requires two antibodies: a primary detection antibody that sticks to the protein of interest and a secondary enzyme-linked antibody complementary to the primary antibody. The primary antibody is added first, followed by a washing step, and then the enzyme-conjugated secondary antibody is added and incubated. After this, the steps are a washing step, the addition of substrate, and detection of a color change. This method has a higher sensitivity when compared to the direct ELISA. It is also less expensive and more flexible due to the many possible primary antibodies that can be used. The only major disadvantage is the risk of cross-reactivity between the secondary detection antibodies and the occurrence of nonspecific signals [28].

Sandwich ELISA (Antibody-Coated Plate; Screening Antigen)

This method appeared to avoid false-positive or false-negative results. Unlike direct and indirect ELISA, the sandwich ELISA begins with a capture antibody coated onto the wells of the plate. The term "sandwich" refers to the way the antigens are "sandwiched" between two layers (capture and detection antibodies). After adding the capture antibody to the plates, the plates are then covered and incubated overnight at 4 °C. Once the coating step is complete, the plates are washed with PBS, then buffered/blocked with BSA, and finally, the plate is washed with PBS before the addition of the antigen. The plate is rewashed, and the primary detection antibody is added, followed by a buffer wash. The secondary enzyme-conjugated antibody is added and incubated, and the plate is rewashed. Finally, the substrate is added to produce a color change.

The sandwich ELISA has the highest sensitivity and specificity among all the ELISA types. It is suitable for complex samples and has more flexibility to quantify antigens between the two layers of antibodies. Its major disadvantages are the time, the use of expensive "matched pair" (divalent/multivalent antigen), and secondary antibodies [24].

Competitive ELISA (Screening Antibody)

This method is based on a competitive binding process between the original antigen in the sample and the add-in antigen;, the more antigen in the sample, the less labeled antigen is retained in the well and the weaker the signal. It utilizes two specific antibodies, an enzyme-conjugated antibody and another antibody present in the test sample (if it is positive). Combining the two antibodies into the wells will allow for a competition for binding to antigen. The presence of a color change means that the test is negative because the enzyme-conjugated antibody binds the antigens, rather than the antibodies of the test sample. The absence of color indicates a positive test and the presence of antibodies in the sample. The method has a low specificity and cannot be used in dilute samples. However, the benefits are that sample purification is less needed, it can measure a large range of antigens in a given sample, and it can be used for small antigens and has low variability [24].

New Methods

In order to improve the ELISA method, in terms of using smaller quantities of samples, shortening the reaction time, avoiding sophisticated reading equipment, and reducing costs side, new methods have been developed:

The *enzyme-linked immunospot assay (ELISpot assay)* is widely used to evaluate a cellular immune response against viral antigens in allergies, autoimmunity, and vaccine development. The method has a relatively wide quantitative range and offers unique sensitivity by revealing cytokine secretion at the single-cell level. This technique, performed on PVDF membranes, has advantages like specificity, sensitivity, and a wide range of detection [24, 29].

The conventional *single-target assays ELISA-Western blot* are suitable for biomarker validation but could be expensive, time-consuming, and sample limiting. While most of the disease conditions may arise when only one single molecule is altered, more often it is the consequence of the interaction between several molecules within the inflammation milieu; therefore, studying the diseases necessitates a comprehensive perspective [24].

The most recent is the ELISA platform with *ELISA on a chip* (ELISA-LOC), which allows the use of only 5 μl of sample on a miniaturized 96-well plate combined with a CCD camera. The system includes three main functional elements: (1) a reagent loading fluidics module, (2) an assay and detection wells plate, and (3) a reagent removal fluidics module. The ELISA-LOC system combines several biosensing elements: (1) carbon nanotube (CNT) technology to enhance primary antibody immobilization, (2) sensitive ECL (electrochemiluminescence) detection, and (3) a charge-coupled device (CCD) detector for measuring the light signal generated by ECL. This method has greater sensitivities than the corresponding standard manual plate-based ELISAs, and that single samples can be assayed in a minor fraction of the time [30].

Clinical Significance

ELISA testing is an important part of medical care and scientific research. ELISAs can be used in many settings, including rapid antibody screening tests for human immunodeficiency virus (HIV), detection of other viruses, bacteria, fungi, autoimmune diseases, cancer biomarkers, food allergens, blood typing, the presence of the pregnancy hormone hCG, laboratory and clinical research, forensic toxicology, and many other diagnostic settings. Some types of ELISAs and their uses are included in the Fig. 2.2 [24, 27].

Western Blot

The immunoblot or Western blot (WB) is one of the analytical and quantitative techniques mostly used in research laboratories throughout the world for identifying specific proteins in many biological samples, liquid or tissue/cellular homogenates [24]. The WB technique was invented by Harry Towbin and co-workers in 1979. The name "Western blot" was given 2 years later by Neal Burnette, inspired in the earlier name of other blotting methods [28].

In this procedure, crude lysates are first separated based on their molecular weight by SDS-PAGE, transferred to a solid membrane surface (usually nitrocellulose or PVDF) and detected with the help of protein-specific antibodies. The membrane is probed by a specific primary antibody, it binds the specific epitope of the protein, and it is labeled by the addition of a secondary antibody recognizing the primary

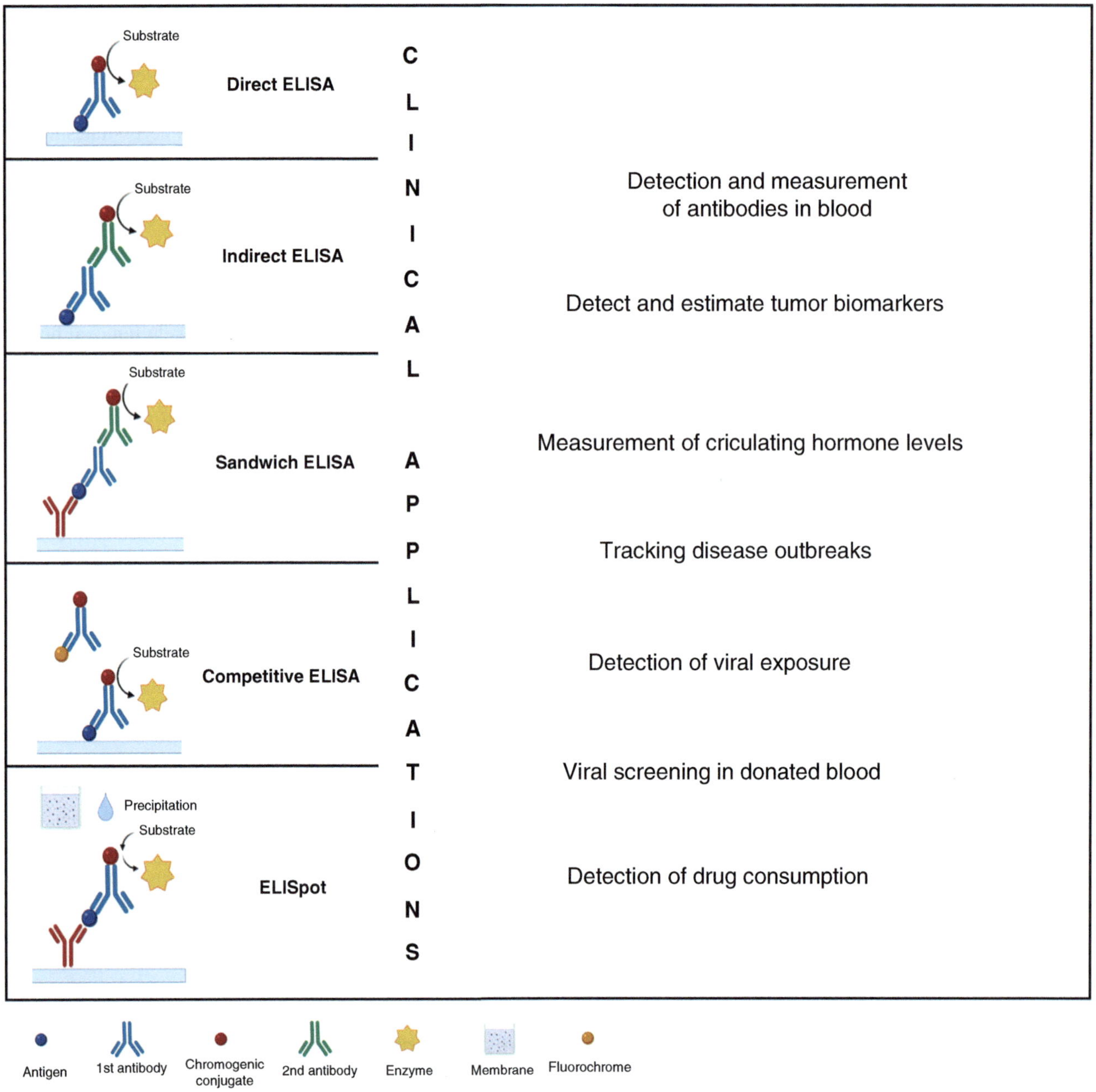

Fig. 2.2 Brief description of ELISA immunoassays and their current clinical applications. In general the working principle of the immunoassays relies on the protein/antigen-antibody reaction, the rest are variations of this principle as depicted in this figure

antibody conjugated with a detection reagent (fluorophore, enzyme, and radioisotope). The visualization is done colorimetric, by chemiluminescence, on X-ray film, or directly in the membrane with the aid of an imaging system [24, 31].

This technique brings concrete and useful information about the amount of protein loaded to independently quantify housekeeping proteins (typically actin, GAPDH, or tubulin). If the target protein present in the sample is altered qualitatively or quantitatively, the band thickness is changed compared to a control being downregulated or overexpressed. The WB results can guide us for a comparison of a target protein expression important in a medical diagnosis or experiment or a genetic investigation in case of partial deletion or duplication in the protein gene [32].

Since WB is a multistep protocol, variations and errors can occur that reduce the reliability and reproducibility of this technique. Also, obtaining maximal sensitivity for the detection of a specific protein remains a fundamental issue, leading to advances in antibody specificity, chemiluminescent formulations, properties of fluorescent molecules and imaging techniques that provide gains in sensitivity, dynamic range, and ease of use. Here we discuss different aspects of methods based on the Western blotting technique and its contemporary application in epidemiology.

Multiplex Western Blot (MWB)

In the last few years, it has become a necessity to analyze multiple target proteins at the same time, in order to compare the expression of proteins involved in a specific pathology. The MWB method revolutionized medical diagnosis and opened new perspectives in biomedical research. The analysis of several proteins involved in different pathologies reduces the cost and time for analysis.

This method was standardized by Anderson and Davison to study different muscle proteins involved in muscular dystrophies. It allows simultaneous screening of multiple proteins in a biphasic polyacrylamide gel system, which enables the corresponding blot to be probed simultaneously with a cocktail of monoclonal antibodies. The gel is optimized so that large proteins of more than 200 Kd can be analyzed in the top part, while smaller proteins under 150 Kd are separated in the lower phase. This basic system allowed establishing a biomarker profile for each patient, providing valuable information for diagnosis as well as for phenotype-genotype correlations [33].

Capillary Electrophoresis (CE) and Capillary Western Blotting (CWB)

This technique was introduced by Nielsen in 1991, with the concept of capillary electrophoresis immunoassay (CEIA), which uses the capillary electrophoresis (CE) technique to visualize the immunocomplex products that form between an antigen and its corresponding antibody. In this method, a mixture of the antigen and antibody is injected into the end of a capillary to quickly separate according to the size of the immune complex from the free antigen (or free antibody), offering a better resolution. This method decreases the time for analysis and requires a smaller volume for samples compared with classical western blotting. Since it is coupling with laser-induced fluorescence (LIF), it enables the highly sensitive detection of fluorescent molecules in a volume as small as nanoliters of the sample, and it can be used to quantify membrane proteins in extracellular vesicles [34].

Microfluidic Western Blotting

This technology reduces even more the amount of the sample required for WB and also the length of the capillaries from centimeters to microns using microfluidic channels. He and Herr developed this automated immunoblotting method, in which proteins are separated by microchip electrophoresis and can be captured on membranes. This process reduces the separation and reduces time to a few minute glass microfluidic chip to in situ immunoblotting, allowing a rapid protein separation, directed electrophoretic transfer, and high-efficiency identification of proteins of interest using antibody-functionalized membranes [24]. Since this system requires only 0.01–0.5 μg of protein, it has been applied to the detection of specific proteins like GAPDH and β-tubulin from A431 cell lysates [24, 35].

Single-Cell Western Blotting

As the most recent proposal technology, single-cell Western blotting is a combination of microfluidics and conventional Western blotting to achieve protein expression analysis at a single-cell resolution. Due to separation by electrophoresis before the antibody probing, it overcomes the issue of cross-reactions. In single-cell Western blotting, a layer of polyacrylamide gel is coated on a glass and patterned with large-array microwells. Single cells are dropped on the thousands of microwells and lysed in situ, and then proteins are separated by gel electrophoresis, immobilized via photoinitiated blotting, and detected by fluorescent labeled antibodies. Although this technique represents a new technology for single-cell protein expression analysis, it has some limitations, since due to cell loss, thousands of cells are required and have limited detection sensitivity because proteins are easily lost during processing procedures such as cell lysing, protein immobilization, and repeated antibody stripping[36].

Dot Blot

In this method, the samples are applied in small dots directly on the membrane and then spotted through circular templates. After membrane drying, the antibodies are applied. The visualization of a target protein is made as in WB, chemiluminescent, or colorimetric [24]. It is used to test the specificity and antibody concentration used for WB or to evaluate the presence of a target protein in the sample before WB. This methodology has been used for detecting *Sarcocystis* spp.'s antibodies in cattle [37] and analyzing conformational changes in herpes simplex virus entry glycoproteins [38].

Far-Western Blotting

It is used to detect a protein-protein interaction in vitro. Instead of the primary antibody for detecting the protein of interest, this method uses a nanoantibody protein that binds to the protein of interest. Far-Western blotting detects proteins on the basis of the presence or the absence of binding sites for the protein probe. This method is important in characterization of protein interactions in biological processes such as signal transductions, receptor-ligand interactions, or screen libraries for interacting proteins [39].

Clinical Significance

Western blotting is frequently used for the confirmatory medical diagnosis of infectious diseases such as Lyme disease, HIV infection, bovine spongiform encephalopathy (BSE), hepatitis C infection, syphilis, inflammatory muscle conditions such as myositis, and certain autoimmune disorders (e.g., paraneoplastic disease). For Lyme disease and HIV infection, these are the only two microbial diseases for which an initial borderline or positive ELISA must be followed by a confirmatory Western blot [24, 31].

Flow Cytometry

Flow cytometry is a multiparametric method which analyzes quantitatively characteristics of individual cells within a heterogeneous population, such as size and granularity simultaneously as the cell flows in suspension. The working principle of this tool relies on the information produced on the light scattering of the cells, which is derived from dyes or antibodies coupled to fluorochromes targeting molecules located on the surface or inside the cells [40], as depicted in Fig. 2.3.

Clinical Applications

Flow cytometry may be a cell-specific identification and quantitative technique with a wide spectrum of applications. Particularly, the main clinical applications of this technique are the disease diagnosis (HIV-infected patients) and monitoring disease progression (cancer, leukemia, and lymphoma). As well, flow cytometry is useful analyzing cell proliferation, phagocytosis, and apoptosis. In the following subsections we dissect the most outstanding examples of the flow cytometry clinical applications.

Phenotypic Characterization of Blood Cells

Immunophenotyping or phenotypic characterization of cells consists of both the identification and quantification of a particular cell group within the mixed population, i.e., blood immune cells (T cells, B cells, NK cells, mast cells, basophils, eosinophils, neutrophils, monocytes, among others). This characterization is possible due to the expression of surface proteins specific for each cell type that can be detected by antibodies [44], for instance, human PBMC. Subsequently, PBMC were labeled with the chosen combination of cell surface antibodies as well as anti-CD3, anti-CD4, and anti-CD14. This cell surface staining section and the labeled cells were analyzed by flow cytometry resulting in the separation of each of the populations [45–47]. Beside the characterization of cell population, the current cytometers can split cell populations (*cell sorting*) for further analyses [48, 49].

Intracellular Antigen Expression

Transcription factors and other intracellular molecules can be stained with fluorochrome-conjugated antibodies after fixation and permeabilization of the cells. Flow cytometry, in contrast to classical microscopy techniques, can provide accurate quantification and high-throughput analysis. Expression levels of a protein in >100,000 individual cells can be measured and visualized within a few minutes. However, this internal staining tends to have higher background, whereas optimal fixation and permeabilization methods vary (such as 0.01% formaldehyde, 1–4% PFA or acetone followed by 0.1–1% NP-40 or ice-cold methanol, etc.) [50].

Characterization of Antigen-Specific Responses

Antigen-specific responses are measured by antigen cell stimulation, and with the following characterization of cellular processes such as proliferation, activation, plasticity, or antigen recognition through major histocompatibility complex (MHC) multimers. In vaccination studies, where the identification of multiple cytokines and surface marker are needed to study in parallel, the most used technique is the *intracellular cytokine staining* or cytokine flow cytometry, since this is a combined technique useful for recognizing the antigen-specific T-cell stimulation in complex cellular samples using more than five fluorescent markers [51].

Another method to measure the antigen-specific responses is using the labeled MHC multimers. Usually, MHC multimers are in a monomeric conformation (MHC-I or MHC-II); these are grouped in multimeric arrangements using a biotinylated fluorescent streptavidin backbone. Then the MHC multimers are *loaded* with the antigen leading to the antigen recognition by the T cells, which indicates the amount of response to a particular application; this method is commonly employed in immunogen studies or in cancer diagnosis and prognosis [52].

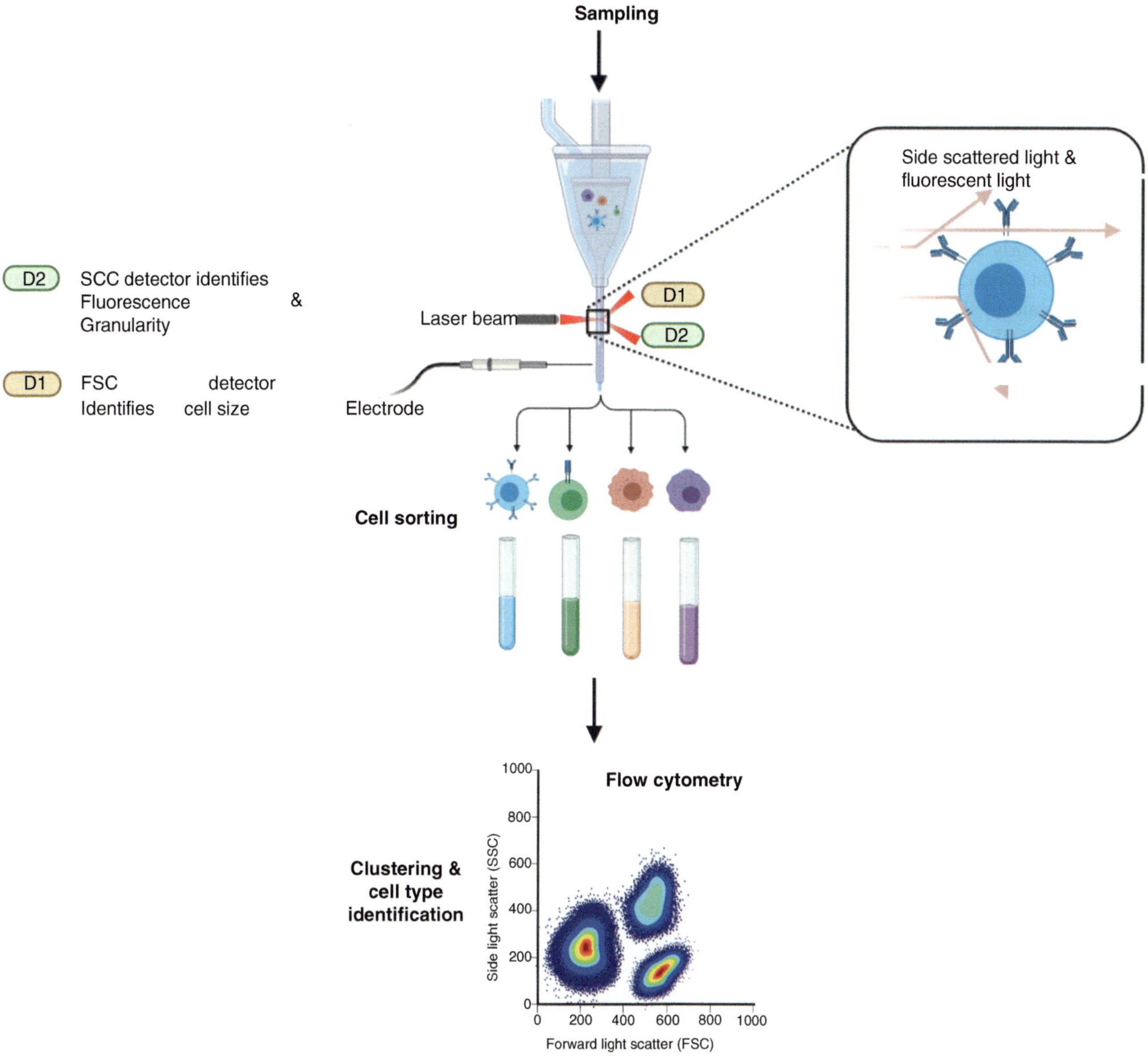

Fig. 2.3 Flow cytometry working principle. Visible light scatter is measured in two different directions, the forward direction (Forward Scatter or FSC) which can indicate the relative size of the cell and at 90° (Side Scatter or SSC) which indicates the internal complexity or granularity of the cell [41]. Light scatter is independent of fluorescence. Samples are prepared for fluorescence measurement through transfection and expression of fluorescent proteins (e.g., Green Fluorescent Protein, GFP), staining with fluorescent dyes (e.g., Propidium Iodide, DNA) or staining with fluorescently conjugated antibodies (e.g., CD3 FITC) [40]. The first flow cytometer was developed to detect the size of the cells; nowadays, these devices are powerful tools capable of detecting up to 14 parameters simultaneously related such as size, shape, complexity, and, of course, any component or cellular function that can be marked with a fluorochrome [42, 43], giving detailed information of cell population in a short period

Cell Cycle Analysis

In cell proliferation, cells may be stained with fluorescent dyes such as succinimidyl carboxyfluorescein ester (CFSE). This dye binds covalently to both intracellular and cell surface proteins and is incorporated equally to the next cell generation (daughter cells) during cell proliferation, with each division CFSE fluorescence decreases twofold leading to identifying up to seven to eight cell divisions accurately [53–55]. Another useful marker to characterize cell proliferation is the thymidine analogs BrdU (5-bromo-21-deoxyuridine) or EdU (ethynyl deoxyuridine), which is similar to the 3H thymidine proliferation assay. The BrdU is a thymidine analog that is incorporated to the newly synthesized DNA and in the subsequent daughter cells; the detection is mediated by the anti-BrdU antibody [53, 56]. Moreover, the use of BrdU is a compatible method is that can be be used simultaneously with other fluorescent markers, and also, both propidium halide (PI) and Hoechst 33342 may be used to quantify DNA content in each phase of the cell cycle, since the cells that are in S phase are going to be brighter than cells in G1 phase, and cells in G2 phase will be just about twofold brighter than the cells in G1[57].

Apoptosis

During the early stage of apoptosis, phosphatidylserine (PS) residues that unremarkably exist only within the plasma membrane relocate to the outer surface making such molecules available for PS binding proteins such as annexin V [42]. Since apoptosis is a cascade of events that occurs at different stages, its detection in flow cytometry utilizes multiple targets that bring a complete overview of events related to this process [57]. For instance, additionally to the annexin V labeling, the endonuclease digestion of polymer is identified by TUNEL (TdT dUTP nick end labeling) assay; the caspase activation is targeted by specific fluorescent-coupled antibodies; mitochondrial uncoupling is targeted with dyes that depend on the mitochondrial membrane potential (JC-1, Rhodamine 123 or Mitotrackers), and chromatin granule condensation within the nucleus detected with Hoescht 33,342 [58, 59].

As seen above flow cytometry is a powerful tool that may be employed to characterize a wide range of cellular and biochemical processes simultaneously. Hence, in the following section we enlist the most outstanding clinical applications in which flow cytometry has demonstrated to be a success:

- Leukemia and lymphomas diagnosis, since leukocyte surface antigens may be identified in neoplastic cells [60–62]
- Detection of minimal residual disease (MDR) in leukemia via CD13, CD19, and CD34 identification in blood and bone marrow [63, 64]
- Hematopoietic progenitor cells count in bone marrow transplantation by CD34 identification [65]
- Histocompatibility cross-matching via IgG measurement after incubating donor's lymphocytes with the recipient's serum [66, 67]
- Posttransplantation monitoring via CD3+ T cell counting [68]
- Immunodeficiencies diagnosis via CD4- and CD8-positive cells counting within the blood and other liquid biopsies [69]
- HIV infection diagnosis via CD4-positive lymphocytes count performed in blood samples [69, 70]
- Detection of fetal red blood cells and maternal F cells detection and quantitation the feto-maternal hemorrhage [71]
- Contaminating leukocytes measurement in blood for transfusion [72]
- DNA content to detect malignancies [73–76]
- Auto−/allo-immune diseases diagnosis via IgG and immune serum globulin detection using antiplatelet antibodies and IgG for antineutrophil antibodies [77–79]

As mentioned above, flow cytometry impacts positively in daily clinical practice, since this tool is widely used in both hematology and immunology; this leads to suggest that flow cytometry is a powerful tool for diagnosing, classifying, and determining the prognosis of assorted diseases. However, to improve the methods and expand the applications of flow cytometry, it is vital to strengthen the collaboration between physicians and biomedical researchers.

Proteomics

The proteome refers to the set of proteins present in a cell or organism at any given time. The DNA contains the needed information for the creation of proteins [80]. However, the relationship between the genome and the proteome is complex, since one single gene may encode for more than a single protein by means of alternative splicing [80]. Furthermore, posttranslational modifications and protein cleavage or modifications give rise to the origin of several protein isoforms for each single gene [80]. One astonishing example is the *DSCAM1* gene in *Drosophila melanogaster* (fruit fly), which has over 30,000 identified isoforms [12]. The human protocadherins, encoded in the *Protocadherin* locus, are essential in neural development, and its locus is thought to have a similar number of isoforms as the *DSCAM1* gene [12]. Thus, studying the proteome provides additional information that otherwise we wouldn't notice only studying the genome or the transcriptome [80].

As a general rule, the genome is the same in every cell of an individual [2]. The proteome, however, varies between cell types and conditions, allowing the classification of cells according to their protein expression patterns and the identification of changes in certain contexts, such as diseases [2]. As mentioned before, the phenotype can be better explained by the proteome than by the genome or the transcriptome due to differences in expression levels and to protein modifications [81]. Furthermore, the proteome is more stable and more easily assessed than the metabolome [81]. Thus, proteomics studies are a powerful tool preferentially used in studying disease, development, aging, among others [81].

Studying the Proteome

Usually, proteomics studies involve mass spectrometry analysis, which requires previous protein purification from tissue or cell samples [81, 82]. Two-dimensional gel electrophoresis, liquid chromatography, and capillary electrophoresis are the most commonly used techniques for this required step [81]. Mass spectrometry subjects proteins to a magnetic and an electric field and calculates a ratio known as *m/z*, which refers to the mass-charge ratio [83]. Finally, the masses of the molecules are plotted as peaks in the mass spectrum. The output of the mass spectrometry is queried against protein databases for the identification and quantification of each specific peptide [83]. Mass spectrometry has the enormous advantage that it is also able to detect modifications in the proteins because these change the behavior of the molecules when exposed to magnetic and electric fields [12].

Protein characterization by mass spectrometry can be coupled with protein-protein interactions and protein structure analysis [84]. Protein-protein interaction analysis reveals proteins that associate with others, and it provides information about its functions, since proteins that interact usually are implicated in similar or interrelated pathways. In accordance to dynamic and context-dependent protein expression, protein-protein interactions adjust to environmental conditions [3]. Therefore, understanding interaction networks — the interactome — in contexts of interest, including diseases, offers a higher level of pathway understanding and effective therapies discovery [85]. Some of the methodologies used for protein-protein interaction assays are yeast two-hybrid (Y2H), bimolecular fluorescence complementation (BiFC),, and fluorescence resonance energy transfer (FRET) [85].

On the other hand, understanding protein structure provides valuable information about protein functions [4]. Proteins acquire their functionality by folding in three-dimensional structures, allowing the formation of channels, binding sites, active sites, among others [86]. However, most peptides are able to fold into millions of different structures, and when misfolding occurs, multiple diseases may arise, by, for example, the formation of toxic aggregates [86]. Cells have a stress response named the unfolded protein response (UPR), which prevents the production of misfolded proteins. However, this response may be altered in diseases, aging, or some viral infections [86]. For this reason, the analysis of misfolded proteins in diseases such as Alzheimer's disease has revealed important molecular hallmarks [86]. Tridimensional structure identification can be done by X-ray diffraction, which is the preferred methodology, combined with others such as nuclear magnetic resonance (NMR) spectroscopy [84]. For a complete review of proteomic, please refer to Chap. 6 in this book.

Concluding Remarks

Routine clinical molecular tools demand rigorous, simple, and most importantly reproducibility procedures that help to characterize accurately biomarkers not only for disease diagnosis but also monitoring the patient's response to clinical interventions. Such characterization may be efficient and performed with the minimal invasion for the patient. However, reaching such a level of success will be only possible if physicians and biomedical researchers collaborate. Hence, with this chapter we aim to encourage such collaboration, since here we bring a brief description of both the most molecular tools employed in clinical screening and the working principle of each one, leading to identify the potential use of such molecular tools in unexplored medicine fields.

Acknowledgments This chapter is part of a registered project at Instituto Nacional de Geriatría DI-PI-002-2020.

Conflict of Interest Authors declare nonconflict of interest.

References

1. Landegren U, Schallmeiner E, Nilsson M, et al. Molecular tools for a molecular medicine: analyzing genes, transcripts and proteins using padlock and proximity probes. J Mol Recognit. 2004;17:194–7.
2. Strachan T, Read A. Human molecular genetics. Garland Science; 2018.
3. Watson JD, Baker TA, Bell SP. Molecular biology of the gene. Benjamin-Cummings Publishing Company; 2014.
4. ND L, Lehninger AL, Nelson DL, Cox MM, University Michael M Cox. Lehninger principles of biochemistry. Macmillan; 2005.
5. Jobling M, Hollox E, Hurles M, Kivisild T, Tyler-Smith C. Human evolutionary genetics. Garland Science: 2nd; 2013.
6. Edwards MC, Gibbs RA. Multiplex PCR: advantages, development, and applications. PCR Methods Appl. 1994;3:S65–75.
7. VanGuilder HD, Vrana KE, Freeman WM. Twenty-five years of quantitative PCR for gene expression analysis. BioTechniques. 2008;44:619–26.
8. Ohan NW, Heikkila JJ. Reverse transcription-polymerase chain reaction: an overview of the technique and its applications. Biotechnol Adv. 1993;11:13–29.
9. Pohl G, Shih I-M. Principle and applications of digital PCR. Expert Rev Mol Diagn. 2004;4:41–7.
10. Becherer L, Borst N, Bakheit M, Frischmann S, Zengerle R, von Stetten F. Loop-mediated isothermal amplification (LAMP) – review and classification of methods for sequence-specific detection. Anal Methods. 2020;12:717–46.
11. Wong Y-P, Othman S, Lau Y-L, Radu S, Chee H-Y. Loop-mediated isothermal amplification (LAMP): a versatile technique for detection of micro-organisms. J Appl Microbiol. 2018;124:626–43.
12. Alberts B, Johnson AD, Lewis J, Morgan D, Raff M, Roberts K, Walter P. Molecular biology of the cell: sixth international. Student ed. W.W. Norton & Company; 2014.
13. Walker JM, Rapley R. Medical BioMethods Handbook. Springer Science & Business Media; 2007.
14. Šestáková Š, Šálek C, Remešová H. DNA methylation validation methods: a coherent review with practical comparison. Biol Proced Online. 2019;21:19.

15. Martin CL, Warburton D. Detection of chromosomal aberrations in clinical practice: from karyotype to genome sequence. Annu Rev Genomics Hum Genet. 2015;16:309–26.
16. Swiger RR, Tucker JD. Fluorescence in situ hybridization: a brief review. Environ Mol Mutagen. 1996;27:245–54.
17. Ratan ZA, Zaman SB, Mehta V, Haidere MF, Runa NJ, Akter N. Application of fluorescence in situ hybridization (FISH) technique for the detection of genetic aberration in medical science. Cureus. 2017;9:e1325.
18. Bishop R. Applications of fluorescence in situ hybridization (FISH) in detecting genetic aberrations of medical significance. Bioscience Horizons. 2010;3:85–95.
19. Borrell A, Grande M, Pauta M, Rodriguez-Revenga L, Figueras F. Chromosomal microarray analysis in Fetuses with growth restriction and Normal karyotype: a systematic review and meta-analysis. Fetal Diagn Ther. 2018;44:1–9.
20. Reddy UM, Page GP, Saade GR. The role of DNA microarrays in the evaluation of fetal death. Prenat Diagn. 2012;32:371–5.
21. Pinkel D, Albertson DG. Comparative genomic hybridization. Annu Rev Genomics Hum Genet. 2005;6:331–54.
22. Pinto IP, da Cruz A, Costa E, Pereira S, Minasi L, da Cruz A. Cytogenetics - past, present and further perspectives. 2018.
23. Mann SP, Treit PV, Geyer PE, Omenn GS, Mann M. Ethical principles, constraints, and opportunities in clinical proteomics. Mol Cell Proteomics. 2021;20:100046.
24. Manole E, Bastian AE, Popescu ID, Constantin C, Mihai S, Gaina GF, Codrici E, Neagu MT. Immunoassay techniques highlighting biomarkers in Immunogenetic diseases. Immunogenetics. 2019; https://doi.org/10.5772/intechopen.75951.
25. Engvall E, Perlmann P. Enzyme-linked immunosorbent assay (ELISA). Quantitative Assay of Immunoglobulin G Immunochemistry. 1971;8:871–4.
26. Van Weemen BK, Schuurs AHWM. Immunoassay using antigen-enzyme conjugates. FEBS Lett. 1971;15:232–6.
27. Alhajj M, Farhana A. Enzyme Linked Immunosorbent Assay. StatPearls; 2021.
28. Slieman TA, Leheste J. Introduction to immunological techniques in the clinical laboratory. Methods Microbiol. 2020:1–16.
29. Yang F, Patton K, Kasprzyk T, Long B, Gupta S, Zoog SJ, Tracy K, Vettermann C. Validation of an IFN-gamma ELISpot assay to measure cellular immune responses against viral antigens in non-human primates. Gene Ther. 2021; https://doi.org/10.1038/s41434-020-00214-w.
30. Rasooly A, Bruck HA, Kostov Y. An ELISA lab-on-a-Chip (ELISA-LOC). Methods Mol Biol. 2013;949:451–71.
31. Heda GD, Shrestha L, Thapa S, Ghimire S, Raut D. Optimization of western blotting for the detection of proteins of different molecular weight. BioTechniques. 2020;68:318–24.
32. Lück C, Haitjema C, Heger C. Simple Western: bringing the Western blot into the twenty-first century. Methods Mol Biol. 2021;2261:481–8.
33. Anderson LV, Davison K. Multiplex Western blotting system for the analysis of muscular dystrophy proteins. Am J Pathol. 1999;154:1017–22.
34. Tani Y, Kaneta T. Indirect capillary electrophoresis immunoassay of membrane protein in extracellular vesicles. J Chromatogr A. 1629;2020:461513.
35. Arvin NE, Dawod M, Lamb DT, Anderson JP, Furtaw MD, Kennedy RT. Fast immunoassay for microfluidic Western blotting by direct deposition of reagents onto capture membrane. Anal Methods. 2020;12:1606–16.
36. Liu L, Chen D, Wang J, Chen J. Advances of single-cell protein analysis. Cell. 2020;9:1271.
37. Sari TK, Gianopulos KA, Nicola AV. Conformational change in herpes simplex virus entry glycoproteins detected by dot blot. Methods Mol Biol. 2060;2020:319–26.
38. Ferreira MST, Fernandes FD, Alves MEM, Bräunig P, Sangioni LA, Vogel FSF. Performance of the dot-blot test method for detecting antibodies to Sarcocystis spp. in cattle. Pesquisa Veterinária Brasileira. 2020;40:385–8.
39. Jadwin JA, Mayer BJ, Machida K. Detection and quantification of protein-protein interactions by far-western blotting. Methods Mol Biol. 2015;1312:379–98.
40. Adan A, Alizada G, Kiraz Y, Baran Y, Nalbant A. Flow cytometry: basic principles and applications. Crit Rev Biotechnol. 2017;37:163–76.
41. Shapiro HM. Lasers for flow cytometry. Current protocols in cytometry. 2004. https://doi.org/10.1002/0471142956.cy0109s27.
42. Wlodkowic D, Telford W, Skommer J, Darzynkiewicz Z. Apoptosis and beyond: cytometry in studies of programmed cell death. Methods Cell Biol. 2011;103:55–98.
43. Wilkerson MJ. Principles and applications of flow cytometry and cell sorting in companion animal medicine. Vet Clin North Am Small Anim Pract. 2012;42:53–71.
44. Biasi SD, De Biasi S, Gibellini L, et al. High speed flow cytometry allows the detection of circulating endothelial cells in hemangioblastoma patients. Methods. 2018;134-135:3–10.
45. Leipold MD, Newell EW, Maecker HT. Multiparameter phenotyping of human PBMCs using mass cytometry. Methods Mol Biol. 2015;1343:81–95.
46. Mei HE, Leipold MD, Maecker HT. Platinum-conjugated antibodies for application in mass cytometry. Cytometry A. 2016;89:292–300.
47. Böyum A. Isolation of mononuclear cells and granulocytes from human blood. Isolation of monuclear cells by one centrifugation, and of granulocytes by combining centrifugation and sedimentation at 1 g. Scand J Clin Lab Invest Suppl. 1968;97:77–89.
48. Barteneva NS, Fasler-Kan E, Vorobjev IA. Imaging flow cytometry. J Histochem Cytochem. 2012;60:723–33.
49. Doan M, Vorobjev I, Rees P, Filby A, Wolkenhauer O, Goldfeld AE, Lieberman J, Barteneva N, Carpenter AE, Hennig H. Diagnostic potential of imaging flow cytometry. Trends Biotechnol. 2018;36:649–52.
50. Chantzoura E, Kaji K. Flow cytometry. Basic Sci Methods Clin Res. 2017:173–89.
51. Gratama JW, Kern F, Manca F, Roederer M. Measuring antigen-specific immune responses, 2008 update. Cytometry A. 2008;73:971–4.
52. Akinfieva O, Nabiev I, Sukhanova A. New directions in quantum dot-based cytometry detection of cancer serum markers and tumor cells. Crit Rev Oncol Hematol. 2013;86:1–14.
53. Han Y, Wang S, Zhang Z, et al. In vivo imaging of protein-protein and RNA-protein interactions using novel far-red fluorescence complementation systems. Nucleic Acids Res. 2014;42:e103.
54. Ansari MJ, Strom TB. Novel diagnostics in transplantation. In: Chronic kidney disease, dialysis, and transplantation. Elsevier; 2010. p. 609–19.
55. Bajgelman MC. Principles and applications of flow cytometry. In: Data processing handbook for complex biological data sources. Elsevier; 2019. p. 119–24.
56. Lai C, Stepniak D, Sias L, Funatake C. A sensitive flow cytometric method for multi-parametric analysis of microRNA, messenger RNA and protein in single cells. Methods. 2018;134-135:136–48.
57. Wlodkowic D, Skommer J, Darzynkiewicz Z. Rapid quantification of cell viability and apoptosis in B-cell lymphoma cultures using cyanine SYTO probes. Methods Mol Biol. 2011;740:81–9.
58. Kwon K, Jang J, Choi W, Ramachandran S, Cho C, Cagle P. Expression of apoptotic nuclei by ultrastructural terminal deoxyribonucleotidyl transferase mediated dUTP nick end labeling and detection of FasL, caspases and PARP protein molecules in cadmium induced acute alveolar cell injury. Toxicology. 2006;218:197–204.

59. Wlodkowic D, Skommer J, Akagi J, Fujimura Y, Takeda K. Multiparameter analysis of apoptosis using lab-on-a-chip flow cytometry. Curr Protoc Cytom. 2013;66:9.42.1–9.42.15.
60. Orfao A, Matarraz S, Pérez-Andrés M, Almeida J, Teodosio C, Berkowska MA, van Dongen JJM, EuroFlow. Immunophenotypic dissection of normal hematopoiesis. J Immunol Methods. 2019;475:112684.
61. DiGiuseppe JA, Wood BL. Applications of flow Cytometric Immunophenotyping in the diagnosis and Posttreatment monitoring of B and T lymphoblastic Leukemia/lymphoma. Cytometry B Clin Cytom. 2019;96:256–65.
62. Debord C, Wuillème S, Eveillard M, Theisen O, Godon C, Le Bris Y, Béné MC. Flow cytometry in the diagnosis of mature B-cell lymphoproliferative disorders. Int J Lab Hematol. 2020;42(Suppl 1):113–20.
63. van Lochem EG, van der Velden VHJ, Wind HK, te Marvelde JG, Westerdaal NAC, van Dongen JJM. Immunophenotypic differentiation patterns of normal hematopoiesis in human bone marrow: reference patterns for age-related changes and disease-induced shifts. Cytometry B Clin Cytom. 2004;60:1–13.
64. Han X, Jorgensen JL, Brahmandam A, Schlette E, Huh YO, Shi Y, Awagu S, Chen W. Immunophenotypic study of basophils by multiparameter flow cytometry. Arch Pathol Lab Med. 2008;132:813–9.
65. Yu H, Yoo J, Hwang JS, et al. Enumeration of CD34-positive stem cells using the ADAMII image-based fluorescence cell counter. Ann Lab Med. 2019;39:388–95.
66. Downing J. The lymphocyte crossmatch by flow cytometry for kidney transplantation. Methods Mol Biol. 2012;882:379–90.
67. McCarthy JF, Cook DJ, Massad MG, et al. Vascular rejection post heart transplantation is associated with positive flow cytometric cross-matching1. Eur J Cardiothorac Surg. 1998;14:197–200.
68. Zhuang Q, Peng B, Wei W, Gong H, Yu M, Yang M, Liu L, Ming Y. The detailed distribution of T cell subpopulations in immune-stable renal allograft recipients: a single center study. PeerJ. 2019;7:e6417.
69. Petkov S, Bekele Y, Lakshmikanth T, Hejdeman B, Zazzi M, Brodin P, Chiodi F. High CD45 expression of CD8+ and CD4+ T cells correlates with the size of HIV-1 reservoir in blood. Sci Rep. 2020;10:20425.
70. Baron U, Werner J, Schildknecht K, et al. Epigenetic immune cell counting in human blood samples for immunodiagnostics. Sci Transl Med. 2018; https://doi.org/10.1126/scitranslmed.aan3508.
71. Farias MG, Dal Bó S, de Castro SM, da Silva AR, Bonazzoni J, Scotti L, Costa SHAM. Flow cytometry in detection of Fetal red blood cells and maternal F cells to identify Fetomaternal Hemorrhage. Fetal Pediatr Pathol. 2016;35:385–91.
72. Gorczyca W, Sun Z-Y, Cronin W, Li X, Mau S, Tugulea S. Immunophenotypic pattern of myeloid populations by flow cytometry analysis. Methods Cell Biol. 2011;103:221–66.
73. Torres-Rendon A, Stewart R, Craig GT, Wells M, Speight PM. DNA ploidy analysis by image cytometry helps to identify oral epithelial dysplasias with a high risk of malignant progression. Oral Oncol. 2009;45:468–73.
74. Tachibana M. Clinical application of flow cytometry to urological malignancies. Keio J Med. 1996;45:73–80.
75. Swerts K, Van Roy N, Benoit Y, Laureys G, Philippé J. DRAQ5: improved flow cytometric DNA content analysis and minimal residual disease detection in childhood malignancies. Clin Chim Acta. 2007;379:154–7.
76. Gerashchenko BI, Huna A, Erenpreisa J. Characterization of breast cancer DNA content profiles as a prognostic tool. Exp Oncol. 2014;36:219–25.
77. Fromm PD, Silveira PA, Hsu JL, et al. Distinguishing human peripheral blood CD16 myeloid cells based on phenotypic characteristics. J Leukoc Biol. 2020;107:323–39.
78. Biran N, Ely S, Chari A. Controversies in the assessment of minimal residual disease in multiple myeloma: clinical significance of minimal residual disease negativity using highly sensitive techniques. Curr Hematol Malig Rep. 2014;9:368–78.
79. Chen B, Vousden KA, Naiman B, et al. Humanised effector-null FcγRIIA antibody inhibits immune complex-mediated proinflammatory responses. Ann Rheum Dis. 2019;78:228–37.
80. Ahmad Y, Lamond AI. A perspective on proteomics in cell biology. Trends Cell Biol. 2014;24:257–64.
81. Analytical tools to assess aging in humans: the rise of geri-omics. Trends Analyt Chem. 2016;80:204–12.
82. Altelaar AFM, Munoz J, Heck AJR. Next-generation proteomics: towards an integrative view of proteome dynamics. Nat Rev Genet. 2013;14:35–48.
83. Timp W, Timp G. Beyond mass spectrometry, the next step in proteomics. Sci Adv. 2020;6:eaax8978.
84. Aslam B, Basit M, Nisar MA, Khurshid M, Rasool MH. Proteomics: technologies and their applications. J Chromatogr Sci. 2017;55:182–96.
85. Snider J, Kotlyar M, Saraon P, Yao Z, Jurisica I, Stagljar I. Fundamentals of protein interaction network mapping. Mol Syst Biol. 2015;11:848.
86. Hartl FU. Protein Misfolding diseases. Annu Rev Biochem. 2017;86:21–6.

Genomic Tools in Clinical Epidemiology

3

Alfredo García-Venzor, Esteban Cruz-Arenas, Victor Takeshi Landero-Yoshioka, and Edna Ayerim Mandujano-Tinoco

Abbreviations

AACR	American Association for Cancer Research
ACE2	Angiotensin-converting enzyme 2
ACPAs	Anti-citrullinated protein antibodies
AFA	Anti-filaggrin antibodies
AKA	Anti-keratin antibodies
AKT1	AKT serine/threonine kinase 1
ANXA1	Annexin A1
APC	Adenomatous polyposis coli tumor suppressor
BCR-VDJ	B-cell receptor variable (V), joining (J) or diversity (D) gene segments
BET	Bromodomain and extraterminal
BRCA1	Breast cancer type 1 susceptibility protein
BRCA2	Breast cancer type 2 susceptibility protein
BTNL2	Butyrophilin-like 2
cagA	Cytotoxin-associated gene A
cagY	Cytotoxin-associated gene Y
CD4	Cluster of differentiation 4
CDKAL1	Cyclin-dependent kinase 5 regulatory-associated protein 1-like 1
DLEU1	Deleted in lymphocytic leukemia 1
DRB1	DRB1 beta chain
EGFR	Epidermal growth factor receptor
ERBB2	Erythroblastic oncogene B
FAPα	Fibroblast Activation Protein Alpha
FBXW7	F-box and WD repeat domain containing 7
FGFR3	Fibroblast growth factor receptor 3
flgE	Flagellar hook protein E
GPRC5A	G protein-coupled receptor class C group 5 member A
gyrA	DNA gyrase subunit A
HERC2	Proteins ubiquitin E3 ligases of the HECT family
HIP-1	Huntingtin-interacting protein 1
HLA	Human leukocyte antigen
HLA-DRAB1	Major histocompatibility complex, class II, DR beta 1
HRAS	Hras proto-oncogene
IFN	Interferons
IgG	Immunoglobulin G
IgM	Immunoglobulin M
IL2RA	Interleukin 2 receptor subunit alpha
IL2RB	Interleukin 2 receptor subunit beta
infB	Translation initiation factor IF-2
infB	Translation initiation factor IF-2
KDM6A	Lysine demethylase 6A
KRT17	Keratin 17
MAGE	Melanoma-associated antigen genes
MHC	Major histocompatibility complex
MMP1	Matrix metalloproteinase-1
MMP3	Matrix metalloproteinase-3
NKκB	Nuclear factor kappa B
NKT	Natural killer T
NOTCH4	Neurogenic locus notch homolog protein 4
NSAIDs	Nonsteroidal anti-inflammatories
OCA2	Oculocutaneous albinism II
PIK3CA	Phosphatidylinositol-4,5-bisphosphate 3-kinase catalytic
PLB1	Phospholipase B1
PSMB10	Proteasome 20S subunit beta 10

A. García-Venzor
Department of Life Sciences, Ben-Gurion University of the Negev, Beer Sheva, Israel

E. Cruz-Arenas
Hospital Epidemiological Surveillance Unit, Instituto Nacional de Rehabilitación "Luis Guillermo Ibarra Ibarra", Mexico City, Mexico

V. T. Landero-Yoshioka
Experimental Surgery Department. Centro Médico Nacional "20 de Noviembre" ISSSTE, Mexico City, Mexico

E. A. Mandujano-Tinoco (✉)
Laboratory of Connective Tissue, Centro Nacional de Investigación y Atención de Quemados, Instituto Nacional de Rehabilitación "Luis Guillermo Ibarra Ibarra", Mexico City, Mexico
e-mail: emandujano@inr.gob.mx

J. C. Gomez-Verjan, N. A. Rivero-Segura (eds.), *Principles of Genetics and Molecular Epidemiology*, https://doi.org/10.1007/978-3-030-89601-0_3

rpl22	Ribosomal protein L22
rpoB	B-subunit of bacterial RNA polymerase
RXRA	Retinoid X receptor alpha
SLC2A1	Solute carrier family 2 member 1
STAT3	Signal transducer and activator of transcription 3
TCR	T-cell antigen receptor
TERT	Telomerase reverse-transcriptase
THY1	Thy-1 cell surface antigen
TMPRSS2	Transmembrane serine 2 protease
TNF	Tumor necrosis factor
TNFi	TNF inhibitor
TP53	Tumor protein P53
TPST1	Tyrosylprotein sulfotransferase 1
ureC	Urease subunit alpha
vacA	Vacuolating cytotoxin A
ZEB1	Zinc finger E-box-binding homeobox 1
ZNF292	Zinc finger protein 292

Introduction

Since the sequencing of the complete human genome was possible, biological and medical sciences have used new and revolutionary technologies to perform accurate and comprehensive analysis about the function of genes, their products, and their interactions. High-throughput (HT) "-omics" technologies enable to generate large-scale biological data and are used for different types of analysis including genomics (DNA level), transcriptomics (RNA level), proteomics (protein level), and other related fields such as epigenomics, metagenomics, metabolomics, interactomics, and microbiome.

Genome HT technologies have evolved from low-throughput Sanger sequencing to complex next-generation sequencing (NGS) and are used to sequence multiple DNA or RNA molecules in a more cost-effective way [1].

Although sequencing-library methods and chemistries vary in each selected platform (such as GS FLX by 454 Life Sciences/Roche diagnostics, Genome analyzer, HiSeq, MiSeq, NextSeq by Illumina, SOLID by Applied Biosystems, Ion Torrent by Life technologies) [2], the basic principle of NGS technologies relies on the in vitro amplification of the input DNA and in the detection of DNA sequences during the synthesis of the complementary DNA strand in a massive manner [2]. For library construction, the template DNA is fragmented into short and double-stranded molecules that will be ligated with synthetic DNA sequences (adapters) in both ends. This mixture of adapters and DNA fragments is known as library, which is then denatured and immobilized on a solid surface (e.g., bead, flow cell). Library fragments are in situ amplified (by Bridge PCR, emulsion PCR, or in situ polonies) to generate a DNA cluster array [3]. Each cluster consists of thousands of copies of the same DNA fragment. DNA clusters will now be massive parallel sequenced by using a DNA polymerase or DNA ligase and following a stepwise reaction series of three steps: (1) nucleotide addition, (2) detection of the incorporated nucleotide on each fragment, and (3) washing of the fluorescent labels [2]. The incorporation of each new nucleotide is recorded as a fluorescent or chemiluminescent optical signal by a charge-coupled device (CCD camera). The construction of each DNA fragment sequence is performed by using the sequential images of each nucleotide addition step. Each sequence fragment is then assembled into a larger sequence until the whole genome or every RNA transcript could be fully sequenced [4]. Finally, the generated data is bioinformatically analyzed and interpreted for each scientific interest. The general workflow of NGS protocols is schematized in Fig. 3.1.

Within the most commonly used NGS technologies are the whole-exome sequencing (WES) and the whole-genome sequencing (WGS), which provide information on variant frequencies in different populations and allow the identification of single-nucleotide variants (SNVs) and mutations of many genetic disorders [5]; RNA sequencing (RNA-seq), which is used to study differentially expressed genes in specific conditions in order to understand phenotypic variation [6]; chromatin immunoprecipitation sequencing (Chip-seq) and methylation sequencing (Methyl-seq), which are helpful in the identification of epigenetic marks to know how the genes are globally regulated [7]; and mass spectrometry (MS), which is the key technology used to quantize thousands of proteins and metabolites in a single sample, but it also serves to detect interactions between nucleic acids (DNA/RNA) and proteins [8]. Table 3.1 shows the basic principle of these NGS technologies and the advantages and disadvantages of their use in this new era of genomic knowledge.

Undoubtedly, clinical epidemiology is a particular area in which HT technologies are having such transformative applications. This science focuses on the understanding of the etiology, distribution, and genetic and nongenetic risk factors of infectious and chronic diseases thus helping in the public health management [9]. In this sense, epidemiological studies are incorporating genomic tools on infection control programs in the detection, treatment efficacy, and recognition of infection persistence, the identification of rare SNPs at the individual and at the population levels, the estimation of disease heritability, the study of environmental risk factors, and the identification of specific biomarkers for the diagnostics, progress, and risk prediction [9–11] (see Table 3.1). In this chapter, we will review how new genomic tools are assisting scientists and epidemiologists in improving public health practices. To this end, we will give specific examples about

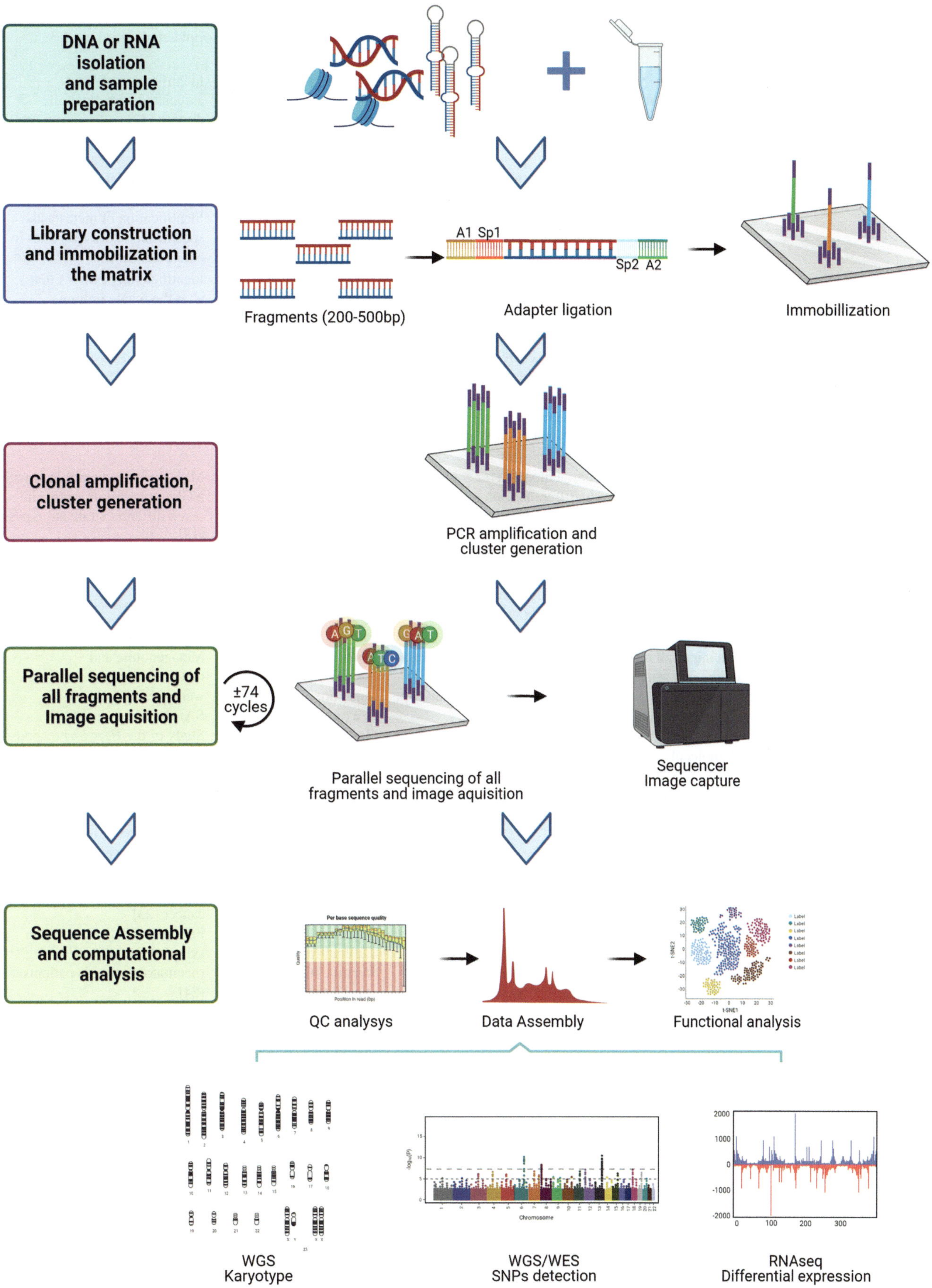

Fig. 3.1 General workflow of next-generation sequencing protocols. (Created with BioRender)

Table 3.1 Basic principle of NGS technologies and the advantages and disadvantages of their use in clinical epidemiology

Technology	Molecule	Test principle	Advantages	Disadvantages	Epidemiological uses
Genomic detection methods					
WGS	DNA	Sequencing of the entire genomic material	Identification of novel SNVs or SNPs Recognition of specific pathogenic or treatment resistant strains Propagation tracking and evolution of outbreaks in real time Identification of novel diagnostic test	Requires high specialized equipment and personnel High cost Specific bioinformatic analysis	Identification of SNPs associated with alcoholic disease [12] Identification of novel mutations associated with antibiotic resistance of *H. pylori* [13] Identification of metastasis-specific signatures of genes in colorectal cancer [14]
WES	DNA	Using of probes to enrich only specific genomic regions, usually exons	Increases the sequencing deepness and coverage of specific genomic Analysis of only coding regions allows the correlation of SNPs with gene function. Increases the number of sequenced samples Easier analysis due to less sequencing data	Loss of information about noncoding regions Does not allow the identification of novel SNPs or SNVs Requires high specialized equipment and personnel	Identification of SNVs that alter the innate immune response in inflammatory diseases [15]. Discovery of novel SNPs that represent risk for rheumatoid arthritis development [16] Analysis of novel mutations in several cancer samples [17]
MLST	DNA	Sequencing of internal fragments of multiple housekeeping genes	Pathogenic bacteria genotyping and differentiation between strains Characterization of pathogenic genes and resistant-associated alleles Sequencing of more locus with higher coverage and deep	Needs the isolation and culture of pathogenic bacteria Requires the use of specialized equipment, and sanger sequencer	MLST has been combined with NGS to classify *H. pylori* in seven different strain subtypes [18]
Transcriptomic detection methods					
RNA-seq	RNA	Sequencing of all the transcripts in a sample. The RNA from a sample must be retrotranscribed into cDNA molecules and amplified by PCR, before its sequencing in a NGS platform	Analysis of differential expressed genes between conditions Identification of novel transcripts and splicing isoforms Identification of noncoding RNA molecules RNA-seq metagenomics allows the identification of mRNAs expressed by pathogens and host	Some genomic features could not be identified, since not all the genome is expressed Require the use of specialized equipment and personnel	Metagenomic and metatranscriptomic analysis were used to genotyping SARS-CoV-2 virus [19] Study of the *H. pylori* response to antibiotics [20] Identification of novel biomarkers for diagnosis in cancer [21] Identification of new drug targets in cancer [22]
scRNA-seq	RNA	Sequencing of the entire transcriptome at the single cell level. The cells of a tissue or culture are isolated by flow cytometry or microfluidics technologies	Transcriptome analysis of the cellular heterogeneity of a population Identification of low represented cellular populations in a tissue Gives information about the components of a complex cell mixture, with proportions and high sequencing deepness	Difficulties with sample preparation Tissues must be processed in fresh Difficulties for the identification of low represented cellular populations High cost and requires specialized equipment and personnel trained	Treatment response of the multicellular ecosystem in lung cancer [23] Fibroblasts heterogeneity in synovial tissue from rheumatoid arthritis patients [24] Peripheral immune cell landscape of COVID-19 patients [25]

Table 3.1 (continued)

Technology	Molecule	Test principle	Advantages	Disadvantages	Epidemiological uses
Epigenomic detection methods					
Methyl-DNAseq	DNA	Genome-wide sequencing to identify methylated cytosines in the human genome. This technology relies in bisulfite reduction of methyl-5-cytosine to uracil, which during sequencing it is read as thymidine. The changes in methylated cytosines can be addressed by its comparison with a genome reference	Assessment of epigenetic anomalous modifications that can highly correlate with specific diseases	Methyl-seq is usually restricted to research laboratories, due to its technical difficulties and high cost	DNA methylation analysis of 18 genes in urine DNA allows the prognosis and risk stratification of patients with non-muscle invasive bladder cancer [26]
ChIP-Seq	DNA	It is used to study the interactions between a specific protein and the DNA. In ChIP-seq experiments, the target protein is immunoprecipitated using a specific antibody. After immunoprecipitation, the isolated genomic material is sequenced with NGS platforms	Allows the identification of all the DNA-binding regions for a specific protein in the genome Profiling of transcription factors and modified histones in the entire genome	ChIP-seq is technically difficult to perform and expensive. This characteristic restricts its use mainly for research laboratories	Epigenetic landscape of fibroblast-like synoviocytes from rheumatoid arthritis patients [27]

WGS whole-genome sequencing, *WES* whole-exome sequencing, *MLST* multi-locus sequencing typing, *CM* clinical metagenomics, *RNA-seq* RNA sequencing, *scRNA-seq* single-cell RNA sequencing

the application of HT technologies in the identification of novel and rare SNPs of different diseases, in guiding viral infection control during the COVID-19 pandemic situation and in the prevention, diagnosis, and treatment of *Helicobacter pylori* bacterial infection and other chronic pathologies such as cancer and rheumatoid arthritis.

NGS in the Study of Relevant SNVs and SNPs for Epidemiological Surveillance

Genomic variation is a naturally occurring phenomenon in all organismal populations, which is evidenced by differences in the DNA sequence among individuals of the same species. These differences in DNA sequences can be changes in a single nucleotide, such as single-nucleotide variations (SNVs) or single-nucleotide polymorphisms (SNPs) [28], but can also compromise several nucleotides (ranging from Kb to Mb), which could be (1) changes in tandem repeats, (2) insertions and deletions of several nucleotides (indels), (3) copy number variations (CNVs) which are additions or deletions that change the copy number of large sequences (even complete genes), and (4) chromosomal rearrangements such as inversions, translocations, or deletions. All these variations could have important outputs in gene and regulatory sequence [29]. In this section, we will focus on the use of NGS to study important SNPs and SNVs for epidemiological surveillance, for example, as prognostic/risk factors for the development of specific diseases in an individual or population and as markers to differentiate between individuals and populations.

SNPs are single-nucleotide changes in a given region of reference genomes and can include insertions, deletions, and substitutions that generate new alleles. They have a frequency of at least 1% of the population, being common SNPs those with minor allele frequencies higher than 5% [30]. The SNPs are the most common type of genetic variation, occurring at a rate of one SNP every DNA kb. The 1000 human genome project found that a typical human genome differs from the reference assembly in 4.1 to 5.0 million sites; importantly, more of the 99.9% of these changes correspond to SNPs and short indels. Also, a typical human genome contains between 2100 and 2500 structural variants (including tandem repeats variations, large deletions, CNVs, inversions, and transposon insertions) which overall affect approximately 20 million bases of sequence [31]. A big proportion of the newly described SNPs in this project has minor allele frequencies below 1% of the population, which categorized them as SNVs instead of SNPs. Since SNPs correspond to the major source of variation between human genomes, they are believed to be responsible of phenotypic differences. The diagnostic and prognostic potential of SNPs comes from studies where all the SNPs of an individual genome are used to make correlations with a specific phenotypic trait, and these studies are called genome-wide association studies (GWAS) and allow the statistical selection of a group of SNPs that together can explain a specific phenotype [32]. For example, the analysis of WGS sequencing datasets and GWAS study of a group of SNPs allowed the identification of 13 SNPs highly associated with alcoholic disease [12]. The authors suggest that the use of specific drugs like magnesium valproate which targets one of the polymorphic proteins could serve in alcoholism treatment. Another study performed a meta-analysis of two previous GWAS studies finding that in European populations, a big number of SNPs are associated with increased weight (3290) and body mass index (941) affecting 610 and 138 genes, respectively [33]. Also, the eye color, which is a very complex phenotypic trait, is the result of polymorphisms in 16 different genes, and SNPs in *HERC2* and *OCA2* are the most highly associated with this phenotype [34]. In these examples, the identification of the SNPs in each genome is usually made by WGS, WES, or both, which makes the implementation of GWAS analysis expensive as routine diagnostic and prognostic tests. This is mainly because of the high deep sequencing needed to identify SNPs (75x in WGS and 140 in WES). However, if the idea is not to identify novel SNPs, there are cheaper alternatives to NGS approaches like SNP microarrays. Microarrays have been used to identify SNPs associated with blood pressure homeostasis and hypertension focusing on only 148 alleles. In total, 874 SNPs have been associated with hypertension, with a minor allele frequency of 11%, being highly specific of each sampled population. Interestingly, half of the associated SNPs caused changes in the protein generating a high protein diversity [35]. Next, we will address some examples regarding the use of NGS and GWAS analysis to identify SNPs of epidemiological relevance.

Diabetes is one of the most important diseases of our days. Several efforts have been made to identify genomic variants associated with diabetes onset and progression. The regulatory subunit of the cyclin-dependent kinase 5 (*CDKAL1*) regulates insulin secretion and has been associated with diabetes development. A GWAS study identified a specific SNP (rs7756992) associated with *CDKAL1* variants in diabetes. rs7756992 is associated with high-fat diets and the risk of developing diabetes in female populations [36]. Also, NGS and GWAS analysis have been used for the prognosis of diabetic retinopathy. A study found 76 SNPs previously associated with diabetes that function as risk factors for diabetic retinopathy, from which 55 SNPs explained a 2.5-fold increase in developing retinopathy [37]. NGS dataset meta-analysis has been used to identify miRNA SNPs. miR-146a rs2910164 is associated with increased diabetes risk in the Latino population, while miR-27a rs895819 and miR-124 rs531564 SNPs are associated with a reduced risk in the Asian population and overall population, respectively [38]. This study is interesting because the relevance of miRNAs gene variations with diabetic risk is controversial, but since miRNAs can be analyzed in blood samples, its genotyping could be an alternative to easily assess diabetogenic risks.

Myocardial infarction is another public health problem that can be studied by SNPs analysis. Genomic variance and SNPs can predict the risk of an individual to suffer myocardial infarction and also serve as prognostic factors for disease progression. A study conducted using a big cohort of patients with early-onset myocardial infarction and controls found that 9 SNPs can help to significantly predict the risk to suffer myocardial infarction in the future [39]. Another GWAS study conducted in an Italian Mediterranean population with myocardial infarction and controls found 4 SNPs located in chromosome 9p21 that were associated with myocardial infarction risk. Interestingly, in this study, these four SNPs were strongly associated with patients with a familial history of myocardial infarction [40].

SNPs analysis and GWAS studies have also been used to determine genomic variations for different cancer risks and prognoses. For colorectal cancer, 30 SNPs can serve as risk assessment factors; interestingly, some of these SNPs are also negatively associated with progression measures such as disease-free survival and overall survival [41]. Similar studies have been conducted in breast cancer patients, where genetic factors account for 60% of the variation in breast density, and different SNPs have been correlated with high breast density and risk of developing breast cancer. Also, some of these SNPs can be used to predict the evolution of the disease and to take clinical decisions [42]. Finally, GWAS analysis has been used to generate novel statistical tools for the risk assessment of 11 different cancers. The genetic risk score (GRS) is a recent statistical tool to measure the cumulative effect of all risk-associated SNPs. GRS tool has been used in the risk prediction of bladder, breast, colorectal, glioma, lung, melanoma, ovarian, pancreatic, prostate, renal, and thyroid cancer with promising results [43]. These studies exemplify the possibility of using NGS and GWAS analysis in patients in order to have a global perspective of the genomic risk factors that can predispose for specific diseases in the future and take prophylactic measures to avoid the onset of these diseases.

The use of NGS and GWAS studies to address risk and prognostic factors has also been used in the epidemiological analysis of infectious diseases. In China, WGS has been used for genotyping *Mycobacterium tuberculosis* strains isolated from patients in urban and rural areas. Using WGS, the authors identified relevant SNPs that allowed a very precise tracking and clustering of the isolated strains thus inferring with great confidence the transmission dynamics from rural to urban areas and their posterior local spreading [44]. In *M. tuberculosis*, some SNPs are associated with multidrug-resistant phenotypes, and 86 SNPs are responsible for multidrug resistance in Chinese strains. The sequencing of these strains has allowed the identification of genomic variations for the design of novel antibiotic molecules that target these specific genotypes [45]. GWAS analysis has been used for the prognosis of tuberculosis infection, based on the presence of different SNPs in *IL-1β*, *TNF-α*, and *IL-6* which affect the host immunological response to the bacteria [46]. Also, the FDA has developed a novel epidemiological surveillance tool for *Escherichia coli*, which consists of a specific microarray (FDA-ECID) with designed probes targeting several genomic characteristics of *E. coli* including genome-wide SNPs information. This novel microarray was applied to strains from food, environmental and clinical samples, outperforming other analytical methods in strain identification, virulence assessment, and phylogenetic reconstruction of each strain. FDA-ECID allows a more precise and rapid analysis of pathogens with a higher quantity of information regarding risks for public health [47]. In viral infections, the study of the complete genome and identification of SNPs is also relevant for the prediction of virulence and reconstruction of infection dynamics by phylogenetic analysis. In the recent outbreak of the SARS-CoV-2 virus, it has been observed that the viral genome shows a tendency to mutate fast, showing different SNPs that correlate with pathogenic traits and can be responsible for the observed pharmacological treatment inefficiency [48].

The use of SNPs for epidemiological studies is a powerful tool for the assessment of relevant variables such as risk, disease progression, prognosis, and in the case of infectious disease the relation between the host and the pathogen, and the phylogenetic evolution of a specific strain. NGSs are important for the identification of novel SNPs in future genotypifications needed for the clinical management of several diseases.

Implication of NGS Technologies on the Control of SARS-CoV-2 Pandemic

After the outbreak of severe acute respiratory syndrome coronavirus (SARS-CoV) in 2002 and Middle East respiratory syndrome coronavirus (MERS-CoV) in 2012, on December of 2019, the World Health Organization (WHO) announced about cases of pneumonia of unknown etiology detected in China in the city of Wuhan, Hubei Province [49]. From December 31, 2019, to January 3, 2020, 44 patients were reported to the WHO, without an identified causative agent [49]. On January 7, a new type of coronavirus (CoV) was identified, and on February 11, the WHO named it as a coronavirus disease (COVID-19); at the same time that the International Virus Classification Commission recognized it as severe acute respiratory syndrome coronavirus 2 (SARS-CoV-2) [50–52]. Until March 12 of 2021, a total of 119,218,587 and 2,642,673 confirmed cases and deaths, respectively, had been reported worldwide due to COVID-19 [49].

As an urgent investigation, NGS was used to ensemble the whole-genome sequence of SARS-CoV-2 from bronchoalveolar lavage fluid and culture isolates of 9 Wuhan patients. The results showed that SARS-CoV-2 genome is 29.8 kb and shares 88% identity with SARS-like coronaviruses, 79% with SARS-CoV, and about 50% with MERS-CoV [19]. Subsequently, phylogenetic analyses on 160 genomes from human samples indicated the existence of 3 SARS-CoV-2 variants (A, B, and C) which differ by few amino acids [53]. Mutation analysis in genome sequences of 95 samples acquired at different times and locations found 116 mutations as indicatives of viral genetic diversity that might guide to greater severity and spread of the SARS-CoV-2 infection [54]. The sequencing of a large number of SARS-CoV-2 strains have generated knowledge about the pathogenesis, the vaccine development, and the antiviral drugs resistance [55].

Clinically, the diagnosis of COVID-19 infection is difficult because it can be confused with other viral infections such as influenza [56]. There are currently numerous commercial SARS-CoV-2 detection kits available that identify (a) specific viral gene regions through nucleic acid amplification technique [RT-PCR] and isothermal nucleic acid amplification, (b) antibodies produced by the immune system in response to the viral infection (serology/immunoglobulin M (IgM)/immunoglobulin G (IgG) tests), and (c) antigen testing by lateral flow assays.

The most common symptoms of COVID-19 are fever, dry cough, and tiredness. The less common symptoms include aches and pains, nasal congestion, headache, conjunctivitis, sore throat, diarrhea, loss of taste or smell, skin rash, and discoloration of fingers and toes. These symptoms are generally mild and begin gradually, while some infected people present only mild symptoms [57].

In the face of the imminent pandemic, several studies performed scRNA-seq to reveal the cellular and molecular pathogenic mechanisms of SARS-CoV-2 [58]. For example, a reconfiguration of the peripheral immune cells landscape was discovered including an interferon-stimulated gene signature, *HLA* class II downregulation, and a neutrophil population related to plasmablasts in those COVID-19 patients with acute respiratory failure [25]. The transcriptional profiles of immune cells showed an intensive expansion of cytotoxic effector T cells (CD8+ effector GNLY (granulysin), NKT CD160, and CD4+ effector-GNLY) in COVID-19 patients with moderate and severe response. Moreover,

B-cell activity is strongly activated in severe patients, and unique B-cell receptor variable (V), joining (J), or diversity (D) gene segment (*BCR-VDJ*) rearrangements are observed in these patients [59]. scRNA-seq has also revealed that the acute respiratory dysfunction that occurs in severe COVID-19 patients is associated with the mucin secretion from club cells which are stimulated by high levels of *IL-1b* and *TNF-α* production [60]. By combining scRNA-seq, mass cytometry, and scATAC-seq, it was found that the COVID-19 vulnerability of the aged people is associated with an upregulation of the genes associated with SARS-CoV-2 susceptibility in certain cell subtypes (effector T cells, NK cells, age-associated B cells, inflammatory monocytes, and age-associated dendritic cells) [61].

COVID-19 transmission can occur by close contact with an infected person through respiratory droplets, or through contact with contaminated objects and surfaces [62, 63]. Evidence has shown that expulsion of the virus is highest in the upper respiratory tract within 3 days prior to the onset of symptoms, indicating that a person can transmit the disease 1–3 days before it presents clinical manifestations [64]; the incubation period of the virus has been described between 5 and 6 days and could be extended up to 14 days [65].

scRNA-seq studies also served to validate that the SARS-CoV-2 entry factors (*ACE2* and *TMPRSS2*) are highly expressed in the nasal epithelial cells but are lowly expressed in the conjunctival epithelium, suggesting that viral infection occurs mainly via the respiratory mucosa but not via the ocular surface [66–68].

From the perspective of public health, it is essential to have plans supported by rigorous epidemiological information, which promote adequate decision-making. The above could be possible with the sufficient knowledge about SARS-CoV-2 infection. In this sense, NGS technologies facilitate the knowledge about the virus biology, the transmission mechanisms, the human response during infection, and the origins and diversification of the pathogen. These have allowed the fast implementation of diagnostic methods, the development of the vaccine, and the better clinical management of the more complicated patients (Fig. 3.2).

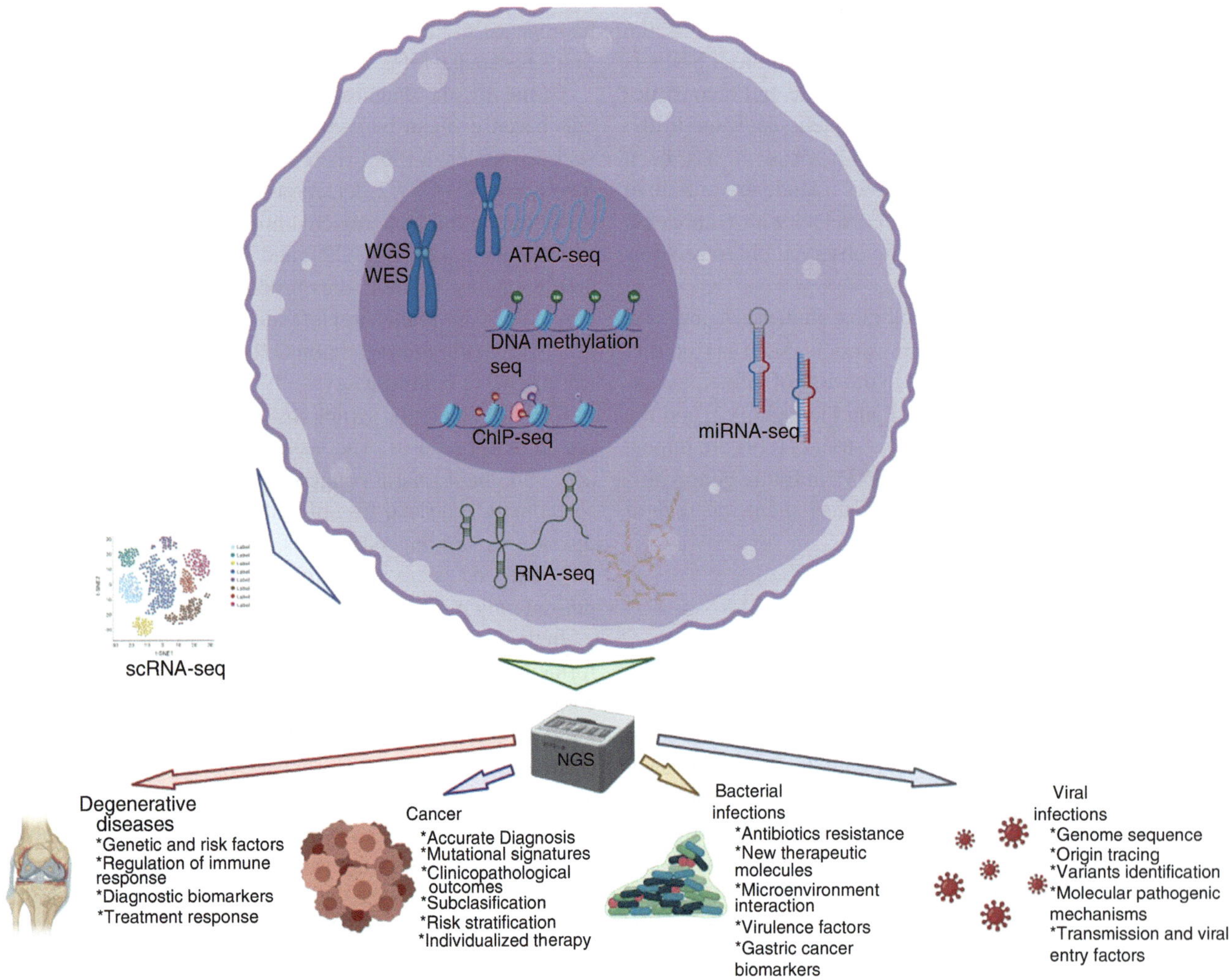

Fig. 3.2 Impact of NGS technologies in different fields of clinical epidemiology. (Created with BioRender)

The Influence of Genomic Tools on the Management of *Helicobacter pylori* Infection

Helicobacter pylori (*H. pylori*) is extremely prevalent in humans showing different clinical settings going from asymptomatic state to even develop gastric adenocarcinoma [69] or mucosa associated tissue lymphoma (MALT) [70]. It is estimated that half of the world population is infected with this bacterium, and it is the main risk factor for stomach cancer [71]. Even the less serious clinical presentations such as gastric and/or duodenal ulcers have a great impact on people economics. The expense of diagnostics test including endoscopy and biopsies and the relatively complicated and costly therapeutic scheme treatments lead patients to stop medication (bringing along antibiotic resistance, reinfection, relapse) or to have a chronic infection with associated complications in which case there is a high risk of developing any form of cancer.

The aforementioned circumstances have prompted all kinds of concerns for the better understanding of the bacterium biological behavior about the mechanisms involved in antibiotic resistance, the biofilm formation for survival, the genotypes associated with cancer development, and the interactions with other gastrointestinal (GI) flora. Thanks to the fact that the complete genome sequence of *H. pylori* was published since 1997; it has been easier to implement the use of genomic technologies for its study [72]. Recently, a group from Switzerland did a quite extensive review about the current studies in *H. pylori* that have used NGSs, highlighting that these tools will play a crucial role in the development of new diagnostic and treatment strategies to tackle antibiotic resistance which is the most alarming trait of the disease [73].

The first-line strategy traditionally used to treat *H. pylori* comprises triple therapy using any proton-pump inhibitor (PPI), amoxicillin, and clarithromycin. Second line of treatment includes use of quadruple scheme using bismuth subcitrate, PPI, metronidazole, and tetracycline [74]. Resistance to clarithromycin is quite a major problem when *H. pylori* eradication is intended. Until 2014, it was known that this resistance was mainly mediated by three point mutations (A2142G, A2143G, and A2142C) in the domain V of the 23S rRNA gene [75]. By using WGS, Binh et al. founded novel mutations in the genes *infB* and *rpl22* in two in vitro *H. pylori*-resistant strains. Mutations in the 23S rRNA gene in combination with *infB* or *rpl22* resulted in higher minimum inhibitory concentration (MIC) values, showing that all mutations have synergic effect on the resistant phenotype of the bacterial strains [13]. Another WGS analysis in 106 metronidazole- and levofloxacin-resistant strains isolated from dyspeptic patients demonstrated that different point mutations (I837V, A2412T/V, Q2079K, and K2068R) in the *rpoB* gene are associated with resistance to the classic but also to the alternative antibiotics rifaximin and garenoxacin [76].

Until now, the usual technique to assess *H. pylori* behavior toward any given antibiotic is to culture and make phenotypic drug susceptibility testing (DST). Until date, the standard method for evaluating *H. pylori* behavior toward a given antibiotic has been to culture the bacteria and do phenotypic drug susceptibility tests (DST). However, DST is not performed before the first-line treatment because it involves intrusive methods (endoscopy) to gather patient samples, which takes time and money. Some authors are proposing the use of WGS as an alternative, less invasive, low-cost, and accurate method to detect multiple antibiotic resistances at early stages of the diagnostic. A comparative study of DST and WGS results in 140 clinical samples showed that resistance to levofloxacin, metronidazole, rifampicin, tetracycline, and clarithromycin was phenotypic and genomic. It was concluded that DST highly correlates with certain SNPs identified in the 23S rRNA (A2146C, A2146G), *gyrA* (N87K, D91G, D91Y), and *rpoB* (H54ON, L525P) genes for clarithromycin, levofloxacin, and rifampicin. Tetracycline susceptibility is related with the absence of double or triple substitutions (A926T, A926G, A928C) in the 16s rRNA gene, while metronidazole resistance does not correlate with the SNPs identified by WGS [77]. Furthermore, using WGS data of common mutations in the 23S rRNA gene, the CRHP webtool has been validated to predict clarithromycin resistance in *H. pylori* strains. This bioinformatic tool can be used in the future to quickly monitor the clinical relevance of novel identified mutations [78].

Transcriptomic (RNA-seq) analysis is a genomic tool that has been used to study how medications influence the *H. pylori* metabolism. For example, through temporal dynamics profiling, it has been demonstrated that treatment with bismuth alters the expression level of 920 genes of *in vitro* strains, most of which are related to the bacterial energy metabolism at the level of tricarboxylic acid cycle (TCA cycle), oxidative phosphorylation, and the generation of oxidative stress [20]. These results explain why bismuth-containing quadruple therapy is more effective for the treatment of *H. pylori*-resistant infections [79] and serve as basis for the design of new therapeutic drugs that target the central carbon metabolism of the bacterium to improve the clinical management. Similar studies have used RNA-seq to analyze how different treatments or environmental conditions could modify the expression of multiple genes thus affecting important clinical traits of *H. pylori* including adherence to host cells, cell shape, and treatment susceptibility [80, 81].

In general, NGS technologies have helped us to better understand the *H. pylori* evolution, structure, metabolism, and its interaction with the surroundings. Thanks to multilocus sequence typing (MLST) results; we now know that this bacterium has coevolved with humans and that classifies in seven different strain subtypes [18, 82]. This knowledge is of medical relevance as each strain is associated with specific virulence factors thus varying the treatment responses [83]. In this regard, genomic tools could have several applications on the identification of genetic differences in *H. pylori* isolates from patients, thus guiding to a new era of individualized therapy.

Other groups are using genomics to study *H. pylori* behaviors during host adaptation. This adaptation requires fast changes at the genomic and transcriptomic level at the early stages of the infection establishment [84]. WGS analysis has showed that clusters of genes of the bacterium have evolved as an advantage to increase genetic recombination events; in this manner, *H. pylori* could promote immune evasion and different strategies to colonize in a dynamic environment [85]. In this regard, RNA-seq results suggested that the bacterium responds to the gastric acidity by changing the expression of several genes involved in protective, acid acclimation, and pathogenic factors (i.e., antioxidant, flagellar, and urease components) [86], and growing as biofilms thus forming microcolonies [87]. With a variation of RNA-seq (dRNA-seq), it was revealed that changes in the bacterial gene expression occur by an active riboregulation conducted by several small regulatory RNAs [88]. Furthermore, single-molecule real-time sequence analyses (SMRT) have revealed that DNA methylation is a key epigenetic factor that contributes in the gene regulation of virulence factors including flagellar components (*flgE*), secretion components (*cagY*), urease biosynthesis components (*ureC*), and the vacuolating cytotoxin A (*vacA*) [89, 90].

Finally, the most fearsome complication of *H. pylori* infection is the induction of cancer. NGS technologies are having a profound impact in the search of prediction factors for gastric adenocarcinoma and MALT lymphoma development. For example, by sequencing the 16S rRNA gene of chronic gastritis and gastric carcinoma specimens, it was discovered that knowing the microbiome composition could serve to discriminate between gastritis and cancer. This is because the gastric carcinoma microbiota is characterized by a dysbiotic microbial community with a severe decrease in the abundance of *Helicobacter* and an increase in other intestinal bacteria such as *non-Helicobacter Proteobacteria and Firmicutes* [91]. The sequencing of virulence genes *cagA* and *vacA* accompanied with MLST analysis had helped to identify novel candidate loci as potential biomarkers of gastric malignancies – one locus in the *cagA* gene to distinguish between gastritis and MALT lymphoma and one locus in the *vacA* to distinguish gastritis from adenocarcinoma [92].

Overall, these studies highlight that using genomic tools can help predict antibiotic resistance, druf efficacy, virulence factors, and cancer risk in bacterial infections with high specificity and sensitivity. In this manner, an accurate diagnosis could be done in an adequate timeframe, and treatment could be personalized according to the characteristics of the *H. pylori* strain that has infected each patient. As NGS technologies continue to develop, genomic determinations will be cost-efficient in comparison with the actual and conventional strategies of diagnostics (Fig. 3.2).

General Applications of NGS Technologies in Oncology Practices

According to the World Health Organization (WHO), cancer disease represents the first up to the fourth leading cause of death in 194 countries. Cancer incidence and mortality are increasing worldwide due to overpopulation, aging, and exposure to carcinogens (occupational or environmental risk factors).

In 2020, the ten most common cancers by incidence in both sexes and all ages are breast, lung, colorectal, prostate, stomach, liver, cervix uteri, esophagus, thyroid, and bladder, whereas the ten most common cancers by mortality are lung, colorectal, liver, stomach, breast, esophagus, pancreas, prostate, cervix uteri, and leukemia [93].

Human lifespan has increased dramatically from 48 years, by the mid-50s, to almost double it at an average of 72.2 years nowadays. Unfortunately, this impressive increase in life expectancy also comes with a higher risk of developing a neoplastic disease. In fact, it is predicted that by the year 2030, there will be a rate of some 22 million new cancer cases each year. For the patients, the society, and the economy of any country, the cost burden of neoplastic diseases can reach astronomical amounts. It was estimated that in 2017, the US cancer healthcare spending was US$161.2 billion, while the European Union healthcare spending was €57.3 billion [94].

Neoplastic disease treatment is costly because it includes not only diagnosis, treatment (surgical, medical, or both), and follow-up, but also dealing with a variety of complications and/or side effects of radiotherapy/chemotherapy, such as taking other medications, hospitalizations, emergency department visits, and relapse. Traditional cancer diagnosis and treatment costs are increasing; therefore, prevention and early detection efforts will be paramount toward more cost-effective treatment, and potentially cost-saving. One of the aims of future management of cancer disease is to try to make it better, more suitable, individualized, less aggressive for the patient as a whole, and ultimately less expensive, and it is here where new genomic technologies come into play.

In the last decades, the advance in NGS technologies has made feasible their use in oncology practices. There are countless examples about their applications on several aspects of clinical management [95]. Here, we will give a general view about how the use of different genomic tools has impacted the diagnosis, risk prediction, classification, and cancer response to treatment.

One of the biggest impacts that genomic tools has on oncology practices is in the field of diagnosis. Gene panels are one of the most used tools for the diagnosis of hereditary cancers including breast, ovarian, and colorectal cancers. These consist in sequencing only a discrete number of genes of interest, which have been already classified as predisposition factors. Ideally, any diagnostic gene panel should be able to maximize clinical sensitivity and minimize analysis of inappropriate/unnecessary genes that may result in variants of uncertain clinical significance [96]. For example, diverse gene panels that classify genes in high, moderate, and low risk are commercially available for diagnosis of hereditary breast cancer (HBC). BROCA is a gene panel which sequence all exons, repeated introns, and promoter regions to detect mutations in 17 high-risk genes (including *APC*, *TP53*, and *BRCA1* and 2), 14 moderate-risk genes, and 17 low-risk genes [97]. There are also commercially available gene panels for ovarian and colorectal cancer [98]. However, it has been reported that a high percentage of patients remain undiagnosed after gene panels testing [99], highlighting that profiling of a restricted number of genes could not be the most adequate approach.

Since 2009, the Cancer Genome Project started to use WES and WGS to decode the mutation landscape in all cancer types [100]. By analyzing 10,952 exomes and 1,048 whole genomes across 40 different human cancer types, the catalogue of somatic mutations (COSMIC) has made available 30 "mutational signatures" which show unique combination of mutation types associated with the development of any specific cancer [101, 102]. Also, the Cancer Genome Atlas (TCGA) uses NGSs to discover molecular aberrations at the genome, epigenome, transcriptome, and proteome levels, thus understanding the complexity of tumors and guiding the diagnostics and clinical decision-making [17, 103]. More recently, the Pan-Cancer Analysis of Whole Genomes Consortium has fully sequenced 2,658 genomes across 38 tumor types. They have found that 705 mutations constantly occurred in all cancer genomes and have evidenced the importance of mutations in noncoding regions as drivers for tumor growth [104]. To integrate the genomic knowledge into clinical applications, there are currently translational research projects trying to join the DNA information with the clinical data of each patient. For example, the GENIE project (Genomics Evidence Neoplasia Information Exchange from the AACR) has integrated their first dataset providing genomic and clinical data of 18,804 sequenced samples from 18,324 patients at 8 different medical centers. From this dataset, they have been able to associate clinicopathologic features and outcomes of breast cancer patients with metastasis, by examining *ERBB2* and *AKT1* rare mutations [105].

In addition to the accuracy and robustness of mutational profiling, one important cue in diagnosis is the possibility to use presurgical biopsies. The genotyping of 143 fine-needle aspiration samples of thyroid nodules with the targeted ThyroSeq v2 NGS panel showed that this approach has a 90–93% sensitivity and specificity with 83% positive predictive value and 92% accuracy with 96% negative predictive value for malignant and benign nodules. ThyroSeq profiles point mutations in 13 genes and 42 types of gene fusions that are frequent in thyroid tumors, and it appears to be an accurate tool for preoperative diagnosis [106]. Furthermore, a recent clinical trial that compares the diagnostic performance between the RNA test "Afirma genomic sequencing classifier" and the new DNA-RNA test "ThyroSeq v3 genomic classifier" showed that both tests are highly specific for malignancy prediction and allowed 50% of patients with indeterminate nodules to avoid diagnostic thyroidectomy [107].

NGS analysis in noninvasive liquid biopsies is often having good results in cancer diagnosis. Liquid biopsy allows real-time biomolecular characterization of a tumor by analyzing human body fluids. The sequencing of a biomarker panel (*TERT*, *FGFR3*, *PIK3CA*, *TP53*, *HRAS*, *KDM6A*, *RXRA*) made on 211 urinary samples from primary or recurrent bladder cancer (BC) and 20 healthy donors showed the high potential of this test to detect non-muscle-invasive bladder cancers (NMIBC) [108]. Also, by associating the *FGFR3* mutation assay with DNA methylation analysis of 18 genes in urine DNA, it was demonstrated that this strategy has a powerful use not only in diagnosis but also in prognosis and in risk stratification in NMIBC patients [26]. This kind of molecular testing is becoming one of the preferred diagnostic strategies by patients instead of repeat biopsy or diagnostic surgery [109, 110] and is now being extended for the diagnosis of several cancer types such as early lung cancer in which the genotyping of circulating tumor cells (CTCs) and cell-free DNA (cfDNA) has already entered clinical practice for detection of *EGFR* mutations [111].

Although less frequently, RNA-seq analysis has also been used for diagnosis and prognosis applications. RNA-seq results from tumor-educated platelets have been analyzed to screen mRNA markers with potential to distinguish several cancer types (breast, colorectal, glioblastoma, hepatobiliar, lung, and pancreas) from healthy patients. Eighteen functional genes resulted to be important for distinguishing cancer including ribosome-associated genes, cell surface proteins, and confirmed tumor-associated genes [21]. Similarly, the expression of 3 mRNA biomarkers (*SLC2A1*, *GPRC5A*, *KRT17*) in urinary extracellular vesicles (EV) resulted to have a good potential to detect early-stage BC [112]. A recent study of integrative transcriptomics performed in TCGA data from 400 patients with BC found a five-RNA signature (three protein coding: *ANXA1*, *TPST1*, and *PSMB10*; one lncRNA: *DLEU1*; and one miRNA: miR-497-5p) that could predict the outcome of patients by classifying them into high- or low-risk groups [113]. The clinical utility of gene expression signature identification is reflected in the broader use of test such as MammaPrint and Oncotype DX (breast cancer), GeneFx (lung cancer), Prolaris (prostate cancer), and ColoPrint (colon cancer) [114].

The possibility to use NGS in the implementation of personalized therapy based on individual genetic profiles has been also studied for several types of cancer. In 2015, the SHIVA clinical trial published the results about the efficacy of molecularly targeted therapy in 741 patients with any kind of metastatic solid tumor refractory to conventional therapy. They found that choosing a therapy only on the basis of molecular profiling but outside its clinical indications does not improve the progression-free survival rates with those obtained in patients treated with conventional therapy [115]. However, the assessment of this kind of directed therapy has been encouraged by diverse large-scale protocols. The Drug Rediscovery Protocol from the Netherlands Cancer Institute has proved the efficacy of targeted therapies in 136 patients who have exhausted standard treatments. They identified a clinical benefit rate of 63% in a group of 30 patients with microsatellite unstable tumors that received nivolumab. Among them, one patient had a complete response while 18 patients had a partial response or stable disease [116]. These results point up the impact that WGS genomic approaches could have on the selection of better therapeutic strategies for small groups of patients with no treatment options.

The potential use of NGSs to identify drug targets or treatment response is well documented. For example, the presence of a mutation in the *TP53* gene on chronic lymphocytic leukemia means that patients won't respond to any kind of chemoimmunotherapy; with this knowledge, physicians could decide early on to choose other kind of treatment such as stem cell transplantation [117, 118]. A research screened 29 cervical cancer samples with a panel of 226 genes finding out that 48% of the participants displayed mutations in genes that can be targeted with approved drugs such as crizotinib, ceritinib, and other tyrosine kinase inhibitors [119]. The WGS analysis of 420 patients with colorectal metastatic cancer allowed the identification of metastasis-specific signatures of genes that then could be studied as novel target molecules. Also, the researchers found that specific mutations in the *FBXW7* gene could be used as a predictive biomarker of poor response to EGFR-targeted treatment [14].

More recently, RNA-seq has been highly useful in the search for molecular drug targets and clinical outcomes biomarkers. To understand transcriptional changes that occur in lung cancers, 3240 RNA-seq data from 23 cell lines treated with several compounds were analyzed. The expression of a module of genes related with stress response resulted to be a potential drug molecule that could be targeted by using bromodomain and extraterminal motif (BET) inhibitors. These are a new class of epigenome drugs that affect transcription by decreasing the chromatin accessibility of gene promoters [22]. Bioinformatic models can be used for the selection of drugs in personalized therapy. By collecting datasets from gene expression profiles associated with chemotherapy responses of 2786 individual cases, a large molecular collection has been created as guide for physicians to know if certain treatment protocols will have positive or negative response depending on the clinical history of each patient [120]. Finally, with the advent of the new single-cell RNA sequencing (scRNA-seq), it is now possible to study the treatment response of the multicellular ecosystem in cancer. The scRNA-seq of 49 biopsies from metastatic lung cancer patients found an alveolar-regenerative cell signature that characterizes a residual disease state (RD). RD samples were taken during treatment and represent those regressing or stable tumors. The increased expression of the alveolar-regenerative signature is associated with a less aggressive malignant state and an improved patient survival. Within the molecular signature associated with RD cells, WNT/β-catenin is a therapeutically targetable pathway for preventing tumor relapse [23].

NGS technologies have enabled an unprecedented understanding about the cancer biology, which has had a profound impact on clinical management. Molecular screening by genome or RNA analysis can aid personalized patient care. While these technologies still have full analysis limitations, they have been spreading in public and private oncological services as accurate and accessible tools to guide physicians in the diagnostics and selection of the best therapy option for each patient (Fig. 3.2).

Advances on the Pathogenesis and Treatment Knowledge of Rheumatoid Arthritis by Using NGS

Rheumatoid arthritis (RA) is a chronic inflammatory autoimmune disease affecting the joints [121, 122]. It is characterized by a progressive symmetric inflammation of affected joints resulting in cartilage destruction, bone erosion, and disability [123].

RA is one of the most frequent chronic inflammatory diseases, with a constant prevalence in many European and North-American populations ranging from 0.4% to 1.3% [123, 124] and a highly variable annual incidence (12–1200 per 100,000 population) depending on gender, race/ethnicity, and calendar year [125]. Some exceptions are the native American-Indian populations, in whom the highest occurrence of RA was recorded, with a prevalence of 5.3% noted for the Pima Indians and of 6.8% for the Chippewa Indians

[125, 126]. Epidemiological studies have not found any RA case in two rural populations in South Africa and Nigeria [127, 128].

RA affects women two to three times more often than men [123]. Female sex hormones may play a protective role in RA [124]. Male sex hormones such as testosterone are lower in RA men. By contrast, levels of female sex hormones are not different between RA cases and controls [129]. The onset of RA seems to be a critical factor in its clinical spectrum and occurs usually in the fourth or fifth decade of life, with a new diagnostic peak in the sixth decade [13]; a patient suffers elderly onset rheumatoid arthritis (EORA) when the disease begins at the age of ≥60 years [130]. Since the general population is aging, beginning of RA in older people is more and more common.

People with RA have a significantly increased risk of death (54%) [131] compared with age- and sex-matched controls without RA from the same community, and it is associated with reduced life expectancy [132]. Mortality rates in patients with RA are 1.5- to 1.6-fold higher than in the general population, with similar patterns over 60 years [133]. The mortality rate in RA cases incidence has declined significantly [134]; however, patients with RA have an increased risk of cardiovascular disease (CVD), in a large international cohort of RA patients, and 30% of CVD events were attributable to RA characteristics [135]. Also, mortality is higher for respiratory, musculoskeletal, and digestive system diseases in patients with RA [131].

Multiple factors have been associated with RA risk. Within genetic factors are familial associations, particularly the first-degree relative's status [136]. Environmental and other risk factors include the presence of alleles containing the shared epitope (SE); female sex; exposure to tobacco smoke; obesity; exposition to UV light; sex hormones; drugs; changes in microbiome of the gut, mouth, and lung; periodontitis; infections; and some dietary factors such as intake of omega-3 fatty acids [136, 137].

NGSs such as deep exon sequencing and GWASs (see SNPs and SNVs section) have been used in order to know which genetic variants could be risk factors for RA [138]. In the Caucasian population, these studies identified several rare variants within the protein-coding portion of genes such as *IL2RA* and *IL2RB* as genetic contributors for RA development [139]. Also, by using WGS, a rare mutation in the *PLB1* gene was identified in RA family members with dominant inheritance, suggesting it as a potent risk for RA [140]. The WES analysis of 19 RA Japanese cases also demonstrated that three SNPs in the *BTNL2* gene confer RA risk independently from the RA-susceptible genes *DRB1* and *NOTCH4* [16]. Also, a recent WES study in Chinese population found five novel and rare variants in genes that alter innate immunity pathways contributing to the risk of RA [15]. More large-scale studies are needed to know which genetic variants are associated with the risk of RA development in other populations.

In RA, an autoimmune tissue destruction is present as synovitis which is an inflammation of the joint capsule consisting of the synovial membrane, synovial fluid, and the respective bones [141]. Several genetic aspects (i.e., major histocompatibility complex (*MHC*) genes, especially *HLA-DRB1* locus), cellular components, soluble mediators (i.e., IL-1β, IFN-γ, TNF cytokines) [142], adhesion molecules and autoantibodies (rheumatoid factor (RF), and anti-citrullinated protein antibodies (ACPAs)) contribute to the development of inflammation and structural changes of joints and internal organs [27].

By analyzing the whole transcriptome of CD4 + T (HLA-DRAB1) cells, it was found that STAT3 and Wnt signaling networks and several transcription factors (i.e., *ZEB1*, *ZNF292*) have an aberrant expression in RA patients, suggesting their pathogenic potential for RA development [143]. More recently, the RNA-seq of 7 CD4 + T cell subtypes isolated from the peripheral blood of both healthy and RA patients showed that several genes involved in GTPase-associated signaling and apoptosis are overexpressed in RA [144]. Also, the roles of synovial fibroblasts have been studied by RNA-seq allowing the identification of new genes and isoforms associated with RA pathogenesis [145]. The fibroblast heterogeneity in RA synovial tissue was studied by scRNA-seq analysis. The authors described a pathogenic fibroblast (FAPα+THY1-) subpopulation in RA patients. These fibroblasts are in the lining layer of the synovial tissue and have an invasive and destructive phenotype, causing damage to cartilage and bone. These results are of relevant importance for the implementation of new cell-based therapies intended to modulate tissue damage [24].

Research at molecular and cellular levels has clarified some mechanisms that regulate the innate and adaptive immune responses (i.e., Th1/Tc1-type immune responses) in RA [24, 26]. For example, a recent study using Chip-seq, ATAC-seq, and RNA-seq revealed the epigenetic landscape of fibroblast-like synoviocytes (FLS) from RA. The genes with different histone modification marks and open chromatin resulted to be enriched for immune pathways and the "Huntington's disease signaling". This pathway is important for RA development as the *HIP-1* gene regulates FLS invasion into matrix [146]. Also, GWAS results showed that immune cells of patients with rheumatic diseases have non-coding variants located at sites with epigenetic modifications [147]. Within them, several miRNAs including miR-4728-5p resulted to have a significant contribution in the RA pathogenesis [148]. Using metagenomic shotgun sequencing, it was detected that functional changes in the mucosal microbiota (dysbiosis) are present in RA. Dysbiosis results in an imbalance of the immune status and epithelial barrier function, diminishes the general gene methylation level, and increases NFκB and TNF-α production, thus resulting in the hyperactivation of T-helper cells (Th1 and Th17) which secrete inflammatory cytokines that finally contribute to the RA hyperinflammatory state [149]. These results highlight the potential of genomic and integrative analysis for the search of new therapeutic targets.

RA diagnosis requires biomarkers that should be detected in blood and/or in the synovial fluid [150]. A number of biomarkers have been discovered and clinically used in RA [151], including anti-keratin antibodies (AKA), anti-cyclic citrullinated peptide (anti-CCP) antibodies, anti-perinuclear factor, anti-filaggrin antibodies (AFA), anti-citrullinated vimentin antibodies (anti-SA), rheumatoid factor (RF), melanoma-associated antigen genes family (*MAGE*) [152], C-reactive protein (CRP), and erythrocyte sedimentation rate (ESR).

However, the clinical activity of RA is not accurately predicted by current laboratory measures. Transcriptome analysis in RA is spreading as a new strategy for the searching of predictive biomarkers of the RA pathogenesis [153–155]. For example, gene expression analysis made on biopsies from RA patients showed that some clusters of genes served to classify RA in three subtypes. The first subtype expressed genes of the adaptive immune response (*MMP1*, *MMP3*, and STAT-induced genes), the second expressed genes involved in the extracellular matrix remodeling, and the third expressed genes related to fibroblast differentiation [156, 157]. Further study of this data could serve to validate the potential of some genes as classification biomarkers for a better diagnosis and treatment selection in each RA presentation.

RA therapy includes minimizing joint pain and swelling, preventing deformity (such as ulnar deviation) and radiographic damage (such as erosions), maintaining quality of life, and controlling extra-articular manifestations [141]. Currently available drugs include nonsteroidal anti-inflammatories (NSAIDs); oral, intramuscular, or intra-articular immunosuppressive glucocorticoids; and disease-modifying antirheumatic drugs DMARDs (biologic or nonbiologic). DMARDs are the mainstay of RA therapy [123].

There is much interest in transcriptome analysis as a strategy for predicting the effect of RA treatment. Although these kinds of analyses have generally focused on the study of whole peripheral blood mononuclear cells (PBMC), studies in particular cell subsets (such as CD4 + T cells or neutrophils) are now emerging, thus increasing our understanding about the disease response. The transcriptome profiling of PBMC from RA patients found a set of 193 genes that are differentially expressed between responders and nonresponders RA patients treated with rituximab (RTX). The set of upregulated genes are involved in the NFκB inflammatory pathway, while the downregulated ones are related to the IFN pathway; these genes could be used as responsiveness markers to identify patients that will not respond to the RTX treatment [158]. The RNA-seq of neutrophils isolated from RA patients also demonstrated that the upregulation of the IFN-response genes is highly correlated with a good response to TNF inhibitor (TNFi) therapy [159].

Undoubtedly, while NGS technologies are better and more implemented on basic research, scientist and physicians will have a better understanding of the complexity of RA disease. This is crucial for the implementation of new diagnostic and therapeutic strategies in the turn toward the individualized management (Fig. 3.2).

Concluding Remarks

The advent of genomic technologies is having a significant impact on clinical epidemiology. SNPs are powerful tools for assessing epidemiological variables such as risk, disease progression, and prognosis. Also, different genomic approaches are now being used in the clinical management of a variety of pathologies, including cancer, where advanced technologies are spreading as an accurate tool guiding physicians in screening, diagnosis, and therapy selection. Moreover, in less-explored areas, genomic tools have potential to transform the current epidemiological methods into strategies with high specificity and sensitivity on the prediction for antibiotic resistance, drug efficacy, and virulence factors depending on each population.

References

1. Hayden EC. Technology: the $1,000 genome. Nature. 2014;507(7492):294–5.
2. Mardis ER. Next-generation sequencing platforms. Annu Rev Anal Chem (Palo Alto, Calif). 2013;6:287–303.
3. Adessi C, Matton G, Ayala G, Turcatti G, Mermod JJ, Mayer P, Kawashima E. Solid phase DNA amplification: characterisation of primer attachment and amplification mechanisms. Nucleic Acids Res. 2000;28(20):E87.
4. Dressman D, Yan H, Traverso G, Kinzler KW, Vogelstein B. Transforming single DNA molecules into fluorescent magnetic particles for detection and enumeration of genetic variations. Proc Natl Acad Sci U S A. 2003;100(15):8817–22.
5. Petersen BS, Fredrich B, Hoeppner MP, Ellinghaus D, Franke A. Opportunities and challenges of whole-genome and -exome sequencing. BMC Genet. 2017;18(1):14.
6. Costa-Silva J, Domingues D, Lopes FM. RNA-Seq differential expression analysis: an extended review and a software tool. PLoS One. 2017;12(12):e0190152.
7. Bogdanovic O, Fernandez-Minan A, Tena JJ, de la Calle-Mustienes E, Gomez-Skarmeta JL. The developmental epigenomics toolbox: ChIP-seq and MethylCap-seq profiling of early zebrafish embryos. Methods. 2013;62(3):207–15.
8. Di Girolamo F, Lante I, Muraca M, Putignani L. The role of mass spectrometry in the "omics" era. Curr Org Chem. 2013;17(23):2891–905.

9. Duggal P, Ladd-Acosta C, Ray D, Beaty TH. The evolving field of genetic epidemiology: from familial aggregation to genomic sequencing. Am J Epidemiol. 2019;188(12):2069–77.
10. Blanton RE. Population genetics and molecular epidemiology of eukaryotes. Microbiol Spectr. 2018;6(6):6.
11. Traynor BJ. The era of genomic epidemiology. Neuroepidemiology. 2009;33(3):276–9.
12. Chen H, Sun J, Jiang H, Wang X, Wu L, Wu W, Wang Q. Inferring alcoholism SNPs and regulatory chemical compounds based on ensemble Bayesian network. Comb Chem High Throughput Screen. 2017;20(2):107–15.
13. Binh TT, Shiota S, Suzuki R, Matsuda M, Trang TT, Kwon DH, Iwatani S, Yamaoka Y. Discovery of novel mutations for clarithromycin resistance in Helicobacter pylori by using next-generation sequencing. J Antimicrob Chemother. 2014;69(7):1796–803.
14. Mendelaar PAJ, Smid M, van Riet J, Angus L, Labots M, Steeghs N, Hendriks MP, Cirkel GA, van Rooijen JM, Ten Tije AJ, et al. Whole genome sequencing of metastatic colorectal cancer reveals prior treatment effects and specific metastasis features. Nat Commun. 2021;12(1):574.
15. Li Y, Lai-Han Leung E, Pan H, Yao X, Huang Q, Wu M, Xu T, Wang Y, Cai J, Li R, et al. Identification of potential genetic causal variants for rheumatoid arthritis by whole-exome sequencing. Oncotarget. 2017;8(67):111119–29.
16. Mitsunaga S, Hosomichi K, Okudaira Y, Nakaoka H, Kunii N, Suzuki Y, Kuwana M, Sato S, Kaneko Y, Homma Y, et al. Exome sequencing identifies novel rheumatoid arthritis-susceptible variants in the BTNL2. J Hum Genet. 2013;58(4):210–5.
17. Giordano TJ. The cancer genome atlas research network: a sight to behold. Endocr Pathol. 2014;25(4):362–5.
18. Lamichhane B, Chua EG, Wise MJ, Laming C, Marshall BJ, Tay CY. The complete genome and methylome of Helicobacter pylori hpNEAfrica strain HP14039. Gut Pathog. 2019;11:7.
19. Lu R, Zhao X, Li J, Niu P, Yang B, Wu H, Wang W, Song H, Huang B, Zhu N, et al. Genomic characterisation and epidemiology of 2019 novel coronavirus: implications for virus origins and receptor binding. Lancet. 2020;395(10224):565–74.
20. Han B, Zhang Z, Xie Y, Hu X, Wang H, Xia W, Wang Y, Li H, Wang Y, Sun H. Multi-omics and temporal dynamics profiling reveal disruption of central metabolism in Helicobacter pylori on bismuth treatment. Chem Sci. 2018;9(38):7488–97.
21. Zhang YH, Huang T, Chen L, Xu Y, Hu Y, Hu LD, Cai Y, Kong X. Identifying and analyzing different cancer subtypes using RNA-seq data of blood platelets. Oncotarget. 2017;8(50):87494–511.
22. Suzuki A, Onodera K, Matsui K, Seki M, Esumi H, Soga T, Sugano S, Kohno T, Suzuki Y, Tsuchihara K. Characterization of cancer omics and drug perturbations in panels of lung cancer cells. Sci Rep. 2019;9(1):19529.
23. Maynard A, McCoach CE, Rotow JK, Harris L, Haderk F, Kerr DL, Yu EA, Schenk EL, Tan W, Zee A, et al. Therapy-induced evolution of human lung cancer revealed by single-cell RNA sequencing. Cell. 2020;182(5):1232–1251 e1222.
24. Croft AP, Campos J, Jansen K, Turner JD, Marshall J, Attar M, Savary L, Wehmeyer C, Naylor AJ, Kemble S, et al. Distinct fibroblast subsets drive inflammation and damage in arthritis. Nature. 2019;570(7760):246–51.
25. Wilk AJ, Rustagi A, Zhao NQ, Roque J, Martínez-Colón GJ, McKechnie JL, Ivison GT, Ranganath T, Vergara R, Hollis T, et al. A single-cell atlas of the peripheral immune response in patients with severe COVID-19. Nat Med. 2020;26(7):1070–6.
26. Roperch JP, Grandchamp B, Desgrandchamps F, Mongiat-Artus P, Ravery V, Ouzaid I, Roupret M, Phe V, Ciofu C, Tubach F, et al. Promoter hypermethylation of HS3ST2, SEPTIN9 and SLIT2 combined with FGFR3 mutations as a sensitive/specific urinary assay for diagnosis and surveillance in patients with low or high-risk non-muscle-invasive bladder cancer. BMC Cancer. 2016;16:704.
27. Croia C, Bursi R, Sutera D, Petrelli F, Alunno A, Puxeddu I. One year in review 2019: pathogenesis of rheumatoid arthritis. Clin Exp Rheumatol. 2019;37(3):347–57.
28. Brookes AJ. The essence of SNPs. Gene. 1999;234(2):177–86.
29. Shastry BS. SNPs in disease gene mapping, medicinal drug development and evolution. J Hum Genet. 2007;52(11):871–80.
30. Frazer KA, Murray SS, Schork NJ, Topol EJ. Human genetic variation and its contribution to complex traits. Nat Rev Genet. 2009;10(4):241–51.
31. Auton A, Brooks LD, Durbin RM, Garrison EP, Kang HM, Korbel JO, Marchini JL, McCarthy S, McVean GA, Abecasis GR. A global reference for human genetic variation. Nature. 2015;526(7571):68–74.
32. Korte A, Farlow A. The advantages and limitations of trait analysis with GWAS: a review. Plant Methods. 2013;9:29.
33. Yengo L, Sidorenko J, Kemper KE, Zheng Z, Wood AR, Weedon MN, Frayling TM, Hirschhorn J, Yang J, Visscher PM, et al. Meta-analysis of genome-wide association studies for height and body mass index in ~700000 individuals of European ancestry. Hum Mol Genet. 2018;27(20):3641–9.
34. White D, Rabago-Smith M. Genotype–phenotype associations and human eye color. J Hum Genet. 2011;56(1):5–7.
35. Halushka MK, Fan JB, Bentley K, Hsie L, Shen N, Weder A, Cooper R, Lipshutz R, Chakravarti A. Patterns of single-nucleotide polymorphisms in candidate genes for blood-pressure homeostasis. Nat Genet. 1999;22(3):239–47.
36. Choi WJ, Jin HS, Kim SS, Shin D. Dietary protein and fat intake affects diabetes risk with CDKAL1 genetic variants in Korean adults. Int J Mol Sci. 2020;21(16):5607.
37. Chong YH, Fan Q, Tham YC, Gan A, Tan SP, Tan G, Wang JJ, Mitchell P, Wong TY, Cheng CY. Type 2 diabetes genetic variants and risk of diabetic retinopathy. Ophthalmology. 2017;124(3):336–42.
38. Chen X, Wang W, Li R, Yu J, Gao L. Association between polymorphisms in microRNAs and susceptibility to diabetes mellitus: a meta-analysis. Medicine. 2019;98(44):e17519.
39. Kathiresan S, Voight BF, Purcell S, Musunuru K, Ardissino D, Mannucci PM, Anand S, Engert JC, Samani NJ, Schunkert H, et al. Genome-wide association of early-onset myocardial infarction with single nucleotide polymorphisms and copy number variants. Nat Genet. 2009;41(3):334–41.
40. Shen GQ, Rao S, Martinelli N, Li L, Olivieri O, Corrocher R, Abdullah KG, Hazen SL, Smith J, Barnard J, et al. Association between four SNPs on chromosome 9p21 and myocardial infarction is replicated in an Italian population. J Hum Genet. 2008;53(2):144–50.
41. Song N, Kim K, Shin A, Park JW, Chang HJ, Shi J, Cai Q, Kim DY, Zheng W, Oh JH. Colorectal cancer susceptibility loci and influence on survival. Genes Chromosomes Cancer. 2018;57(12):630–7.
42. Sartor H, Brandt J, Grassmann F, Eriksson M, Czene K, Melander O, Zackrisson S. The association of single nucleotide polymorphisms (SNPs) with breast density and breast cancer survival: the Malmö Diet and Cancer Study. Acta Radiologica (Stockholm, Sweden: 1987). 2020;61(10):1326–34.
43. Shi Z, Yu H, Wu Y, Lin X, Bao Q, Jia H, Perschon C, Duggan D, Helfand BT, Zheng SL, et al. Systematic evaluation of cancer-specific genetic risk score for 11 types of cancer in The Cancer Genome Atlas and Electronic Medical Records and Genomics cohorts. Cancer Med. 2019;8(6):3196–205.
44. Yang C, Lu L, Warren JL, Wu J, Jiang Q, Zuo T, Gan M, Liu M, Liu Q, DeRiemer K, et al. Internal migration and transmission dynamics of tuberculosis in Shanghai, China: an epidemiological, spatial, genomic analysis. Lancet Infect Dis. 2018;18(7):788–95.

45. Huang H, Ding N, Yang T, Li C, Jia X, Wang G, Zhong J, Zhang J, Jiang G, Wang S, et al. Cross-sectional whole-genome sequencing and epidemiological study of multidrug-resistant mycobacterium tuberculosis in China. Clin Infect Dis. 2019;69(3):405–13.
46. Wu S, Wang M-G, Wang Y, He J-Q. Polymorphisms of cytokine genes and tuberculosis in two independent studies. Sci Rep. 2019;9(1):2507.
47. Patel IR, Gangiredla J, Lacher DW, Mammel MK, Jackson SA, Lampel KA, Elkins CA. FDA Escherichia coli identification (FDA-ECID) microarray: a Pangenome molecular toolbox for serotyping, virulence profiling, molecular epidemiology, and phylogeny. Appl Environ Microbiol. 2016;82(11):3384–94.
48. Vankadari N. Overwhelming mutations or SNPs of SARS-CoV-2: a point of caution. Gene. 2020;752:144792.
49. World Health Organization https://www.who.int/docs/default--source/coronaviruse/situation-reports/20200121-sitrep-1-2019-ncov.pdf?sfvrsn=20a99c10_4. Accessed 29 Mar 2021.
50. Coronaviridae Study Group of the International Committee on Taxonomy of V. The species severe acute respiratory syndrome-related coronavirus: classifying 2019-nCoV and naming it SARS-CoV-2. Nat Microbiol. 2020;5(4):536–44.
51. Wassenaar TM, Zou Y. 2019_nCoV/SARS-CoV-2: rapid classification of betacoronaviruses and identification of Traditional Chinese Medicine as potential origin of zoonotic coronaviruses. Lett Appl Microbiol. 2020;70(5):342–8.
52. Yüce M, Filiztekin E, Özkaya KG. COVID-19 diagnosis -a review of current methods. Biosens Bioelectron. 2021;172:112752.
53. Forster P, Forster L, Renfrew C, Forster M. Phylogenetic network analysis of SARS-CoV-2 genomes. Proc Natl Acad Sci U S A. 2020;117(17):9241–3.
54. Khailany RA, Safdar M, Ozaslan M. Genomic characterization of a novel SARS-CoV-2. Gene Rep. 2020;19:100682.
55. Islam MR, Hoque MN, Rahman MS, Alam A, Akther M, Puspo JA, Akter S, Sultana M, Crandall KA, Hossain MA. Genome-wide analysis of SARS-CoV-2 virus strains circulating worldwide implicates heterogeneity. Sci Rep. 2020;10(1):14004.
56. Wang C, Yu H, Horby PW, Cao B, Wu P, Yang S, Gao H, Li H, Tsang TK, Liao Q, et al. Comparison of patients hospitalized with influenza A subtypes H7N9, H5N1, and 2009 pandemic H1N1. Clin Infect Dis. 2014;58(8):1095–103.
57. World Health Organization https://www.who.int/emergencies/diseases/novel-coronavirus-2019/question-and-answers-hub/q-a-detail/q-a-coronaviruses. Acce.sed 20 Mar 2021.
58. Luo G, Gao Q, Zhang S, Yan B. Probing infectious disease by single-cell RNA sequencing: progresses and perspectives. Comput Struct Biotechnol J. 2020;18:2962–71.
59. Zhang JY, Wang XM, Xing X, Xu Z, Zhang C, Song JW, Fan X, Xia P, Fu JL, Wang SY, et al. Single-cell landscape of immunological responses in patients with COVID-19. Nat Immunol. 2020;21(9):1107–18.
60. He J, Cai S, Feng H, Cai B, Lin L, Mai Y, Fan Y, Zhu A, Huang H, Shi J, et al. Single-cell analysis reveals bronchoalveolar epithelial dysfunction in COVID-19 patients. Protein Cell. 2020;11(9):680–7.
61. Zheng Y, Liu X, Le W, Xie L, Li H, Wen W, Wang S, Ma S, Huang Z, Ye J, et al. A human circulating immune cell landscape in aging and COVID-19. Protein Cell. 2020;11(10):740–70.
62. Liu J, Liao X, Qian S, Yuan J, Wang F, Liu Y, Wang Z, Wang FS, Liu L, Zhang Z. Community transmission of severe acute respiratory syndrome coronavirus 2, Shenzhen, China, 2020. Emerg Infect Dis. 2020;26(6):1320–3.
63. Li Q, Guan X, Wu P, Wang X, Zhou L, Tong Y, Ren R, Leung KSM, Lau EHY, Wong JY, et al. Early transmission dynamics in Wuhan, China, of novel coronavirus-infected pneumonia. N Engl J Med. 2020;382(13):1199–207.
64. Wei WE, Li Z, Chiew CJ, Yong SE, Toh MP, Lee VJ. Presymptomatic transmission of SARS-CoV-2 - Singapore, January 23-March 16, 2020. MMWR Morb Mortal Wkly Rep. 2020;69(14):411–5.
65. Malik YA. Properties of coronavirus and SARS-CoV-2. Malays J Pathol. 2020;42(1):3–11.
66. Sungnak W, Huang N, Bécavin C, Berg M, Queen R, Litvinukova M, Talavera-López C, Maatz H, Reichart D, Sampaziotis F, et al. SARS-CoV-2 entry factors are highly expressed in nasal epithelial cells together with innate immune genes. Nat Med. 2020;26(5):681–7.
67. Lange C, Wolf J, Auw-Haedrich C, Schlecht A, Boneva S, Lapp T, Horres R, Agostini H, Martin G, Reinhard T, et al. Expression of the COVID-19 receptor ACE2 in the human conjunctiva. J Med Virol. 2020;92(10):2081–6.
68. Collin J, Queen R, Zerti D, Dorgau B, Georgiou M, Djidrovski I, Hussain R, Coxhead JM, Joseph A, Rooney P, et al. Co-expression of SARS-CoV-2 entry genes in the superficial adult human conjunctival, limbal and corneal epithelium suggests an additional route of entry via the ocular surface. Ocul Surf. 2021;19:190–200.
69. Graham DY. Helicobacter pylori update: gastric cancer, reliable therapy, and possible benefits. Gastroenterology. 2015;148(4):719–731 e713.
70. Wotherspoon AC, Ortiz-Hidalgo C, Falzon MR, Isaacson PG. Helicobacter pylori-associated gastritis and primary B-cell gastric lymphoma. Lancet. 1991;338(8776):1175–6.
71. Covacci A, Telford JL, Del Giudice G, Parsonnet J, Rappuoli R. Helicobacter pylori virulence and genetic geography. Science. 1999;284(5418):1328–33.
72. Tomb JF, White O, Kerlavage AR, Clayton RA, Sutton GG, Fleischmann RD, Ketchum KA, Klenk HP, Gill S, Dougherty BA, et al. The complete genome sequence of the gastric pathogen Helicobacter pylori. Nature. 1997;388(6642):539–47.
73. Pohl D, Keller PM, Bordier V, Wagner K. Review of current diagnostic methods and advances in Helicobacter pylori diagnostics in the era of next generation sequencing. World J Gastroenterol. 2019;25(32):4629–60.
74. Safavi M, Sabourian R, Foroumadi A. Treatment of Helicobacter pylori infection: current and future insights. World J Clin Cases. 2016;4(1):5–19.
75. Versalovic J, Osato MS, Spakovsky K, Dore MP, Reddy R, Stone GG, Shortridge D, Flamm RK, Tanaka SK, Graham DY. Point mutations in the 23S rRNA gene of Helicobacter pylori associated with different levels of clarithromycin resistance. J Antimicrob Chemother. 1997;40(2):283–6.
76. Miftahussurur M, Waskito LA, Syam AF, Nusi IA, Siregar G, Richardo M, Bakry AF, Rezkitha YAA, Wibawa IDN, Yamaoka Y. Alternative eradication regimens for Helicobacter pylori infection in Indonesian regions with high metronidazole and levofloxacin resistance. Infect Drug Resist. 2019;12:345–58.
77. Lauener FN, Imkamp F, Lehours P, Buissonniere A, Benejat L, Zbinden R, Keller PM, Wagner K. Genetic determinants and prediction of antibiotic resistance phenotypes in Helicobacter pylori. J Clin Med. 2019;8(1):53.
78. Yusibova M, Hasman H, Clausen P, Imkamp F, Wagner K, Andersen LP. CRHP finder, a webtool for the detection of clarithromycin resistance in Helicobacter pylori from whole-genome sequencing data. Helicobacter. 2020;25(6):e12752.
79. Dore MP, Lu H, Graham DY. Role of bismuth in improving Helicobacter pylori eradication with triple therapy. Gut. 2016;65(5):870–8.
80. Estibariz I, Overmann A, Ailloud F, Krebes J, Josenhans C, Suerbaum S. The core genome m5C methyltransferase JHP1050 (M.Hpy99III) plays an important role in orchestrating gene expression in Helicobacter pylori. Nucleic Acids Res. 2019;47(5):2336–48.

81. Loh JT, Beckett AC, Scholz MB, Cover TL. High-salt conditions alter transcription of Helicobacter pylori genes encoding outer membrane proteins. Infect Immun. 2018;86(3):e00626–17.
82. Linz B, Balloux F, Moodley Y, Manica A, Liu H, Roumagnac P, Falush D, Stamer C, Prugnolle F, van der Merwe SW, et al. An African origin for the intimate association between humans and Helicobacter pylori. Nature. 2007;445(7130):915–8.
83. Seo JW, Park JY, Shin TS, Kim JG. The analysis of virulence factors and antibiotic resistance between Helicobacter pylori strains isolated from gastric antrum and body. BMC Gastroenterol. 2019;19(1):140.
84. Haley KP, Gaddy JA. Helicobacter pylori: genomic insight into the host-pathogen interaction. Int J Genomics. 2015;2015:386905.
85. Fischer W, Breithaupt U, Kern B, Smith SI, Spicher C, Haas R. A comprehensive analysis of Helicobacter pylori plasticity zones reveals that they are integrating conjugative elements with intermediate integration specificity. BMC Genomics. 2014;15:310.
86. Marcus EA, Sachs G, Scott DR. Acid-regulated gene expression of Helicobacter pylori: insight into acid protection and gastric colonization. Helicobacter. 2018;23(3):e12490.
87. Hathroubi S, Zerebinski J, Ottemann KM. Helicobacter pylori biofilm involves a multigene stress-biased response, including a structural role for flagella. MBio. 2018;9(5):e01973–18.
88. Sharma CM, Hoffmann S, Darfeuille F, Reignier J, Findeiß S, Sittka A, Chabas S, Reiche K, Hackermüller J, Reinhardt R, et al. The primary transcriptome of the major human pathogen Helicobacter pylori. Nature. 2010;464(7286):250–5.
89. Khosravi Y, Rehvathy V, Wee WY, Wang S, Baybayan P, Singh S, Ashby M, Ong J, Amoyo AA, Seow SW, et al. Comparing the genomes of Helicobacter pylori clinical strain UM032 and mice-adapted derivatives. Gut Pathog. 2013;5:25.
90. Furuta Y, Namba-Fukuyo H, Shibata TF, Nishiyama T, Shigenobu S, Suzuki Y, Sugano S, Hasebe M, Kobayashi I. Methylome diversification through changes in DNA methyltransferase sequence specificity. PLoS Genet. 2014;10(4):e1004272.
91. Ferreira RM, Pereira-Marques J, Pinto-Ribeiro I, Costa JL, Carneiro F, Machado JC, Figueiredo C. Gastric microbial community profiling reveals a dysbiotic cancer-associated microbiota. Gut. 2018;67(2):226–36.
92. Hashinaga M, Suzuki R, Akada J, Matsumoto T, Kido Y, Okimoto T, Kodama M, Murakami K, Yamaoka Y. Differences in amino acid frequency in CagA and VacA sequences of Helicobacter pylori distinguish gastric cancer from gastric MALT lymphoma. Gut Pathog. 2016;8:54.
93. World Health Organization, GLOBOCAN 2020: https://gco.iarc.fr/today/online-analysis-pie?v=2020&mode=cancer&mode_population=continents&population=900&populations=900&key=total&sex=0&cancer=39&type=0&statistic=5&prevalence=0&population_group=0&ages_group%5B%5D=0&ages_group%5B%5D=17&nb_items=7&group_cancer=1&include_nmsc=1&include_nmsc_other=1&half_pie=0&donut=0. Accessed 16 Mar 2021.
94. The Cancer Atlas https://canceratlas.cancer.org/taking-action/economic-burden/. Accessed 23 Mar 2021.
95. Kamps R, Brandao RD, Bosch BJ, Paulussen AD, Xanthoulea S, Blok MJ, Romano A. Next-generation sequencing in oncology: genetic diagnosis, risk prediction and cancer classification. Int J Mol Sci. 2017;18(2):308.
96. Bean LJH, Funke B, Carlston CM, Gannon JL, Kantarci S, Krock BL, Zhang S, Bayrak-Toydemir P, Committee ALQA. Diagnostic gene sequencing panels: from design to report-a technical standard of the American College of Medical Genetics and Genomics (ACMG). Genet Med. 2020;22(3):453–61.
97. Available online: http://www.tests.labmed.washington.edu/BROCA. Accessed 19 Mar 2021.
98. Available online: https://www.paragongenomics.com/applications/oncology/hereditary-cancer-risk-assessment/. Accessed 05 Mar 2021.
99. Okur V, Chung WK. The impact of hereditary cancer gene panels on clinical care and lessons learned. Cold Spring Harb Mol Case Stud. 2017;3(6):a002154.
100. Available online: http://www.sanger.ac.uk/science/groups/cancer-genomeproject. Accessed on 19 Mar 2021.
101. Available online: https://cancer.sanger.ac.uk/signatures/signatures_v2/. Accessed on 20 Mar 2021.
102. Alexandrov LB, Kim J, Haradhvala NJ, Huang MN, Tian Ng AW, Wu Y, Boot A, Covington KR, Gordenin DA, Bergstrom EN, et al. The repertoire of mutational signatures in human cancer. Nature. 2020;578(7793):94–101.
103. Cancer Genome Atlas Research N, Weinstein JN, Collisson EA, Mills GB, Shaw KR, Ozenberger BA, Ellrott K, Shmulevich I, Sander C, Stuart JM. The cancer genome atlas pan-cancer analysis project. Nat Genet. 2013;45(10):1113–20.
104. Consortium ITP-CAoWG. Pan-cancer analysis of whole genomes. Nature. 2020;578(7793):82–93.
105. AACR Project GENIE. Powering precision medicine through an international consortium. Cancer Discov. 2017;7(8):818–31.
106. Nikiforov YE, Carty SE, Chiosea SI, Coyne C, Duvvuri U, Ferris RL, Gooding WE, Hodak SP, LeBeau SO, Ohori NP, et al. Highly accurate diagnosis of cancer in thyroid nodules with follicular neoplasm/suspicious for a follicular neoplasm cytology by ThyroSeq v2 next-generation sequencing assay. Cancer. 2014;120(23):3627–34.
107. Livhits MJ, Zhu CY, Kuo EJ, Nguyen DT, Kim J, Tseng CH, Leung AM, Rao J, Levin M, Douek ML, et al. Effectiveness of molecular testing techniques for diagnosis of indeterminate thyroid nodules: a randomized clinical trial. JAMA Oncol. 2021;7(1):70–7.
108. Ward DG, Baxter L, Gordon NS, Ott S, Savage RS, Beggs AD, James JD, Lickiss J, Green S, Wallis Y, et al. Multiplex PCR and next generation sequencing for the non-invasive detection of bladder cancer. PLoS One. 2016;11(2):e0149756.
109. Witt RL. Targeted next generation sequencing with ThyroSeq v2.1 for indeterminate thyroid nodules in clinical practice. Del Med J. 2016;88(12):366–72.
110. Tse RT, Zhao H, Wong CY, Cheng CK, Kong AW, Peng Q, Chiu PK, Ng CF, Teoh JY. Urinary cell-free DNA in bladder cancer detection. Diagnostics (Basel). 2021;11(2):306.
111. Russo A, De Miguel PD, Gunasekaran M, Scilla K, Lapidus R, Cooper B, Mehra R, Adamo V, Malapelle U, Rolfo C. Liquid biopsy tracking of lung tumor evolutions over time. Expert Rev Mol Diagn. 2019;19(12):1099–108.
112. Murakami T, Yamamoto CM, Akino T, Tanaka H, Fukuzawa N, Suzuki H, Osawa T, Tsuji T, Seki T, Harada H. Bladder cancer detection by urinary extracellular vesicle mRNA analysis. Oncotarget. 2018;9(67):32810–21.
113. Liu D, Zhou B, Liu R. An RNA-sequencing-based transcriptome for a significantly prognostic novel driver signature identification in bladder urothelial carcinoma. PeerJ. 2020;8:e9422.
114. Wang Y, Mashock M, Tong Z, Mu X, Chen H, Zhou X, Zhang H, Zhao G, Liu B, Li X. Changing technologies of RNA sequencing and their applications in clinical oncology. Front Oncol. 2020;10:447.
115. Le Tourneau C, Delord JP, Gonçalves A, Gavoille C, Dubot C, Isambert N, Campone M, Trédan O, Massiani MA, Mauborgne C, et al. Molecularly targeted therapy based on tumour molecular profiling versus conventional therapy for advanced cancer (SHIVA): a multicentre, open-label, proof-of-concept, randomised, controlled phase 2 trial. Lancet Oncol. 2015;16(13):1324–34.
116. van der Velden DL, Hoes LR, van der Wijngaart H, van Berge Henegouwen JM, van Werkhoven E, Roepman P, Schilsky RL, de Leng WWJ, Huitema ADR, Nuijen B, et al. The Drug Rediscovery

protocol facilitates the expanded use of existing anticancer drugs. Nature. 2019;574(7776):127–31.
117. Monti P, Lionetti M, De Luca G, Menichini P, Recchia AG, Matis S, Colombo M, Fabris S, Speciale A, Barbieri M, et al. Time to first treatment and P53 dysfunction in chronic lymphocytic leukaemia: results of the O-CLL1 study in early stage patients. Sci Rep. 2020;10(1):18427.
118. Nogrady B. How cancer genomics is transforming diagnosis and treatment. Nature. 2020;579(7800):S10–s11.
119. Muller E, Brault B, Holmes A, Legros A, Jeannot E, Campitelli M, Rousselin A, Goardon N, Frebourg T, Krieger S, et al. Genetic profiles of cervical tumors by high-throughput sequencing for personalized medical care. Cancer Med. 2015;4(10):1484–93.
120. Borisov N, Sorokin M, Tkachev V, Garazha A, Buzdin A. Cancer gene expression profiles associated with clinical outcomes to chemotherapy treatments. BMC Med Genet. 2020;13(Suppl 8):111.
121. Liu CM, Chang SL, Yeh YH, Chung FP, Hu YF, Chou CC, Hung KC, Chang PC, Liao JN, Chan YH, et al. Enhanced detection of cardiac arrhythmias utilizing 14-day continuous ECG patch monitoring. Int J Cardiol. 2021;332:78–84.
122. Smolen JS, Aletaha D, McInnes IB. Rheumatoid arthritis. Lancet. 2016;388(10055):2023–38.
123. Lin YJ, Anzaghe M, Schülke S. Update on the pathomechanism, diagnosis, and treatment options for rheumatoid arthritis. Cell. 2020;9(4):880.
124. Silman AJ, Pearson JE. Epidemiology and genetics of rheumatoid arthritis. Arthritis Res. 4 Suppl 3(Suppl 3):S265–72.
125. Otón T, Carmona L. The epidemiology of established rheumatoid arthritis. Best Pract Res Clin Rheumatol. 2019;33(5):101477.
126. Harvey J, Lotze M, Stevens MB, Lambert G, Jacobson D. Rheumatoid arthritis in a Chippewa Band. I. Pilot screening study of disease prevalence. Arthritis Rheum. 1981;24(5):717–21.
127. Brighton SW, de la Harpe AL, van Staden DJ, Badenhorst JH, Myers OL. The prevalence of rheumatoid arthritis in a rural African population. J Rheumatol. 1988;15(3):405–8.
128. Silman AJ, Ollier W, Holligan S, Birrell F, Adebajo A, Asuzu MC, Thomson W, Pepper L. Absence of rheumatoid arthritis in a rural Nigerian population. J Rheumatol. 1993;20(4):618–22.
129. Heikkilä R, Aho K, Heliövaara M, Knekt P, Reunanen A, Aromaa A, Leino A, Palosuo T. Serum androgen-anabolic hormones and the risk of rheumatoid arthritis. Ann Rheum Dis. 1998;57(5):281–5.
130. Targońska-Stępniak B. Rheumatoid arthritis in elderly people. Wiadomosci lekarskie (Warsaw, Poland: 1960). 2019;72(9 cz 1):1676–82.
131. van den Hoek J, Boshuizen HC, Roorda LD, Tijhuis GJ, Nurmohamed MT, van den Bos GA, Dekker J. Mortality in patients with rheumatoid arthritis: a 15-year prospective cohort study. Rheumatol Int. 2017;37(4):487–93.
132. Naz SM, Symmons DP. Mortality in established rheumatoid arthritis. Best Pract Res Clin Rheumatol. 2007;21(5):871–83.
133. Sokka T, Abelson B, Pincus T. Mortality in rheumatoid arthritis: 2008 update. Clin Exp Rheumatol. 2008;26(5 Suppl 51):S35–61.
134. Abhishek A, Nakafero G, Kuo CF, Mallen C, Zhang W, Grainge MJ, Doherty M. Rheumatoid arthritis and excess mortality: down but not out. A primary care cohort study using data from Clinical Practice Research Datalink. Rheumatology (Oxford). 2018;57(6):977–81.
135. Crowson CS, Rollefstad S, Ikdahl E, Kitas GD, van Riel P, Gabriel SE, Matteson EL, Kvien TK, Douglas K, Sandoo A, et al. Impact of risk factors associated with cardiovascular outcomes in patients with rheumatoid arthritis. Ann Rheum Dis. 2018;77(1):48–54.
136. Deane KD, Demoruelle MK, Kelmenson LB, Kuhn KA, Norris JM, Holers VM. Genetic and environmental risk factors for rheumatoid arthritis. Best Pract Res Clin Rheumatol. 2017;31(1):3–18.
137. Damgaard D, Friberg Bruun Nielsen M, Quisgaard Gaunsbaek M, Palarasah Y, Svane-Knudsen V, Nielsen CH. Smoking is associated with increased levels of extracellular peptidylarginine deiminase 2 (PAD2) in the lungs. Clin Exp Rheumatol. 2015;33(3):405–8.
138. Wiley GB, Kelly JA, Gaffney PM. Use of next-generation DNA sequencing to analyze genetic variants in rheumatic disease. Arthritis Res Ther. 2014;16(6):490.
139. Diogo D, Kurreeman F, Stahl EA, Liao KP, Gupta N, Greenberg JD, Rivas MA, Hickey B, Flannick J, Thomson B, et al. Rare, low-frequency, and common variants in the protein-coding sequence of biological candidate genes from GWASs contribute to risk of rheumatoid arthritis. Am J Hum Genet. 2013;92(1):15–27.
140. Okada Y, Diogo D, Greenberg JD, Mouassess F, Achkar WA, Fulton RS, Denny JC, Gupta N, Mirel D, Gabriel S, et al. Integration of sequence data from a consanguineous family with genetic data from an outbred population identifies PLB1 as a candidate rheumatoid arthritis risk gene. PLoS One. 2014;9(2):e87645.
141. Wasserman AM. Diagnosis and management of rheumatoid arthritis. Am Fam Physician. 2011;84(11):1245–52.
142. Khani-Hanjani A, Lacaille D, Hoar D, Chalmers A, Horsman D, Anderson M, Balshaw R, Keown PA. Association between dinucleotide repeat in non-coding region of interferon-gamma gene and susceptibility to, and severity of, rheumatoid arthritis. Lancet. 2000;356(9232):820–5.
143. Ye H, Zhang J, Wang J, Gao Y, Du Y, Li C, Deng M, Guo J, Li Z. CD4 T-cell transcriptome analysis reveals aberrant regulation of STAT3 and Wnt signaling pathways in rheumatoid arthritis: evidence from a case-control study. Arthritis Res Ther. 2015;17(1):76.
144. Sumitomo S, Nagafuchi Y, Tsuchida Y, Tsuchiya H, Ota M, Ishigaki K, Nakachi S, Kato R, Sakurai K, Hanata N, et al. A gene module associated with dysregulated TCR signaling pathways in CD4(+) T cell subsets in rheumatoid arthritis. J Autoimmun. 2018;89:21–9.
145. Heruth DP, Gibson M, Grigoryev DN, Zhang LQ, Ye SQ. RNA-seq analysis of synovial fibroblasts brings new insights into rheumatoid arthritis. Cell Biosci. 2012;2(1):43.
146. Ai R, Laragione T, Hammaker D, Boyle DL, Wildberg A, Maeshima K, Palescandolo E, Krishna V, Pocalyko D, Whitaker JW, et al. Comprehensive epigenetic landscape of rheumatoid arthritis fibroblast-like synoviocytes. Nat Commun. 2018;9(1):1921.
147. Okada Y, Kishikawa T, Sakaue S, Hirata J. Future directions of genomics research in rheumatic diseases. Rheum Dis Clin North Am. 2017;43(3):481–7.
148. Okada Y, Muramatsu T, Suita N, Kanai M, Kawakami E, Iotchkova V, Soranzo N, Inazawa J, Tanaka T. Significant impact of miRNA-target gene networks on genetics of human complex traits. Sci Rep. 2016;6:22223.
149. Chen B, Sun L, Zhang X. Integration of microbiome and epigenome to decipher the pathogenesis of autoimmune diseases. J Autoimmun. 2017;83:31–42.
150. Aletaha D, Neogi T, Silman AJ, Funovits J, Felson DT, Bingham CO 3rd, Birnbaum NS, Burmester GR, Bykerk VP, Cohen MD, et al. Rheumatoid arthritis classification criteria: an American College of Rheumatology/European League Against Rheumatism collaborative initiative. Ann Rheum Dis. 2010;69(9):1580–8.
151. Niu X, Chen G. Clinical biomarkers and pathogenic-related cytokines in rheumatoid arthritis. J Immunol Res. 2014;2014:698192.
152. Aiman AQ, Nesrin M, Amal A, Nassar AD. A new tool for early diagnosis of rheumatoid arthritis using combined biomarkers; synovial MAGE-1 mRNA and serum anti-CCP and RF. Pan Afr Med J. 2020;36:270.
153. Giannopoulou EG, Elemento O, Ivashkiv LB. Use of RNA sequencing to evaluate rheumatic disease patients. Arthritis Res Ther. 2015;17(1):167.
154. Sumitomo S, Nagafuchi Y, Tsuchida Y, Tsuchiya H, Ota M, Ishigaki K, Suzuki A, Kochi Y, Fujio K, Yamamoto K. Transcriptome anal-

ysis of peripheral blood from patients with rheumatoid arthritis: a systematic review. Inflammation and Regeneration. 2018;38:21.

155. Glocker MO, Guthke R, Kekow J, Thiesen HJ. Rheumatoid arthritis, a complex multifactorial disease: on the way toward individualized medicine. Med Res Rev. 2006;26(1):63–87.

156. van der Pouw Kraan TC, van Gaalen FA, Huizinga TW, Pieterman E, Breedveld FC, Verweij CL. Discovery of distinctive gene expression profiles in rheumatoid synovium using cDNA microarray technology: evidence for the existence of multiple pathways of tissue destruction and repair. Genes Immun. 2003;4(3):187–96.

157. van de Sande MG, Baeten DL. Immunopathology of synovitis: from histology to molecular pathways. Rheumatology (Oxford). 2016;55(4):599–606.

158. Sellam J, Marion-Thore S, Dumont F, Jacques S, Garchon HJ, Rouanet S, Taoufik Y, Hendel-Chavez H, Sibilia J, Tebib J, et al. Use of whole-blood transcriptomic profiling to highlight several pathophysiologic pathways associated with response to rituximab in patients with rheumatoid arthritis: data from a randomized, controlled, open-label trial. Arthritis Rheumatol (Hoboken, NJ). 2014;66(8):2015–25.

159. Wright HL, Thomas HB, Moots RJ, Edwards SW. Interferon gene expression signature in rheumatoid arthritis neutrophils correlates with a good response to TNFi therapy. Rheumatology (Oxford). 2015;54(1):188–93.

Epigenetics in Epidemiology

4

Humberto Nicolini, Alma Delia Genis-Mendoza, and José Jaime Martínez-Magaña

Abbreviations

AHRR	Aryl-hydrocarbon receptor repressor
ChIP-chip	Chromatin immunoprecipitation coupled to microarrays
ChIP-seq	Chromatin immunoprecipitation coupled to sequencing
CpG	Cytosine nucleotide followed by a guanine
DNA	Deoxyribonucleic acid
EWAS	Epigenome-wide association study
FDA	Food and Drugs Administration
PCR	Polymerase chain reaction
PIWI	P-element-induced wimpy testis
RNA	Ribonucleic acid

Introduction

Epigenetics is defined in several ways and is not a unique definition. Armstrong defines epigenetics as "the study of changes in gene function that are mitotically and/or meiotically heritable and that do not entail a change in the sequence of DNA" [1]. Meanwhile, Conrad Waddington, who coined the term, defines epigenetics as "the interaction of genes with their environment, which bring the phenotype into being" [2]. The National Human Genome Research Institute defines epigenetics as *an emerging field of science that studies heritable changes caused by the activation and deactivation of genes without any change in the underlying DNA sequence of the organism*. The word epigenetics is of Greek origin and means *over and above (epi) the genome*. The latest definitions are more related to the organism-environment, but others see epigenetics to a more cellular level [3, 4]. Nanney noted that cells with the same genotype might manifest different phenotypes, and differences in the expression of these could persist during cellular division [5]. Independently of the definition, all suggested that the focus of epigenetics is not the gene's sequence itself, but the surroundings. A higher focus in human epigenetics began after the completion of the human genome sequence, to the opportunity to link the environment to the genome [6], and the field has emerged in different areas like biological development, public health, cell differentiation, and epigenomic epidemiology.

H. Nicolini (✉) · A. D. Genis-Mendoza · J. J. Martínez-Magaña
Laboratorio de Genómica de Enfermedades Psiquiátricas y Neurodegenerativas, Instituto Nacional de Medicina Genómica, CDMX, Mexico
e-mail: hnicolini@inmegen.gob.mx; adgenis@inmegen.gob.mx

Epigenetic Process

The epigenetic process includes DNA methylation, histone modification (methylation, acetylation, phosphorylation, and ubiquitination), and those modulated by RNA (microRNAs and long noncoding RNA) [7]. These epigenetic processes are also called epigenetic marks [8].

DNA methylation is defined as adding a methyl group to a cytosine; the cytosine has to be linked to continuous guanine and is also known as a CpG site. The addition of the methyl groups is catalyzed by enzymes, known as DNA methyltransferases (DNMTs). The main DNMTs are DNMT1, DNMT3A, and DNMT3B, which catalyze the transfer of methyl groups from the S-adenosyl-L-methionine (SAM) to the 5′ position of the cytosine [8]. DNMT1 is essential for maintenance, and DNMT3A/DNMT3B are de novo methyltransferases [9, 10]. DNA methylation is recognized by DNA methyl-binding proteins that interfere with the union of transcription factors or promoters the recruitment of chromatin remodeling proteins (SWI/SNF, ISW1, CHD, and INO80) [11]. One of the first mechanisms identified between DNA methylation and gene expression regulation is the CpG islands' effect present in gene promoters

J. C. Gomez-Verjan, N. A. Rivero-Segura (eds.), *Principles of Genetics and Molecular Epidemiology*, https://doi.org/10.1007/978-3-030-89601-0_4

[12]. CpG islands are sequences of more than 1000 base pairs with an elevated guanine and cytosine content, which could be present in high density in gene promoters. The methylation of CpG sites on CpG islands associated with promoters could recruit proteins that repress transcription of the gene, suppressing the expression. Nevertheless, the relationship between DNA methylation, principally outside CpG islands, in the control of gene expression is still under study [13].

DNA sequence and histone proteins conformed to the basic unit of chromatin, known as a nucleosome. The nucleosome is conformed to a core of two copies of histones H2A, H2B, H3, and H4 [14]. The structure of chromatin controls gene expression by altering the compaction of DNA and, consequently, the accessibility of transcription factors to regions of the DNA controlling gene expression [15]. The compaction of the DNA is regulated by adding functional groups (acetyl, methyl, and phosphoryl) to the histone proteins, principally in the tails of the histones [15]. These functional groups are produced as posttranslational modifications created by specific enzymes acting mainly at lysine (K) and arginine (R) amino acids. These modifications could be present in multiple but specific sites of the histones. The main effect of these modifications is to activate or repress gene expression depending on the environment stimulus, acting at genes that coded transcription factors and signaling pathways of the cells [16]. Histone acetylation is principally added in the lysine residues; at the difference, methylation occurs at both residues (lysine and arginine). The enzymes that catalyzed the addition or removal of acetyl groups are known as histone acetyltransferases/histone deacetylases. In homology, the modifications by methyl groups are catalyzed by histone methyltransferases/histone demethylases [17, 18]. Even when the dynamics of the histone modifications are complex, some marks have been associated with gene expression regulation [19, 20]. For example, trimethylation of the 27-lysine residue (H3K27me3) and trimethylation of the nine-lysine residue of the histone 3 (H3K9me3) are repressive marks. The same histone, trimethylation of 4 and 36 residues of histone 3 (H3K4me3 and H3K36me3), is an activation mark that leads to a permissive state (i.e., allows binding of transcription factors) of chromatin.

The other epigenetic process is based on non-protein-coding RNAs. Non-protein-coding RNAs (also called noncoding RNAs) are functional transcripts that are not translated to proteins but could impact gene expression regulation [21]. Noncoding RNAs are highly diverse, but one characteristic that could classify the former is the base pair size. Mainly, we could find two different noncoding RNAs: small (<200 bp) including microRNAs silencing RNAs and PIWI-interacting) and long noncoding (>200 bp) [22]. MicroRNAs are nucleotide guides (21–24 base pairs) that regulate the expression of messenger RNAs (mainly at 3′ untranslated regions) that contain complementary sequences for the particular microRNA [23]. In humans, the pairing of the messenger RNA (protein-coding) and microRNAs reduced the expression of proteins by activating diverse mechanisms for messenger RNA degradation [24, 25]. Meanwhile, the effect of long noncoding RNAs is more diverse; in general, these RNAs could interact with DNA, RNA, and proteins, where the interaction modulates chromatin structure, and also RNA splicing, stability, and translation [26].

Epigenetic Technology Analysis

The evaluation of the epigenetic process, in the laboratory setting, has three main essential points to be considered: (i) biological sample (tissues or liquid biopsies, blood, chemical fixed tissues) where the evaluation is going to be performed, (ii) the epigenetic process to be analyzed (DNA methylation, RNA-based mechanisms), (iii) the extension of the genes evaluated, a limited number of genes (candidates or pathways) or genome-wide analysis (all the genome evaluated, mainly refer as epigenome-wide analysis). Concerning the biological sample, the epigenetic marks are stable in fluids, like plasma, serum, or urine [27–29]. These kinds of marks had an advantage compared to other sources of molecular marks (like metabolites). Also, some authors report that the stability of these epigenetic marks is seen on tissue preparations (frozen, dried blood spots, or formalin-fixed paraffin-embedded tissues) [30, 31]. A disadvantage of the epigenetic marks related to the biological sample is the tissue-specific and temporal changes of these marks [32–34]. The epigenetic marks depend on cell differentiation, making these marks fluctuate according to the development and tissue microenvironment [35–37]. The tissue specific epigenetic marks outpoint the need to be evaluated in the tissue under interest or perform the analysis in proxy biological samples (i.e., tissues with high correlation between these marks). The use of these proxy tissues is exemplified in brain-related diseases, where having a biological brain sample is complicated, and blood samples are extensively used [38–40]. Another point to consider is the extension of genes to be evaluated, some or all the genome [41–43]. This extension of analysis will be crucial for the epigenetic technology to be applied during the study.

The most studied subfield of epigenetics is DNA methylation. The improvement of DNA methylation analysis became possible with the development of bisulfite transformation of DNA [44]. In bisulfite transformation, DNA is treated with bisulfite, which converts cytosine to uracil, while methylated cytosine is not converted [45]. Next, this bisulfite-treated DNA is amplified by PCR, whereby complementarity, nonmethylated cytosines are recognized as thymine, and methylated cytosines remain as cytosines. Once amplified, the DNA could be evaluated by sequencing (whole-genome bisulfite sequencing) or by array-based analysis [46, 47]. Depending on the regions and number of samples targeted to be analyzed, the amount of data generated increases and now is the point to be processed by bioinformatic techniques to obtain meaningful biological information [48–50].

Histone modification is evaluated by a technique known as chromatin immunoprecipitation (ChIP) [51–53]. The basic steps in this technique are: (i) fixation of proteins [2], sonication [3], immunoprecipitation, and [4] analysis of the DNA. The fixation is mainly performed by cross-linking proteins with formaldehyde; once this cross-linking is performed, the proteins that interact with DNA act as a shield for the next step. In the next step, all the DNAs are sonicated to reduce the size of the fragments (200–1000 base pairs). Once sonicated, the DNA fragments are exposed to an antibody to precipitate all the complex (protein plus DNA) using a specific antibody. The antibody could target all the proteins that interact with DNA, transcription factors, enhancers, or histones, making this technique very versatile for identifying DNA sequences associated with these proteins. Posterior to antibody precipitation, the DNA that interacts with the protein (sequences co-precipitated with protein) is analyzed by DNA microarray (ChIP-chip) or sequencing techniques (ChIP-seq). The ChIP techniques allow us to identify all the DNA sequences associated with the specific protein.

The previously described techniques could be named now as high-resolution techniques. These high-resolution techniques are characterized by generating large amounts of information (epigenome-wide), but the cost and infrastructure required for these techniques are not under every clinical laboratory. For these reasons, the use of these technologies in clinical settings is reduced. Some techniques with lower resolution have been developed to analyze the epigenetics under clinical laboratories, like pyrosequencing, methylation-sensitive single-strand conformation analysis, single nucleotide primer extension, photo-crosslinking hybridization assays, or PCR in situ hybridization, and others [54, 55].

Overview of Bioinformatic Analysis of Epigenome

As previously mentioned, the most studied epigenome-wide mark is DNA methylation [32, 56]. Consequently, we will center on exposing some bioinformatic tools and pipelines to analyze this data with a focus on array-based analysis. The most commonly used arrays for DNA methylation analysis are the commercial Infinium 450 K (Illumina, USA) and recently updated to the Infinium Human Methylation Epic (Illumina, USA), being the first approximations to epigenome-wide analysis [57]. The previous arrays are a

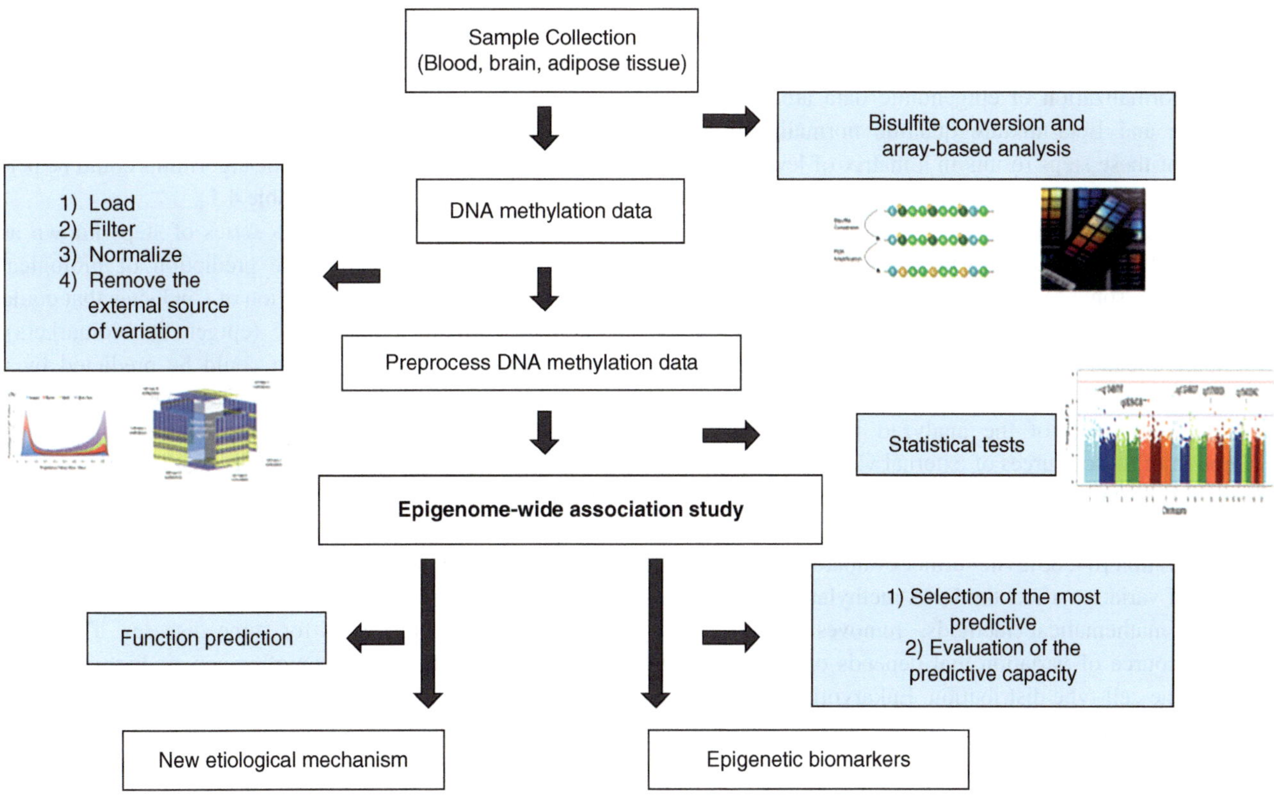

Fig. 4.1 Overview of the steps for an epigenome-wide association study (EWAS)

fixed collection of know sequence small nucleotide probes associated with silica beads and measured by the fluorescence intensity, depending on the hybridization to the target sequence [58]. Even when different software and algorithms perform the analysis of these arrays, most of them follow the next steps: (1) load, (2) filter, (3) normalize, (4) remove the source external variations, and (5) statistical tests (Fig. 4.1). Load or import the data is the action to store the information in the computer memory [59]. The load of the data could be a limitation state in hardware with low random-access memory, depending on sample size the need for computing power could increase exponentially, showing the necessity for high-performance cluster computing [60]. The former computing clusters increase the cost of the bioinformatic analysis of epigenomic data and, consequently, have the same effect on the study budget.

Epigenomic data has a high level of complexity [61]. The complexity behind the epigenomic data adds the challenge to identify robust signals from noise; to overcome this, the filter, normalization, and removal of the source of external variations by bioinformatic and statistical algorithms are fundamental. The algorithms of the previous steps depend on the experiment design and the tissue explored, but we will expose some steps that could be a guide. Filtering array-based epigenome data follow the next steps: (i) removal of probes and samples with low detection rates, (ii) remove of non-CpG probes, SNP-related, multi-hit, and sexual chromosome-associated probes. Next to filtering, the normalization process reduces the systematic errors behind the differential distribution of probes. The most effective methods for normalization of epigenomic data are peak-based correction and Beta-mixture quantile normalization [62]. The end of these steps results in a matrix of levels of DNA methylation (Beta-values or M-values) by each probe and each analyzed sample [63].

The next step, remotion of the external sources of variation, requires some technical knowledge of the experiment design like tissue, type of cell distribution, experiment facilities (experimental staff, number, and position of arrays), and covariates (i.e., species, age, gender, developmental stage, comorbidities, etc.) of the analyzed biological samples. Remotion of these sources of external variations is based on unsupervised statistical methods, like single-value deconvolution, principal component analysis, mean centering, or standardization [64, 65]. The former evaluates if any of these sources of variation affect the DNA methylation data matrix and, by mathematical methods, removes these effects. Another source of variation that depends on the evaluated tissue is the cell-type distribution. Eukaryotic cells follow a process of differentiation and changes in DNA methylation patterns, promoting a source of experimental noise. The removal of cell-dependent DNA methylation patterns is based on supervised and unsupervised methods. The supervised methods are based on the development of a catalog of cell-specific DNA methylation signatures by sorting (Cell reference generation), posterior evaluation of DNA methylation, and mathematical computation (deconvolution) of the experimental data with this reference, to infer the probability distribution of the cell types in the samples [66–68]. Meanwhile, the unsupervised methods find from the data the probable distribution of the cells and adjust the data based on these distributions [69, 70]. The previous steps are known as preprocessing of data to allow us to have a reduction in the noise and increase the replicability of the results generated by the next step (statistical tests), the *epigenome-wide association study* (EWAS).

EWAS is the mathematical contrast of the preprocessed DNA methylation data with a phenotypic variable. The variables could be categorical or continuous depending on the study design. EWAS is explored in four primary study designs: case-control, families, cohorts, and longitudinal studies [71]. The most explored study design is the case-control study, where cases are affected by a disease (named phenotype). However, this design has the limitation of requiring large sample sizes to get well-powered EWAS. The case-control statistical contrast is based on logistic regression models adjusted by different covariates (gender, age, ancestry, etc.) and with a p-value cutoff of 5e-8 to be considered as statistically significant [72]. Once generated the statistical analysis, the CpG sites (known as differentially methylated probes or sites) with lower p-values are the ones with the higher difference between both groups, suggesting an effect of disease on the CpG site or vice versa. The case-control study could not establish causality between the CpG site and the disease, pointing to the need for functional studies [71]. The previous bioinformatic algorithms could be performed in numerous software (Table 4.1).

Once performed the EWAS, a series of steps known as post-EWAS starts, including the prediction of biological functional impact and the generation of CpG sites that could predict the evaluated phenotype (epigenetic biomarkers). The biological functional impact could be predicted by a series of diverse prediction algorithms and databases access, like CpG site annotation to gene expression regulatory elements [50, 73, 74], enrichment analysis [75–77], protein-protein interaction networks [78], and cell-specific epigenetic marks [79]. The former allows us to identify new etiologic mechanisms behind the analyzed disorder; Table 4.2 presents a summary of some EWAS that have generated some new etiological mechanisms for some diseases. The other post-EWAS step, the prediction of epigenetic biomarkers, is a more clinical and epidemiological application of the differentially methylated sites generated by EWAS. The identification of epigenetic biomarkers follows the next two steps: (1) selection of the most predictive CpG sites and (2) evaluation of the prediction capacity of the selected sites.

Table 4.1 Example of pipelines for epigenome-wide analysis data

Software	Filter	Normalization	External variation remotion	EWAS	Website
wateRmelon	Yes	Yes	Yes	No	https://bioconductor.org/packages/release/bioc/html/wateRmelon.html
bigmelon	Yes	Yes	Yes	No	http://bioconductor.org/packages/release/bioc/html/bigmelon.html
minfi	Yes	Yes	Yes	Yes	https://www.bioconductor.org/help/course-materials/2015/BioC2015/methylation450k.html
ChAMP (The Chip Analysis Methylation Pipeline)	Yes	Yes	Yes	Yes	https://bioconductor.org/packages/release/bioc/vignettes/ChAMP/inst/doc/ChAMP.html
missMethyl	Yes	Yes	Yes	Yes	https://www.bioconductor.org/packages/release/bioc/vignettes/missMethyl/inst/doc/missMethyl.html
OSCA (OmicS-data-based Complex trait Analysis)	No	Yes	Yes	Yes	https://cnsgenomics.com/software/osca/#Overview
EWAS	No	No	No	Yes	http://www.bioapp.org/ewas./service.html
GLINT	No	No	Yes	Yes	https://glint-epigenetics.readthedocs.io/en/latest/
Partek ®	Yes	Yes	No	Yes	https://www.partek.com/application-page/methylation/
GenomeStudio ®	Yes	Yes	No	Yes	https://www.illumina.com/techniques/microarrays/array-data-analysis-experimental-design.html

Table 4.2 Examples of important epigenome-wide association studies (EWAS)

Phenotype	Sample size	Main findings	Ethnicity	Reference
Lifetime estrogen exposure	31,864 unrelated	DNA methylation score associated with breast cancer risk	European	[85]
COVID-19 severity	406 unrelated	Genes of the inflammasome and HLA-C	European	[86]
Coffee and tea consumption	15,789 unrelated	Differential methylation of AHRR, F2RL3, FLJ43663, HDAC4, GFI1, and PHGDH genes	European and African-American	[87]
Kidney function	13,537 unrelated	CpG sites enriched in kidney development	Trans-ethnic	[88]
Cigarette smoking-related lung diseases	6000 unrelated	Hypomethylation and lower expression of CHRNA5 are causally linked to increased risk of COPD and lung cancer	European	[89]
Alzheimer's disease	1453 unrelated	Differentially methylation on 121 genes associated with neuropathology	European	[90]
Depression	724 twins	Genes implicated in response to stress	European	[91]
Schizophrenia	1831 unrelated	Genes in neuronal function, genes previously associated with schizophrenia, and genes also involved in T-cell development	European	[92]
Educational attainment	10,767 unrelated	Association with smoking	European	[93]
Birth weight	8825 unrelated	CpG sites associated with maternal smoking and BMI	European	[94]

Well-powered EWAS generates many associated CpG sites, making it difficult to use these sites in clinical settings; different statistical techniques have been applied to reduce these CpG sites. The applied techniques are based on machine learning algorithms, like support vector machines, Gaussian-mixture models, random forest, hierarchical clustering [80–82], methylation risk scores [83], or recently a combination of both [84]. The algorithms select the "best" conjunct of CpG sites and a mathematical model to predict the phenotype. Once this conjunct and model are generated, it must be evaluated on a different sample to know the replication capacity of the marker. If the conjunct has an acceptable predictive capacity on the replication sample, then it could be called a possible *epigenetic biomarker* and ready for testing under other clinical settings. The most outstanding EWAS appear depicted in Table 4.2.

Epigenetic Biomarkers

The epigenetic alterations could be a significant source of knowledge behind the etiology of the diseases [95–97], but a more clinical application is developing epigenetic disease biomarkers [55]. The development of these epigenetic biomarkers could be of great interest for epigenetic epidemiology to characterize the utility of these biomarkers in different human diseases, mainly to characterize the relationship between the exposures to different environments in clinical data [98–100]. An epigenetic biomarker could be any mark that could be applied in risk measurement, diagnosis, prediction, and monitoring of any disease. Nevertheless, to be used in a daily clinical setting, an epigenetic biomarker must be sensible and specific, and inexpensive [101–103]. The amount of information generated in the last years regarding epigenetic biomarkers is large. A search in the PubMed database (April 2021) using the term "epigenetic biomarkers" led to a total of 13,489 results. Even when the amount of information is large, very few epigenetic markers have been approved for use under clinical diagnostic settings, like Epi proColon, Cologuard, AssureMDx, GynTect, PredictMDx, and others [104]. Most of these biomarkers focus on different types of cancer (colorectal, hepatocellular carcinoma, lung, prostate, bladder, cervical, and glioblastoma). Most of the claimed biomarkers are still under preclinical evaluation or still under development. The gap between the number of epigenetic biomarkers and the information generated could be explained under the clinical laboratory settings. Most clinical laboratories do not have the required infrastructure to perform routine analysis of epigenetic information [105].

Epigenetic biomarkers had been applied more advanced as risk prediction and proposed to be used in epidemiologic studies. Some of these areas are applied in aging, substance use, and nutrition [106–112]. One of these applications is the development of epigenetic markers for cigarette smoking. Cigarette smoking is one of the leading causes of preventable mortality [113]. The accurate detection of smoking is of high priority for targeting those individuals for treatment purposes [114, 115]. To detect smoking behavior in individuals is based on two measurements: self-reported and cotinine (nicotine metabolite) levels, but these indicators had some limitations. On the one hand, in epidemiological studies, the self-reported smoking behavior is reported to be accurate [116–118]; nevertheless, under some circumstances (adolescents or pregnant women), the self-reported have elevated rates of disagreement [117, 119, 120]. On the other hand, the cotinine levels had a short half-life, promoting difficulty in being measured under epidemiological analysis [121–123]. To overcome these limitations, some authors proposed using epigenetic biomarkers, like the CpG site (cg05575921), located in the aryl hydrocarbon receptor repressor (*AHRR*). The demethylation of this site has been consistently associated with smoking behavior [123, 124] and has a high correlation with other smoking biomarkers [107]. In a recent study, this CpG site is proposed to be a valuable marker to predict the risk of heavy smoking behavior.

Another example of an epigenetic biomarker that could be used in an epidemiological perspective is the epigenetic clocks. In 2013, the first epigenetic age, using DNA methylation data, estimation method with accuracy in different tissues was published by Horvath [125]. The method proposed by Horvath estimates a statistical parameter based on an elastic net regression, which he called DNA methylation age. Comparing DNA methylation age (a proxy variable of biological aging) with chronological age allows us to calculate the acceleration or non-acceleration of aging and how the environment could alter this process [65, 108, 126, 127]. Since the publishing of this clock, many authors have applied this method to evaluate the effects of environmental exposures on the aging process. In a recent meta-analysis, Ryan et al. report that body mass index increased DNA methylation age and was associated with frailty index [128]. Even when Horvath started a milestone in the epigenetic age, other authors had improved this method by adding other variables to the model to perform more accurate predictions, like PhenoAge [129] or GrimAge [130]. These clocks are highly associated with all causes of mortality (cancer, cardiovascular disorders, and cognitive impairment) and reduction in lifespan [131]. The accelerations on epigenetic age could capture different sources of stress promoting a shorting of life expectation in response to these events.

Epigenetics in Pharmacology

Besides the development of epigenetic biomarkers and disease-associated epigenetic changes, these changes could also affect the response to environmental factors [132]. Some environmental factors that have demonstrated epigenetic changes are toxins, diet, stress, and xenobiotics [133, 134]. Inside the xenobiotics, we could find pharmacological drugs; moreover, some have been reported to promote epigenetic modification, known as epigenetic drugs. The use of these epigenetic drugs is a field under development. Some authors suggested that these drugs could have higher therapeutic response rates with lower adverse reactions, pointing to their use under a personalized medicine approach [135–138]. In deficiency, the use of these epigenetic drugs requires knowledge about the epigenetic alterations that are consistent in different disorders to target the most precise epigenetic alteration [138, 139]. These drugs are azacitidine and decitabine, which the FDA approved to treat chronic leukemia and myelodysplastic syndrome. These drugs promote a hypomethylation of some genes that had been silent under cancer development [140, 141]. Other epigenetic drugs under development are oriented to treating multiple sclerosis, pain, and memory deficits [142–144].

Conclusions

The epigenetics biomarkers are becoming more regularly analyzed in investigation settings, finding more associations with different disorders, and possibly will generate new diagnostic strategies focusing on precision medicine. The perspective is that the development of new technologies and clinical guidelines using epigenetic biomarkers could help in the diagnosis, screening, and treatment of several human diseases.

References

1. Pinel C, Prainsack B, McKevitt C. Markers as mediators: a review and synthesis of epigenetics literature. BioSocieties. 2018;13(1):276–303.
2. Waddington CH. The epigenotype. Int J Epidemiol. 2012;41(1):10–3.
3. Haig D. Commentary: the epidemiology of epigenetics: Figure 1. Int J Epidemiol. 2012;41(1):13–6.
4. Dolinoy D, Weidman J, Jirtle R. Epigenetic gene regulation: linking early developmental environment to adult disease. Reprod Toxicol. 2007;23(3):297–307.
5. Nanney DL. Epigenetic control systems. Proc Natl Acad Sci U S A. 1958;44(7):712–7.
6. Meloni M. The social brain meets the reactive genome: neuroscience, epigenetics and the new social biology. Front Hum Neurosci. 2014;8:309.
7. Kim JK, Samaranayake M, Pradhan S. Epigenetic mechanisms in mammals. Cell Mol Life Sci CMLS. 2009;66(4):596–612.
8. Jin B, Robertson KD. DNA methyltransferases, DNA damage repair, and cancer. Adv Exp Med Biol. 2013;754:3–29.
9. Okano M, Xie S, Li E. Cloning and characterization of a family of novel mammalian DNA (cytosine-5) methyltransferases. Nat Genet. 1998;19(3):219–20.
10. Okano M, Bell DW, Haber DA, Li E. DNA methyltransferases Dnmt3a and Dnmt3b are essential for de novo methylation and mammalian development. Cell. 1999;99(3):247–57.
11. Zhang P, Torres K, Liu X, Liu C-G, Pollock RE. An overview of chromatin-regulating proteins in cells. Curr Protein Pept Sci. 2016;17(5):401–10.
12. Deaton AM, Bird A. CpG islands and the regulation of transcription. Genes Dev. 2011;25(10):1010–22.
13. Siegfried Z, Simon I. DNA methylation and gene expression. Wiley Interdiscip Rev Syst Biol Med. 2010;2(3):362–71.
14. Luger K, Mäder AW, Richmond RK, Sargent DF, Richmond TJ. Crystal structure of the nucleosome core particle at 2.8 A resolution. Nature. 1997;389(6648):251–60.
15. Bannister AJ, Kouzarides T. Regulation of chromatin by histone modifications. Cell Res. 2011;21(3):381–95.
16. Kouzarides T. Chromatin modifications and their function. Cell. 2007;128(4):693–705.
17. Marmorstein R, Trievel RC. Histone modifying enzymes: structures, mechanisms, and specificities. Biochim Biophys Acta. 2009;1789(1):58–68.
18. Butler JS, Koutelou E, Schibler AC, Dent SYR. Histone-modifying enzymes: regulators of developmental decisions and drivers of human disease. Epigenomics. 2012;4(2):163–77.
19. Gibcus JH, Dekker J. The context of gene expression regulation. F1000 Biol Rep. 2012;4:8.
20. Pfluger J, Wagner D. Histone modifications and dynamic regulation of genome accessibility in plants. Curr Opin Plant Biol. 2007;10(6):645–52.
21. Hubé F, Francastel C. Coding and non-coding RNAs, the frontier has never been so blurred. Front Genet. 2018;9:140.
22. Boon RA, Jaé N, Holdt L, Dimmeler S. Long noncoding RNAs: from clinical genetics to therapeutic targets? J Am Coll Cardiol. 2016;67(10):1214–26.
23. Pasquinelli AE. MicroRNAs and their targets: recognition, regulation and an emerging reciprocal relationship. Nat Rev Genet. 2012;13(4):271–82.
24. Krol J, Loedige I, Filipowicz W. The widespread regulation of microRNA biogenesis, function and decay. Nat Rev Genet. 2010;11(9):597–610.
25. Huntzinger E, Izaurralde E. Gene silencing by microRNAs: contributions of translational repression and mRNA decay. Nat Rev Genet. 2011;12(2):99–110.
26. Statello L, Guo C-J, Chen L-L, Huarte M. Gene regulation by long non-coding RNAs and its biological functions. Nat Rev Mol Cell Biol. 2021;22(2):96–118.
27. Glinge C, Clauss S, Boddum K, Jabbari R, Jabbari J, Risgaard B, et al. Stability of circulating blood-based MicroRNAs - pre-analytic methodological considerations. PLoS One. 2017;12(2):e0167969.
28. Park NJ, Zhou H, Elashoff D, Henson BS, Kastratovic DA, Abemayor E, et al. Salivary microRNA: discovery, characterization, and clinical utility for oral cancer detection. Clin Cancer Res Off J Am Assoc Cancer Res. 2009;15(17):5473–7.
29. Zubakov D, Boersma AWM, Choi Y, van Kuijk PF, Wiemer EAC, Kayser M. MicroRNA markers for forensic body fluid identification obtained from microarray screening and quantitative RT-PCR confirmation. Int J Leg Med. 2010;124(3):217–26.
30. Peiró-Chova L, Peña-Chilet M, López-Guerrero JA, García-Giménez JL, Alonso-Yuste E, Burgues O, et al. High stability of microRNAs in tissue samples of compromised quality. Virchows Arch Int J Pathol. 2013;463(6):765–74.
31. Patnaik SK, Mallick R, Yendamuri S. Detection of microRNAs in dried serum blots. Anal Biochem. 2010;407(1):147–9.
32. Barros SP, Offenbacher S. Epigenetics: connecting environment and genotype to phenotype and disease. J Dent Res. 2009;88(5):400–8.
33. Yamada L, Chong S. Epigenetic studies in Developmental Origins of Health and Disease: pitfalls and key considerations for study design and interpretation. J Dev Orig Health Dis. 2017;8(1):30–43.
34. Jacques M, Hiam D, Craig J, Barrès R, Eynon N, Voisin S. Epigenetic changes in healthy human skeletal muscle following exercise– a systematic review. Epigenetics. 2019;14(7):633–48.
35. Wilson CB, Rowell E, Sekimata M. Epigenetic control of T-helper-cell differentiation. Nat Rev Immunol. 2009;9(2):91–105.
36. Henning AN, Roychoudhuri R, Restifo NP. Epigenetic control of CD8+ T cell differentiation. Nat Rev Immunol. 2018;18(5):340–56.
37. Atkinson S, Armstrong L. Epigenetics in embryonic stem cells: regulation of pluripotency and differentiation. Cell Tissue Res. 2008;331(1):23–9.
38. Svoboda LK, Neier K, Wang K, Cavalcante RG, Rygiel CA, Tsai Z, et al. Tissue and sex-specific programming of DNA methylation by perinatal lead exposure: implications for environmental epigenetics studies. Epigenetics. 2020;8:1–21.
39. Walton E, Hass J, Liu J, Roffman JL, Bernardoni F, Roessner V, et al. Correspondence of DNA methylation between blood and brain tissue and its application to schizophrenia research. Schizophr Bull. 2016;42(2):406–14.
40. Bakulski KM, Halladay A, Hu VW, Mill J, Fallin MD. Epigenetic research in neuropsychiatric disorders: the "tissue issue". Curr Behav Neurosci Rep. 2016;3(3):264–74.
41. Qureshi IA, Mehler MF. Advances in epigenetics and epigenomics for neurodegenerative diseases. Curr Neurol Neurosci Rep. 2011;11(5):464–73.
42. Agarwal G, Kudapa H, Ramalingam A, Choudhary D, Sinha P, Garg V, et al. Epigenetics and epigenomics: underlying mechanisms, relevance, and implications in crop improvement. Funct Integr Genomics. 2020;20(6):739–61.
43. Rosen ED, Kaestner KH, Natarajan R, Patti M-E, Sallari R, Sander M, et al. Epigenetics and epigenomics: implications for diabetes and obesity. Diabetes. 2018;67(10):1923–31.

44. Tollefsbol TO. Advances in epigenetic technology. Methods Mol Biol Clifton NJ. 2011;791:1–10.
45. Zhang Y, Rohde C, Tierling S, Stamerjohanns H, Reinhardt R, Walter J, et al. DNA methylation analysis by bisulfite conversion, cloning, and sequencing of individual clones. Methods Mol Biol Clifton NJ. 2009;507:177–87.
46. Leti F, Llaci L, Malenica I, DiStefano JK. Methods for CpG methylation array profiling via bisulfite conversion. Methods Mol Biol Clifton NJ. 1706;2018:233–54.
47. Rauluseviciute I, Drabløs F, Rye MB. DNA methylation data by sequencing: experimental approaches and recommendations for tools and pipelines for data analysis. Clin Epigenetics. 2019;11(1):193.
48. Kurdyukov S, Bullock M. DNA methylation analysis: choosing the right method. Biology. 2016;5(1):E3.
49. Asselman J. Bioinformatic analysis of methylation patterns using bisulfite sequencing data. Methods Mol Biol Clifton NJ. 2019;1858:157–75.
50. Medvedeva YA, Lennartsson A, Ehsani R, Kulakovskiy IV, Vorontsov IE, Panahandeh P, et al. EpiFactors: a comprehensive database of human epigenetic factors and complexes. Database J Biol Databases Curation. 2015;2015:bav067.
51. Gade P, Kalvakolanu DV. Chromatin immunoprecipitation assay as a tool for analyzing transcription factor activity. Methods Mol Biol Clifton NJ. 2012;809:85–104.
52. Milne TA, Zhao K, Hess JL. Chromatin immunoprecipitation (ChIP) for analysis of histone modifications and chromatin-associated proteins. Methods Mol Biol Clifton NJ. 2009;538:409–23.
53. Das PM, Ramachandran K, van Wert J, Singal R. Chromatin immunoprecipitation assay. Biotechniques. 2004;37(6):961–9.
54. Tollefsbol TO. Methods of epigenetic analysis. Methods Mol Biol Clifton NJ. 2004;287:1–8.
55. García-Giménez JL, Seco-Cervera M, Tollefsbol TO, Romá-Mateo C, Peiró-Chova L, Lapunzina P, et al. Epigenetic biomarkers: current strategies and future challenges for their use in the clinical laboratory. Crit Rev Clin Lab Sci. 2017;54(7–8):529–50.
56. DeAngelis JT, Farrington WJ, Tollefsbol TO. An overview of epigenetic assays. Mol Biotechnol. 2008;38(2):179–83.
57. McEwen LM, Jones MJ, Lin DTS, Edgar RD, Husquin LT, MacIsaac JL, et al. Systematic evaluation of DNA methylation age estimation with common preprocessing methods and the Infinium MethylationEPIC BeadChip array. Clin Epigenetics [Internet]. 2018 Dec [cited 2021 Jul 17];10(1). Available from: https://clinicalepigeneticsjournal.biomedcentral.com/articles/10.1186/s13148-018-0556-2.
58. Bibikova M. High-throughput DNA methylation profiling using universal bead arrays. Genome Res. 2006;16(3):383–93.
59. Lim SJ, Tan TW, Tong JC. Computational epigenetics: the new scientific paradigm. Bioinformation. 2010;4(7):331–7.
60. Becker M, Worlikar U, Agrawal S, Schultze H, Ulas T, Singhal S, et al. Scaling genomics data processing with memory-driven computing to accelerate computational biology. In: Sadayappan P, Chamberlain BL, Juckeland G, Ltaief H, editors. High performance computing [Internet]. Cham: Springer International Publishing; 2020 [cited 2021 Jul 17]. p. 328–44. Available from: http://link.springer.com/10.1007/978-3-030-50743-5_17.
61. Michels KB, Binder AM, Dedeurwaerder S, Epstein CB, Greally JM, Gut I, et al. Recommendations for the design and analysis of epigenome-wide association studies. Nat Methods. 2013;10(10):949–55.
62. Wang T, Guan W, Lin J, Boutaoui N, Canino G, Luo J, et al. A systematic study of normalization methods for Infinium 450K methylation data using whole-genome bisulfite sequencing data. Epigenetics. 2015;10(7):662–9.
63. Du P, Zhang X, Huang C-C, Jafari N, Kibbe WA, Hou L, et al. Comparison of Beta-value and M-value methods for quantifying methylation levels by microarray analysis. BMC Bioinformatics [Internet]. 2010 Dec [cited 2021 Jul 17];11(1). Available from: https://bmcbioinformatics.biomedcentral.com/articles/10.1186/1471-2105-11-587.
64. Luo J, Schumacher M, Scherer A, Sanoudou D, Megherbi D, Davison T, et al. A comparison of batch effect removal methods for enhancement of prediction performance using MAQC-II microarray gene expression data. Pharmacogenomics J. 2010;10(4):278–91.
65. Chen C, Grennan K, Badner J, Zhang D, Gershon E, Jin L, et al. Removing batch effects in analysis of expression microarray data: an evaluation of six batch adjustment methods. Kliebenstein D, editor. PLoS One. 2011;6(2):e17238.
66. Titus AJ, Gallimore RM, Salas LA, Christensen BC. Cell-type deconvolution from DNA methylation: a review of recent applications. Hum Mol Genet. 2017;26(R2):R216–24.
67. Houseman EA, Kelsey KT, Wiencke JK, Marsit CJ. Cell-composition effects in the analysis of DNA methylation array data: a mathematical perspective. BMC Bioinformatics [Internet]. 2015 Dec [cited 2021 Jul 17];16(1). Available from: https://bmcbioinformatics.biomedcentral.com/articles/10.1186/s12859-015-0527-y.
68. Houseman EA, Accomando WP, Koestler DC, Christensen BC, Marsit CJ, Nelson HH, et al. DNA methylation arrays as surrogate measures of cell mixture distribution. BMC Bioinformatics [Internet]. 2012 Dec [cited 2021 Jul 17];13(1). Available from: https://bmcbioinformatics.biomedcentral.com/articles/10.1186/1471-2105-13-86.
69. Rahmani E, Schweiger R, Rhead B, Criswell LA, Barcellos LF, Eskin E, et al. Cell-type-specific resolution epigenetics without the need for cell sorting or single-cell biology. Nat Commun [Internet]. 2019 Dec [cited 2021 Jul 17];10(1). Available from: http://www.nature.com/articles/s41467-019-11052-9.
70. Houseman EA, Kile ML, Christiani DC, Ince TA, Kelsey KT, Marsit CJ. Reference-free deconvolution of DNA methylation data and mediation by cell composition effects. BMC Bioinformatics [Internet]. 2016 Dec [cited 2021 Jul 17];17(1). Available from: https://bmcbioinformatics.biomedcentral.com/articles/10.1186/s12859-016-1140-4.
71. Rakyan VK, Down TA, Balding DJ, Beck S. Epigenome-wide association studies for common human diseases. Nat Rev Genet. 2011;12(8):529–41.
72. Xu J, Zhao L, Liu D, Hu S, Song X, Li J, et al. EWAS: epigenome-wide association study software 2.0. Valencia A, editor. Bioinformatics. 2018;34(15):2657–8.
73. Davis CA, Hitz BC, Sloan CA, Chan ET, Davidson JM, Gabdank I, et al. The Encyclopedia of DNA elements (ENCODE): data portal update. Nucleic Acids Res. 2018;46(Database issue):D794–801.
74. Lizio M, Abugessaisa I, Noguchi S, Kondo A, Hasegawa A, Hon CC, et al. Update of the FANTOM web resource: expansion to provide additional transcriptome atlases. Nucleic Acids Res. 2019;47(Database issue):D752–8.
75. Liao Y, Wang J, Jaehnig EJ, Shi Z, Zhang B. WebGestalt 2019: gene set analysis toolkit with revamped UIs and APIs. Nucleic Acids Res. 2019;47(W1):W199–205.
76. Kuleshov MV, Jones MR, Rouillard AD, Fernandez NF, Duan Q, Wang Z, et al. Enrichr: a comprehensive gene set enrichment analysis web server 2016 update. Nucleic Acids Res. 2016;44(W1):W90–7.
77. Watanabe K, Taskesen E, van Bochoven A, Posthuma D. Functional mapping and annotation of genetic associations with FUMA. Nat Commun. 2017;8(1):1826.

78. von Mering C, Huynen M, Jaeggi D, Schmidt S, Bork P, Snel B. STRING: a database of predicted functional associations between proteins. Nucleic Acids Res. 2003;31(1):258–61.
79. Breeze CE, Reynolds AP, van Dongen J, Dunham I, Lazar J, Neph S, et al. eFORGE v2.0: updated analysis of cell type-specific signal in epigenomic data. Bioinforma Oxf Engl. 2019;35(22):4767–9.
80. Holder LB, Haque MM, Skinner MK. Machine learning for epigenetics and future medical applications. Epigenetics. 2017;12(7):505–14.
81. Fan S, Chen Y, Luo C, Meng F. Machine learning methods in precision medicine targeting epigenetic diseases. Curr Pharm Des. 2018;24(34):3998–4006.
82. Rauschert S, Raubenheimer K, Melton PE, Huang RC. Machine learning and clinical epigenetics: a review of challenges for diagnosis and classification. Clin Epigenetics. 2020;12:51.
83. Hüls A, Czamara D. Methodological challenges in constructing DNA methylation risk scores. Epigenetics. 2019;15(1–2):1–11.
84. Rauschert S, Melton PE, Heiskala A, Karhunen V, Burdge G, Craig JM, et al. Machine learning-based DNA methylation score for fetal exposure to maternal smoking: development and validation in samples collected from adolescents and adults. Environ Health Perspect. 2020;128(9):097003.
85. Johansson A, Palli D, Masala G, Grioni S, Agnoli C, Tumino R, et al. Epigenome-wide association study for lifetime estrogen exposure identifies an epigenetic signature associated with breast cancer risk. Clin Epigenetics [Internet]. 2019 Dec [cited 2021 Jul 19];11(1). Available from: https://clinicalepigeneticsjournal.biomedcentral.com/articles/10.1186/s13148-019-0664-7.
86. Castro de Moura M, Davalos V, Planas-Serra L, Alvarez-Errico D, Arribas C, Ruiz M, et al. Epigenome-wide association study of COVID-19 severity with respiratory failure. EBioMedicine. 2021;66:103339.
87. Karabegović I, Portilla-Fernandez E, Li Y, Ma J, Maas SCE, Sun D, et al. Epigenome-wide association meta-analysis of DNA methylation with coffee and tea consumption. Nat Commun. 2021;12(1):2830.
88. Breeze CE, Batorsky A, Lee MK, Szeto MD, Xu X, McCartney DL, et al. Epigenome-wide association study of kidney function identifies trans-ethnic and ethnic-specific loci. Genome Med. 2021;13(1):74.
89. Yao C, Joehanes R, Wilson R, Tanaka T, Ferrucci L, Kretschmer A, et al. Epigenome-wide association study of whole blood gene expression in Framingham Heart Study participants provides molecular insight into the potential role of CHRNA5 in cigarette smoking-related lung diseases. Clin Epigenetics. 2021;13(1):60.
90. Smith RG, Pishva E, Shireby G, Smith AR, Roubroeks JAY, Hannon E, et al. A meta-analysis of epigenome-wide association studies in Alzheimer's disease highlights novel differentially methylated loci across cortex. Nat Commun. 2021;12(1):3517.
91. Starnawska A, Tan Q, Soerensen M, McGue M, Mors O, Børglum AD, et al. Epigenome-wide association study of depression symptomatology in elderly monozygotic twins. Transl Psychiatry. 2019;9(1):1–14.
92. Montano C, Taub MA, Jaffe A, Briem E, Feinberg JI, Trygvadottir R, et al. Association of DNA methylation differences with schizophrenia in an epigenome-wide association study. JAMA Psychiat. 2016;73(5):506.
93. BIOS Consortium, Karlsson Linnér R, Marioni RE, Rietveld CA, Simpkin AJ, Davies NM, et al. An epigenome-wide association study meta-analysis of educational attainment. Mol Psychiatry. 2017;22(12):1680–90.
94. Küpers LK, Monnereau C, Sharp GC, Yousefi P, Salas LA, Ghantous A, et al. Meta-analysis of epigenome-wide association studies in neonates reveals widespread differential DNA methylation associated with birthweight. Nat Commun [Internet]. 2019 Dec [cited 2021 Jul 19];10(1). Available from: http://www.nature.com/articles/s41467-019-09671-3.
95. Zoghbi HY, Beaudet AL. Epigenetics and Human Disease. Cold Spring Harb Perspect Biol. 2016;8(2):a019497.
96. Moosavi A, Motevalizadeh AA. Role of epigenetics in biology and human diseases. Iran Biomed J. 2016;20(5):246–58.
97. Tzika E, Dreker T, Imhof A. Epigenetics and metabolism in health and disease. Front Genet. 2018;9:361.
98. Bakulski KM, Fallin MD. Epigenetic epidemiology: promises for public health research. Environ Mol Mutagen. 2014;55(3):171–83.
99. Motta V, Bonzini M, Grevendonk L, Iodice S, Bollati V. Epigenetics applied to epidemiology: investigating environmental factors and lifestyle influence on human health. Med Lav. 2017;108(1):10–23.
100. Rozek LS, Dolinoy DC, Sartor MA, Omenn GS. Epigenetics: relevance and implications for public health. Annu Rev Public Health. 2014;35:105–22.
101. Lech G, Słotwiński R, Słodkowski M, Krasnodębski IW. Colorectal cancer tumour markers and biomarkers: Recent therapeutic advances. World J Gastroenterol. 2016;22(5):1745–55.
102. Andersen AM, Dogan MV, Beach SRH, Philibert RA. Current and future prospects for epigenetic biomarkers of substance use disorders. Genes. 2015;6(4):991–1022.
103. Wong CC, Li W, Chan B, Yu J. Epigenomic biomarkers for prognostication and diagnosis of gastrointestinal cancers. Semin Cancer Biol. 2019;55:90–105.
104. Taryma-Leśniak O, Sokolowska KE, Wojdacz TK. Current status of development of methylation biomarkers for in vitro diagnostic IVD applications. Clin Epigenetics. 2020;12(1):100.
105. Haghshenas S, Bhai P, Aref-Eshghi E, Sadikovic B. Diagnostic utility of genome-wide DNA methylation analysis in mendelian neurodevelopmental disorders. Int J Mol Sci. 2020;21(23):E9303.
106. Kaur G, Begum R, Thota S, Batra S. A systematic review of smoking-related epigenetic alterations. Arch Toxicol. 2019;93(10):2715–40.
107. Andersen AM, Philibert RA, Gibbons FX, Simons RL, Long J. Accuracy and utility of an epigenetic biomarker for smoking in populations with varying rates of false self-report. Am J Med Genet Part B Neuropsychiatr Genet Off Publ Int Soc Psychiatr Genet. 2017;174(6):641–50.
108. Gibson J, Russ TC, Clarke T-K, Howard DM, Hillary RF, Evans KL, et al. A meta-analysis of genome-wide association studies of epigenetic age acceleration. PLoS Genet. 2019;15(11):e1008104.
109. Levine ME. Assessment of epigenetic clocks as biomarkers of aging in basic and population research. J Gerontol A Biol Sci Med Sci. 2020;75(3):463–5.
110. Masoro EJ. Overview of caloric restriction and ageing. Mech Ageing Dev. 2005;126(9):913–22.
111. Rohde K, Keller M, la Cour PL, Blüher M, Kovacs P, Böttcher Y. Genetics and epigenetics in obesity. Metabolism. 2019;92:37–50.
112. Samblas M, Milagro FI, Martínez A. DNA methylation markers in obesity, metabolic syndrome, and weight loss. Epigenetics. 2019;14(5):421–44.
113. Samet JM. Tobacco smoking: the leading cause of preventable disease worldwide. Thorac Surg Clin. 2013;23(2):103–12.
114. Jatlow P, Toll BA, Leary V, Krishnan-Sarin S, O'Malley SS. Comparison of expired carbon monoxide and plasma cotinine as markers of cigarette abstinence. Drug Alcohol Depend. 2008;98(3):203–9.
115. Shadel WG, Shiffman S, Niaura R, Nichter M, Abrams DB. Current models of nicotine dependence: what is known and what is needed to advance understanding of tobacco etiology among youth. Drug Alcohol Depend. 2000;59 Suppl 1:S9–22.
116. Vartiainen E, Seppälä T, Lillsunde P, Puska P. Validation of self reported smoking by serum cotinine measurement in a

community-based study. J Epidemiol Community Health. 2002;56(3):167–70.
117. Caraballo RS, Giovino GA, Pechacek TF, Mowery PD. Factors associated with discrepancies between self-reports on cigarette smoking and measured serum cotinine levels among persons aged 17 years or older: Third National Health and Nutrition Examination Survey, 1988–1994. Am J Epidemiol. 2001;153(8):807–14.
118. Larzelere MM, Williams DE. Promoting smoking cessation. Am Fam Physician. 2012;85(6):591–8.
119. Hilberink SR, Jacobs JE, van Opstal S, van der Weijden T, Keegstra J, Kempers PL, et al. Validation of smoking cessation self-reported by patients with chronic obstructive pulmonary disease. Int J Gen Med. 2011;4:85–90.
120. Britton GRA, Brinthaupt J, Stehle JM, James GD. Comparison of self-reported smoking and urinary cotinine levels in a rural pregnant population. J Obstet Gynecol Neonatal Nurs JOGNN. 2004;33(3):306–11.
121. Onor IO, Stirling DL, Williams SR, Bediako D, Borghol A, Harris MB, et al. Clinical effects of cigarette smoking: epidemiologic impact and review of pharmacotherapy options. Int J Environ Res Public Health. 2017;14(10):E1147.
122. Hsieh SJ, Ware LB, Eisner MD, Yu L, Jacob P, Havel C, et al. Biomarkers increase detection of active smoking and second-hand smoke exposure in critically ill patients. Crit Care Med. 2011;39(1):40–5.
123. Joehanes R, Just AC, Marioni RE, Pilling LC, Reynolds LM, Mandaviya PR, et al. Epigenetic signatures of cigarette smoking. Circ Cardiovasc Genet. 2016;9(5):436–47.
124. Philibert R, Hollenbeck N, Andersen E, Osborn T, Gerrard M, Gibbons FX, et al. A quantitative epigenetic approach for the assessment of cigarette consumption. Front Psychol. 2015;6:656.
125. Horvath S. DNA methylation age of human tissues and cell types. Genome Biol. 2013;14(10):R115.
126. Okazaki S, Numata S, Otsuka I, Horai T, Kinoshita M, Sora I, et al. Decelerated epigenetic aging associated with mood stabilizers in the blood of patients with bipolar disorder. Transl Psychiatry. 2020;10(1):129.
127. Perna L, Zhang Y, Mons U, Holleczek B, Saum K-U, Brenner H. Epigenetic age acceleration predicts cancer, cardiovascular, and all-cause mortality in a German case cohort. Clin Epigenetics. 2016;8:64.
128. Ryan J, Wrigglesworth J, Loong J, Fransquet PD, Woods RL. A systematic review and meta-analysis of environmental, lifestyle, and health factors associated with DNA methylation age. J Gerontol A Biol Sci Med Sci. 2020;75(3):481–94.
129. Levine ME, Lu AT, Quach A, Chen BH, Assimes TL, Bandinelli S, et al. An epigenetic biomarker of aging for lifespan and healthspan. Aging. 2018;10(4):573–91.
130. Lu AT, Quach A, Wilson JG, Reiner AP, Aviv A, Raj K, et al. DNA methylation GrimAge strongly predicts lifespan and healthspan. Aging. 2019;11(2):303–27.
131. Ecker S, Beck S. The epigenetic clock: a molecular crystal ball for human aging? Aging. 2019;11(2):833–5.
132. Fraga MF, Ballestar E, Paz MF, Ropero S, Setien F, Ballestar ML, et al. Epigenetic differences arise during the lifetime of monozygotic twins. Proc Natl Acad Sci U S A. 2005;102(30):10604–9.
133. Handy DE, Castro R, Loscalzo J. Epigenetic modifications: basic mechanisms and role in cardiovascular disease. Circulation. 2011;123(19):2145–56.
134. McKay JA, Mathers JC. Diet induced epigenetic changes and their implications for health. Acta Physiol Oxf Engl. 2011;202(2):103–18.
135. Zheng Y-C, Feng S-Q. Epigenetic modifications as therapeutic targets. Curr Drug Targets. 2020;21(11):1046.
136. Szyf M. Epigenetics, DNA methylation, and chromatin modifying drugs. Annu Rev Pharmacol Toxicol. 2009;49:243–63.
137. Hunter P. The second coming of epigenetic drugs: a more strategic and broader research framework could boost the development of new drugs to modify epigenetic factors and gene expression. EMBO Rep. 2015;16(3):276–9.
138. Kronfol MM, Dozmorov MG, Huang R, Slattum PW, McClay JL. The role of epigenomics in personalized medicine. Expert Rev Precis Med Drug Dev. 2017;2(1):33–45.
139. Feinberg AP, Irizarry RA, Fradin D, Aryee MJ, Murakami P, Aspelund T, et al. Personalized epigenomic signatures that are stable over time and covary with body mass index. Sci Transl Med. 2010;2(49):49ra67.
140. Liang G, Gonzales FA, Jones PA, Orntoft TF, Thykjaer T. Analysis of gene induction in human fibroblasts and bladder cancer cells exposed to the methylation inhibitor 5-aza-2'-deoxycytidine. Cancer Res. 2002;62(4):961–6.
141. Cheishvili D, Boureau L, Szyf M. DNA demethylation and invasive cancer: implications for therapeutics. Br J Pharmacol. 2015;172(11):2705–15.
142. Singh P, Konar A, Kumar A, Srivas S, Thakur MK. Hippocampal chromatin-modifying enzymes are pivotal for scopolamine-induced synaptic plasticity gene expression changes and memory impairment. J Neurochem. 2015;134(4):642–51.
143. Sun Y, Sahbaie P, Liang D, Li W, Shi X, Kingery P, et al. DNA methylation modulates nociceptive sensitization after incision. PLoS One. 2015;10(11):e0142046.
144. Peedicayil J. Epigenetic drugs for multiple sclerosis. Curr Neuropharmacol. 2016;14(1):3–9.

Principles of Clinical Transcriptomics and Splicing

5

Juan Carlos Gomez-Verjan, Juan Carlos Yustis-Rubio, and Elizabeth Sulvaran-Guel

Abbreviations

AS	Alternative splicing
BPS	Branch point site
cDNA	Complementary DNA
ERCB	European Renal cDNA Bank
ESE	Exonic splicing enhancer
ESS	Exonic splicing silencer
EST	Expressed sequence tags
GTEx	Genotype-Tissue Expression Project
GWAS	Genome-wide association studies
HCA	Human Cell Atlas
HGMD	Human Gene Mutation Database
hnRNP	Heterogeneous nuclear ribonucleoprotein
ISE	Intronic splicing enhancer
ISS	Intronic splicing silencer
miRNA	MicroRNA
mRNA	Messenger RNA
NGS	Next-generation sequencing
NMD	Nonsense-mediated mRNA decay
PCR	Polymerase chain reaction
pre-mRNA	Precursor mRNA
PTC	Premature termination codon
qPCR	Quantitative PCR
RNA-seq	RNA sequencing
rRNA	Ribosomal RNA
RT-PCR	Reverse transcription PCR
RUST	Regulated unproductive splicing and translation
SAGE	Serial analysis of gene expression
sc-RNA-seq	Single-cell RNA sequencing
snRNP	Small nuclear ribonucleoprotein
SRE	Splicing regulatory element
SS	Splice site
TCGA	The Cancer Genome Atlas Program
TWAS	Transcription-wide association studies

J. C. Gomez-Verjan (✉)
Dirección de Investigación, Instituto Nacional de Geriatría (INGER), Ciudad de México, Mexico
e-mail: jverjan@inger.gob.mx

J. C. Yustis-Rubio
Departamento de Ecología Funcional, Instituto de Ecología, UNAM, Mexico City, Mexico
e-mail: jcyustis@ciencias.unam.mx

E. Sulvaran-Guel
Licenciatura en Ciencias Genómicas, Universidad Nacional Autónoma de México, Mexico City, Mexico
e-mail: sulvaran@lcg.unam.mx

Introduction

With the introduction of molecular biology into the clinical framework as a toolbox for diagnosis and possible pathology treatment, a great effort has been put into linking diseases with their underlying genetic causes. The advent of the Human Genome Project in the early 1990s and other similar initiatives proved to be an invaluable tool for clinical researchers, as it opened human genetics entirely for dissection to find the origins of a wide variety of diseases and other pathological conditions [1]. Initially, this meant studying the entirety of an organism's DNA sequences (the genome) and finding variations in these sequences (mutations) that could be linked to specific ailments. At this stage, aside from studying individual genes to discover specific mutations that could alter their function or finding their expression patterns, genome-wide association studies (GWAS) became a widely used tool for assessing genotype-phenotype relationships [2]. GWAS in medical research essentially works by searching the genome for common and recurring genetic variations that could be associated with the development of certain diseases under the "common disease – common variant" hypothesis [3]. Although GWAS proved to be a valuable tool in discovering novel pathogenic candidate genes and potential therapeutic molecular targets, its predictive power falls short as it fails to provide a direct causal link between genetic variants and disease susceptibility [4]. Also, to provide more

J. C. Gomez-Verjan, N. A. Rivero-Segura (eds.), *Principles of Genetics and Molecular Epidemiology*, https://doi.org/10.1007/978-3-030-89601-0_5

comprehensive results, GWAS require other types of information such as functional gene annotation, genetic interaction data, and other molecular analyses [5]. In this sense, one of the most striking discoveries provided by the sequencing of the human genome was that although over 90% of the human genome is transcribed into RNA, only between 1% and 2% of transcribed human genes are translated into proteins [6, 7]. Such results suggest that there are other levels of complexity and functionality beyond the genome, such as the transcriptome.

Transcriptome refers to the set of all the RNA molecules (coding and noncoding) transcribed from an entire organism's genes, tissue, or cell type in a particular moment and condition. The term was initially coined by Charles Auffray in 1996 and was first used on a research paper in 1997 [8, 9]. Before the term, there were several attempts to analyze gene expression since the early 1990s. For instance, in 1991, an effort led by J. Craig Venter compiled a set of 609 Expressed Sequence Tags (ESTs) from the human brain. This work was one of the first attempts to use the newly developed automated Sanger sequencing technology [10].

Unlike other fields that have recently acquired the "-ome" appendage (whether deservingly of it or not) [11], the transcriptome, and therefore transcriptomics, has rapidly developed, thanks to intense development in both genomic technologies (that have rendered previous molecular techniques obsolete). The diverse bioinformatic applications developed only for transcriptome analyses [12, 13]. Transcriptomics can generate massive amounts of information, including gene expression levels, analysis of expression patterns, splicing and alternative splicing information, and the prevalence of transcript isoforms, among others. Nevertheless, the massive amounts of information generated by transcriptomic technologies represent quite a challenge for storage and analysis, taking into account that artifacts and biases still exist and need to be identified to generate concrete results and conclusions. In this context, transcriptomics has significantly benefited from big data technologies and novel statistical protocols [14].

Interestingly, transcriptomics could be used with other "omics" technologies as its results have molecular implications upstream or downstream on signaling pathways. Not all the classes of RNAs (coding and noncoding) are transcribed in every tissue or cell. Hence, understanding genetic expression under different contexts, such as environmental, chemical, genetic, and social, could provide valuable information applicable to diverse fields on human health. In this sense, the recently named transcriptome-wide association studies (TWAS) have demonstrated to be quite powerful tools to identify candidate genes whose genetically regulated expression is associated with traits of interest. TWAS methods are highly complex at statistical tests associating genetic expression and disease risk; they promise to prioritize candidate causal genes (genes mediating the phenotypic effects of disease) and tissues [15].

The Leap From Classic Gene Expression Studies to Transcriptomics

Before transcriptomics was consolidated as the biological research powerhouse that it is today, efforts were made in previous decades to analyze individual or even reduced groups of transcripts. In 1979, a library of cDNA was constructed from silk moth (*Bombyx mori*) chorion (eggshell) enriched mRNA to analyze the evolution, chromosomal organization, and regulated developmental expression of the chorion multigene families [16]. When low-throughput Sanger sequencing became more commonplace in the 1980s, expression sequencing tags (ESTs) began to be sequenced from cDNA libraries. During the 1990s, EST sequencing was a commonplace and efficient method to determine an organism's gene content without the need to sequence the entire genome [17]. Alongside these more "in-bulk" techniques to study genes and genetic expression, molecular biology techniques to study individual transcripts were also being developed and widely used. Northern blots (and its reverse northern blot variant) and reverse transcriptase-polymerase chain reaction (RT-PCR) became popular methods to study gene expression. However, these methods were time-consuming and required a high degree of expertise and only allowed the capture of a transcriptome's minimal subset [18, 19]. Despite the rapid technological and computational advancements until this point in time, how transcriptomes as a whole were expressed and regulated remained a scientific mystery. The Sanger method dominated sequencing technologies in the field until more high-throughput methods became more commonplace, and these methods were themselves overcome by the advent of next-generation sequencing (NGS) technologies, which have considerably surpassed Moore's law expectations even more so that a new interpretation of Moore's law was needed [20]. Interestingly, by 2008 two human transcriptomes were firstly published, and by 2015 hundreds of individuals' transcriptomes were published [21–24]. Currently, transcriptomes of individual tissues, specific diseases, and even single cells are generated regularly. In the following sections, we will focus on the most used transcriptomic technologies.

General Overview of Transcriptomic Technologies

As previously stated, knowledge of gene expression patterns can supply an overview of different active genetic pathways at a specific time or context [25]. In this sense, there are different methodologies for studying gene expression.

SAGE-Derived Sequencing Technologies

Serial analysis of gene expression (SAGE) was introduced in 1995 to identify and quantify expressed genes [26]. SAGE technology is based on selecting, concatenating, and sequencing short but representative tags in expressed mRNAs. The first step in SAGE methodology is the mRNA extraction from the cell [27]. Once extracted, the mRNAs are retro-transcribed to cDNA with biotinylated oligo(dT) molecules which bind the poly-A tail and cleaved ~256 base pairs from the primer with a restriction enzyme called anchoring enzyme [26]. The biotin molecules attached to the oligo(dT) allow for binding the cDNA to a streptavidin bead [26]. The mixture is divided into two. Each division is now added with two different linkers, a restriction site for another restriction enzyme called tagging enzyme and a restriction site for another anchoring enzyme [28]. The tagging enzyme is then supplemented in the mixture, cutting the fragment 14 base pairs from its recognition site [28]. Two fragments, one from each mixture, are ligated on their free ends (the ones that do not contain the linker), forming di-tags, and finally, the anchoring enzyme cleaves the linker [28]. All the di-tags are concatenated and, lastly, sequenced [27]. The concatenation of all the tags allows the simultaneous sequencing of several mRNAs [27]. However, it must be noted that this methodology surged before next-generation sequencing techniques (see below); thus, other sequencing technologies such as Sanger were available, which generates ~600 base pair reads, allowing the sequencing of ~50 mRNA tags [27]. Tools such as the SAGE Software Suite are needed to quantify each tag and for its matching to reference sequences in databases [28].

Examples of modifications of SAGE are LongSAGE, microSAGE, miniSAGE, and DeepSAGE [29]. LongSAGE is similar to the original methodology, but the tagging enzyme cuts longer fragments, allowing a higher probability of single tags coming from a unique mRNA [29]. MicroSAGE, on the other side, adds a PCR amplification step to the original methodology, allowing the quantification of RNA, beginning with very little material; nonetheless, it has the disadvantage that PCR can introduce bias into the final quantification of transcripts [29]. Similarly, miniSAGE also reduces the initial amount of cells, but it does not require PCR amplification. Instead, it has additional mRNA purification steps and uses a single tube from mRNA extraction until tagging to avoid genetic material loss [30]. Conversely, DeepSAGE uses a different sequencing methodology, known as 454 (see below) [30]. DeepSAGE simplifies the protocol and makes it better to detect low abundance transcripts, but 454 sequencing was introduced until 1999, so for a long time, this SAGE modification was not available.

Since its creation and until the beginning of the millennium, SAGE allowed the analysis of thousands of transcripts, including studies on cancer and immunology [27]. Although it has the advantage that it generally does not require PCR amplification, and thus, it provides highly accurate quantification of transcripts, its methodology is extremely time-consuming and complex [27]

Microarrays

Microarrays are pretty important when we talk about transcriptomics since such technologies allow the analysis of hundreds of thousands of transcripts simultaneously. They are mainly based on three different technologies: PCR, DNA libraries, and hybridization [31]. Microarrays were initially thought for DNA sequencing [32, 33]; however, they were introduced as a hybridization technique for measuring transcription levels in the 1990s [32]. This first approach consisted of a glass slide where cDNA probes were placed, a still widely used technique [25]. Library clones need to be generated; thus, mRNA sequences for different genes in an organism are inserted into identical bacterial DNA vectors [34]. Subsequently, robotic devices place each distinct clone corresponding to a single gene in the glass array at a known coordinate, and they are immobilized with DNA-binding chemicals [25]. Afterward, mRNA sequences from samples to be analyzed are retro-transcribed to cDNA. The latter are labeled with fluorescent dyes of different colors, for example, green for control and red for test [25]. Robotic devices also place these new molecules over the same slide, but instead of placing them in a known coordinate, they are placed throughout the whole array; therefore, all of the cDNA from the samples analyzed will hybridize with the clones adhered to the slide in their corresponding coordinate [31]. Unhybridized fragments are washed away, and the slide is analyzed in a laser scanner [32]. The resulting colors in digital imaging can be interpreted either in both samples, present in none samples, present only in control samples and present only in test samples [32].

To date, two prominent companies produce microarray technologies: Thermo-Affymetrix and Illumina. The first one, Thermo-Affymetrix technology, is based on the synthesis of the DNA probes (25 base pairs) directly over the glass slide using photolithography [31]. Once oligonucleotides are synthesized, the slide is provided with the cDNA for analysis [32]. Illumina technology is based on the assembly of pre-synthesized oligonucleotides probes with a known code coupled to a batch of silica microspheres called beads [32]. These beads are placed over a particular microarray with wells where they are immobilized and decoded, thus identifying each gene sequence's specific location [32]. The posterior protocol and analysis are the same (Table 5.1).

Microarrays' first application was the identification of transcripts' presence in different samples. However, they have been used to detect splice variants and diagnose several

Table 5.1 Available microarray brands with underlying working mechanisms

Microarray	Principle	Reference
Thermo-Affymetrix	Photolithography synthesis of 25 base pairs DNA probes directly over the slide	[31, 32]
Illumina Inc.	Coded probes coupled to beads	[32]
Cy5-Cy3	Hybridization of DNA with two colors (Cy3 and Cy5) cDNA from samples. Also called home microarrays	[36]
Agilent Inc.	Hybridization of DNA with different fluorescent samples. 60 base pairs of oligonucleotides. Allows multicolor hybridization	[37, 38]
Roche-Nimblegen	Photolithography synthesis of 70 base pairs DNA probes directly over the slide. Allows multicolor hybridization	[38]

diseases, such as cancer, where the expression of different genes can help in the distinction of tumors [32]. It is essential to be noted that although microarrays are a powerful technology for the detection of differentially expressed genes, their main limitation is that they may be insensitive for transcripts of low abundance, which may represent an essential loss for transcripts whose normal expression levels are low [35].

RNA-seq

In contrast with microarrays and SAGE technologies, RNA sequencing (commonly known as RNA-seq) is a relatively recent field, which starts with the emergence of next-generation sequencing (NGS) at the beginning of the last decade [39]. NGS methodologies allow the determination of base pairs order in a nucleotide sequence in an efficient and low-cost manner [40]. There are several sequencing methodologies, but most of them can only sequence DNA; thus, the generation of cDNA is a required step for most RNA-seq protocols [41]. Depending on the scope of the project, generally, the first step is the removal of cellular rRNAs for the selection of remaining RNA classes that might be differentially expressed, or, in contrast, the selection of specific RNA classes, such as mRNAs and lncRNAs by poly-A selection with oligo(dT) molecules, or miRNAs by size selection, among others [39]. Subsequently, selected RNAs are retro-transcribed to cDNA and fragmented by sonication or with enzymes such as endonucleases or transposases. Most methodologies can only sequence hundreds or a few thousands of base pairs [39]. It must be noted that the construction of cDNA libraries generates DNA sequences for the mRNA and its complementary sequence; thus, the maintenance of the strand identity, that is, which one proceeds from the mRNA, is required [39, 41]. The most common approach used for this issue is the addition of labels during synthesis that can be recognized for strand degradation, but many others exist [41].

The most common sequencing technology is Illumina, but SOLiD and Roche/454 are also widely used [42]. Illumina sequencing is done by synthesis: the fragment to be sequenced is adhered to a glass flow cell containing nanowells. A polymerase synthesizes the complementary sequence with fluorescently labeled nucleotides – each nucleotide with a different label [43]. The cell is then excited with a light source, and the fluorescence emission is recorded, allowing the identification of the nucleotide added [43]. Like Illumina, SOLiD sequences by ligation, and the first step is emulsion PCR, where the fragments adhere to silica beads and these to a glass plaque [43]. SOLiD sequences by ligating an eight base pair probe adjacent to a primer (named n primer) and identifying the group to which the first two base pairs correspond, thanks to a fluorescent molecule added to the probe (there are 4 groups, each with 4 of the 16 possible combinations for 2 base pairs) [43]. Then, the last three nucleotides of the probe are removed, the remaining only five, and the process is repeated for the whole fragment [43]. Finally, the whole last process is repeated for a one base pair shorter primer (named n-1 primer), a two-base pair shorter primer (named n-2), and so on, until n-4, allowing the identification of the groups for the remaining three nucleotides of the probes, and with this, the identification of every single nucleotide [43]. Lastly, Roche/454 is based on pyrosequencing, which takes advantage of the pyrophosphate released in the nucleotide incorporation during DNA synthesis [43]. When the pyrophosphate reacts with the luciferin molecule and with ATP sulfurylase and luciferase enzymes, it generates oxyluciferin, which emits light [43]. The first step of Roche/454 also involves emulsion PCR [43]. Then, the nucleotides are added one by one, and when a polymerase incorporates the correct one, the light is detected, allowing the identification of the base pair added [43].

Once the sequence is available, reads must be mapped to the reference genome or an annotated transcriptome [44]. Mapping to the transcriptome is faster and computationally less demanding, but it cannot find new transcripts not previously described [44]. Bowtie is one of the most famous mapping algorithms, but it has the downside that its performance in identifying alternative splice sites is lacking [41]. Thus, other mapping algorithms, such as TopHat, are more useful for RNA mapping [44]. Finally, reads must be quantified to identify genomic regions under- or overexpressed considering biases, such as transcript length, which might affect the total fragments mapped to a locus [44]. Cufflinks, FluxCapacitor, and MISO are well-known tools that quantify reads and normalize them [41]. Quantified and normalized reads are now ready for differential gene expression analysis, which can be done with several R libraries, such as edgeR, NOISeq, and EDASeq, depending on the project's scope [44].

RNA-seq has almost surpassed other transcriptomic analysis methodologies because only small amounts of material are needed and have better transcript calls, even for genes whose expression is low [44]. However, it still has limitations. One of them is the inaccurate long transcripts reconstruction, given the restriction that most methodologies can only sequence short reads [41]. Even though long read sequencing methodologies exist, such as PacBio and Nanopore, there is a lack of efficient and practical tools for mapping these kinds of reads [44]. Additionally, few reproducibility across different tools have been reported, which indicates that there is still the need for different parameter standardization [44]. However, the future of RNA sequencing is promising with new emerging protocols that exhibit vast opportunities, such as single-cell RNA-sequencing (scRNA-seq) [44]. The complementation of RNA-seq technologies with other emerging tools for analyzing chromatin accessibility, transcription factor binding sites, nucleosome occupancy, histone modifications, DNA methylation [45], SNPs, and regulatory elements will undoubtedly provide new insights about human genomics, including promising clinical fields.

Bulk Transcriptomics Profiling and Single-Cell Transcriptomics

Transcriptomic technologies such as microarrays and RNA-seq utilize samples from bulk tissues and assume that all cells present in that material represent a homogenous population with, at least, reasonably similar gene expression patterns [46]. However, even among genetically identical cells of the same population, variations exist in their gene expression patterns; this happens mainly because of the stochastic nature of gene expression and randomness coming from both transcription and translation generating the cell-to-cell variability observed [47]. Even though cell-to-cell gene expression variations are something to account for, bulk transcriptomics has been a beneficial and successful approach in medical research when searching for novel disease biomarkers, genetic mechanisms of specific pathologies, as well as potential therapeutic targets [46]. Single-cell transcriptomics is a reasonably novel approach that, apart from taking into account the stochasticity of gene expression, can look past the biases of bulk transcriptomics by analyzing the gene expression patterns of a single-cell type in its tissue context. Single-cell transcriptomics uses a wide variety of techniques and protocols to isolate specific cell types, and they differ in the number of cells they isolate and how the cells are selected. Single-cell RNA-seq (scRNA-seq) has become a potent tool to analyze the transcriptome of a wide variety of cell types and coupled with spatial transcriptomic methods. It can investigate single-cell transcriptomes together with their physical location and tissue context [48, 49].

Single-cell transcriptomic technologies can be scaled to the entire human body. Efforts to generate cell atlases of the entire human body have already begun, and a result of this is the Human Cell Atlas (HCA) global consortium [50, 51]. Its mission states that the HCA wishes "to create comprehensive reference maps of all human cells—the fundamental units of life—as a basis for both understanding human health and diagnosing, monitoring, and treating disease" [https://www.humancellatlas.org]. Using high-throughput technologies at the single-cell resolution, the consortium aims to generate an analogue to a "Google Maps" of the human body. Efforts have been made to generate cell atlases of organs such as the brain, heart, liver, thymus, gut, and kidney, using single-cell technologies. Single-cell transcriptomics has also been used to study a wide variety of diseases such as cancer, chronic kidney disease, neurodevelopmental disorders, as well as an autoinflammatory disease [49].

Splicing and Alternative Splicing

Aside from all the complexity that arises when trying to analyze how transcriptomes are expressed and regulated, the processing of mRNA once transcribed adds yet another layer of complexity that makes transcriptomic analyses an even more daunting enterprise. Eukaryotic genes are arranged in a discontinuous fashion where protein-coding segments, known as exons, are interspaced by noncoding sequences, known as introns. Once transcribed, precursor mRNA (pre-mRNA) carries both exons and introns in its sequence. Through a highly regulated process known as splicing, introns are removed, and exons joined together, and this process, in turn, forms the mature form of mRNA. While some genes have only one splicing isoform (those constitutively spliced), others can produce multiple mRNA isoforms from a single pre-mRNA molecule through a process known as alternative splicing (AS). The process of AS has gained notoriety in recent years as an essential regulator of organism development. It has been proposed that AS has been used in evolutionary history as a mechanism to overcome the relatively low number of genes compared to the total size of the genome while attaining a higher level of genetic complexity [52, 53]. Recent RNA-seq data provides evidence that >95–100% of human protein-coding genes undergo a process of AS, with at least two isoforms *per* gene [21, 54]. Compared to other organisms, humans are the ones that present a higher degree of occurrence of AS events, and together with transcriptome data, these observations seem to suggest that a higher occurrence of AS is directly correlated with an increase in organism complexity [55–57].

Even though the process of splicing (and AS) is commonly referred to as a "cut and paste" event, the actual biochemical reactions that take place are two consecutive S_N2-type transesterification reactions involving functional groups from three reactive regions present in the pre-mRNA. Two of these regions are found at the 5′ or 3′ ends of introns are known as 5′ or 3′ splice sites (SS). The third region involved in the splicing process is known as the branch point site (BPS) and is located near the 3′ end of the intron, around 15–50 nucleotides upstream of the 3′ SS [58, 59].

The splicing process is carried out by a piece of complex molecular machinery known as the spliceosome. The spliceosome is composed of RNA and proteins, and it is commonly known as a ribonucleoprotein, with its catalytic core being composed of RNA [60]. The spliceosome is a highly regulated molecular complex that acts in a stepwise cycle, and with every splice reaction, it is assembled and disassembled. The spliceosome is composed of different subunits of small nuclear ribonucleoproteins (snRNPs) at each of the splicing steps. In turn, each of these snRNPs is composed of a specific small nuclear RNA (snRNA) and other accompanying proteins. These snRNAs are the ones that, through a base-pairing mechanism, can identify the sequences in the DNA (5′ and 3′ SS, BPS) that allow for differentiating introns from exons [61, 62].

As mentioned before, AS can give rise to several transcript isoforms from a single transcriptional event. AS can be regulated by many factors, including splicing sequences in the pre-mRNA, trans-acting splicing factors that can either promote or repress AS, the chromatin environment, and transcription elongation activity [63]. AS can generate different splicing events depending on the organization of the sequences within the pre-mRNA. The AS events can occur exon skipping, intron retention, alternative 3′ and 5′ SS selection, mutually exclusive exons, alternative promoter, and alternative polyadenylation (Fig. 5.1). In humans, the most common AS event is exon skipping. All of these events can happen either individually or simultaneously, and this can, in turn, generate a wide variety of AS isoforms from a single transcript [22, 56].

Implications of Alternative Splicing in Clinics

Evidence has surfaced linking mRNA splicing (both canonical and alternative) as an essential developmental regulator in recent years. Therefore, alterations of the splicing process have gained relevance as the source of many diseases [64, 65]. Around 95% of human genes are subject to a process of AS. While AS has been described as a mechanism through which the transcriptome and proteome can be expanded without genome expansion, the many ways in which this highly complex and regulated process can be altered have spawned, an emerging field in which these numerous splicing alterations are studied to uncover their role in the onset and severity of many diseases. Together with this, research has focused on how splicing factors and AS products can be used and targeted as therapeutic targets and on developing novel treatments of certain diseases [66, 67].

As mentioned before, mRNA splicing is a highly regulated process modulated by cis- and trans-acting factors. Both of these can be subject to alterations that can lead to pathological conditions. Given that SS consensus sequences are poorly conserved, splicing efficiency must rely on other factors [63]. Cis-acting splicing factors, also known as splicing regulatory elements (SREs), are sequences found within exons and introns that can act as either splicing enhancers or silencers. Therefore, these SREs can be intronic splicing enhancers (ISE) and silencers (ISS), as well as exonic splicing enhancers (ESE) and silencers (ESS). SREs act as binding sites for trans-acting splicing regulators that recruit the different spliceosome subunits to a specific site within the genome. The trans-acting splicing factors are RNA-binding proteins that form multiprotein complexes that either favor or repress spliceosome components' recruitment. Although many RNA-binding proteins can act as trans-acting splicing factors, two groups of these proteins are the most common and well-studied. In this sense, serine-arginine (SR)-rich proteins are generally considered to act as splicing promoters; meanwhile, the heterogeneous nuclear ribonucleoprotein (hnRNP) family of proteins are considered splicing repressors. However, recent evidence suggests that the function of both of these types of proteins depends on the sequence to which they bind, meaning that they can act as both promoters and repressors of splicing [68–70].

Data from the Human Gene Mutation Database (HGMD) indicates that more than one-third of all disease-causing mutations in the human genome are related to mRNA splicing [71–73]. These data only considers mutations located at established SSs, but not those present at other SREs or mutations at loci of trans-acting splicing factors, meaning that the incidence of alterations of splicing and AS as a source of pathological conditions might be even higher than previously anticipated.

Splicing alterations can be grouped into these categories: mutations in the core splicing consensus sequences that include alterations of SS sequences and the BPS, mutations in additional cis-acting splicing elements (exonic and intronic splicing enhancers and silencers), mutations of trans-acting splicing factors (SR proteins and hnRNPs), and mutations of core spliceosome elements. In recent years, a great deal of research effort has been put into dissecting the causes of splicing alterations-related pathologies and looking for therapeutic targets in both cis- and trans-acting splicing factors [66].

mRNA splicing and AS can generate transcripts that harbor premature termination codons (PTC) that, if translated, these proteins can cause deleterious effects that can be detrimental to the development of an organism. In regards to this, eukaryotes have developed an RNA surveillance mechanism that recognizes and degrades faulty transcripts called nonsense-mediated mRNA decay (NMD). NMD was first described as a posttranscriptional surveillance and quality control mechanism that focused on degrading faulty PTC-containing transcripts. However, more recent evidence has shown that the NMD pathway is also responsible for regulating the abundance of 10–20% of naturally occurring eukaryotic mRNAs [74–76]. To make matters even more

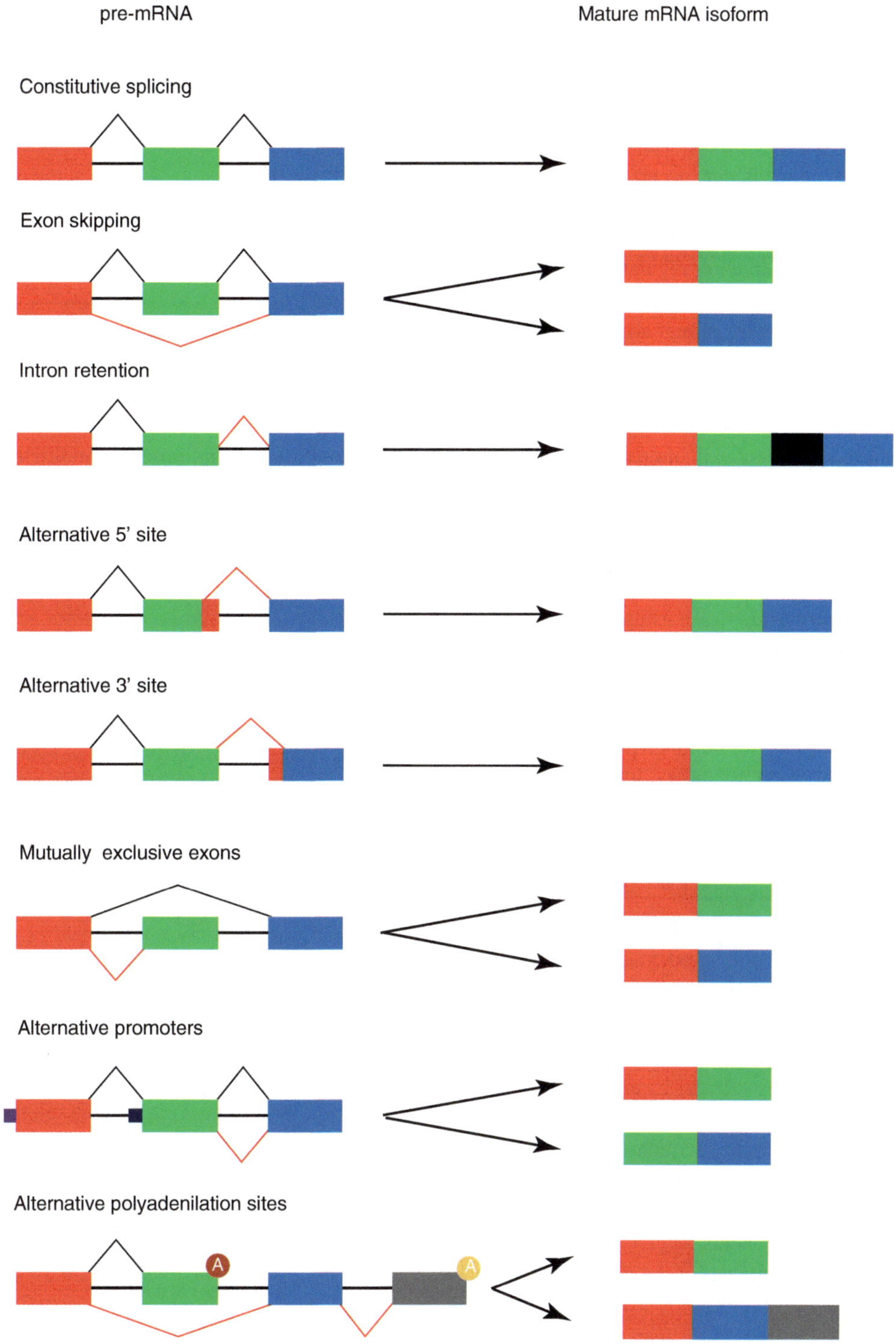

Fig. 5.1 Canonical and noncanonical splicing events. Different splicing events can give rise to several transcript isoforms. These splicing events can happen individually or together within the same transcript

complicated, it is known that not all PTC-containing transcripts trigger the NMD pathway and that some transcripts that do not carry a PTC are also targeted by NMD [77]. In terms of medical interest, since its discovery, NMD has been implicated in human disease. The role of NMD during development and neural development has also been shown. It has been reviewed how NMD can impact several physiological responses such as stress response, immune response, and viral replication [77, 78]. It is estimated that up to 30% of genetic-related diseases are caused by alterations that lead to PTC-carrying transcript, affected by NMD [78].

AS is considered a significant source of PTC-containing transcripts because alternatively spliced transcripts may contain altered reading frames that can introduce a PTC. This process in which NMD regulates AS-derived transcripts has been termed AS coupled to NMD (AS-NMD) [78]. It has been predicted that around one-third of alternatively spliced transcripts can contain a PTC. Therefore AS-NMD could function as a widely used mechanism of transcript regulation [79]. Some transcripts harboring PTCs or other sequence signals known to trigger NMD seem to escape NMD-mediated degradation and, in turn, have managed to utilize AS-NMD as a mechanism to regulate their transcript abundance, and this can occur in a wide variety of transcripts, both naturally occurring and disease-related [80]. Specifically, AS-NMD seems to be relatively frequent in members of splicing regulator families, such as SR proteins and hnRNPs [81, 82]. These splicing regulators, as well as some core spliceosomal components, seem to use this type of regulation as a way to self-limit their range of protein expression by binding their transcripts and catalyzing the splicing of the isoform targeted by NMD, in a process also known as regulated unproductive splicing and translation (RUST) [81, 82]. Even though AS-NMD may not be a widely spread regulatory mechanism as initially thought, it is known that it contributes to the diversity and abundance of transcripts involved in physiological processes such as tissue and brain development and the splicing regulation process [78]. It is also known that AS-NMD deregulation can lead to disease, and evidence suggests that such deregulation is involved to some degree in cancer development and some neurological, neurodevelopmental, and neurodegenerative diseases [77].

Applications of Transcriptomics in Clinical Disease

Table 5.2 is a very brief resume of different works in which transcriptomic technologies have been used in medical research in various tissues, organs, and the study of certain diseases.

Table 5.2 Examples of critical medical research studies where transcriptomics technologies were used to obtain important information from patients

Tissue/Disease	Technology / Cohort	Observations	Reference
Transcriptomic age	7074 human peripheral blood samples from six independent cohort studies, including EGCUT, FHS, IN CHIANTI, KORA, ROTTERDAM STUDY, and SHIP-TREND.	Identification of 1497 differentially expressed genes associated with chronological age used the gene expression profiles to calculate the *transcriptomic age* associated with biological features related to aging's molecular pillars	[83]
G-TEX	54 non-diseased tissue sites across nearly 1000 individuals/Genotype-Tissue Expression (GTEx) project	An ongoing effort to build a comprehensive public resource to study tissue-specific gene expression and regulation (https://www.gtexportal.org/home/)	[84]
Chronic kidney disease	Microarrays/(ERCB) European Renal cDNA Bank	Transcript levels of the epidermal growth factor (EGF) mRNA in urine significantly correlate with the Glomerular Filtration Rate (GFR) and work as a chronic kidney disease risk indicator	[85]
Obesity	RNA-seq	RNA-seq performed on healthy and obese patients' placenta revealed 288 differentially expressed genes enriched in lipid metabolism, angiogenesis, hormone activity, and cytokine activity	[86, 87]
Postmortem human occipital cortex – Rett syndrome	Single-cell RNA sequencing (sc-RNAseq)	Rett syndrome is caused by mutations in the methyl-DNA-binding protein (MECP2) gene, located on the X chromosome. The disease's severity correlates with the fraction of mutant alleles present in brain cells after X chromosome inactivation	[88]
Whole peripheral blood	Microarray/The LIFE-Adult-Study	Illumina Expression Bead-Chips were used for gene expression analysis in blood. The assays were associated with medical, physical, and cognitive examinations, together with interviews and questionnaires for ~10,000 40–70-year-old adults	[89]
Diabetes	Single-cell transcriptomics/human diabetic kidney samples three control and three early diabetic nephropathy samples	Cell-type-specific changes in gene expression that are important for ion transport, angiogenesis, and immune cell activation and increased potassium secretion and angiogenic signaling represent early kidney responses in human diabetic nephropathy	[90]
Alzheimer's Disease	Microarray/1440 of blood-based microarray gene expression profiles	Identified and replicated five genes (CREB5, CD46, TMBIM6, IRAK3, and RPAIN) as significantly dysregulated and that CREB5 was also associated with brain atrophy and increased amyloid-beta (Aβ) accumulation	[91]
Breast cancer	Data obtained from the TCGA breast invasive carcinoma (BRCA) dataset, mainly from RNA-seq	The usage of an artificial intelligence algorithm for the integration of expression level and alternative splicing transcriptome data allowed the classification of breast cancer patients in cancer subtypes required for the appropriate diagnosis and treatment	[92]

Concluding Remarks

Although transcriptomics represents an excellent opportunity due to the amount of information that we could generate from clinical studies, it is also quite a challenge at a technological, statistical, and computational level due to the amount of commercial platforms available and of bioinformatic approaches that need to be generated, including those concerning data warehousing. Nevertheless, there is no doubt that transcriptomic studies will continue to improve our knowledge on the different fields that such technologies have permeated (oncology, aging, pharmacology, chronic diseases, neurobiology, etc.) and that the number of articles generated in this field will only increase in the following years. Therefore, it becomes necessary to start to educate all medical researchers in using this technology, so we could accelerate its implementation and use accelerating the *bench-to-bedside* process.

Acknowledgments This chapter is part of a registered project at the Instituto Nacional de Geriatria with the number DI-PI-003/2018.

References

1. Samuelsson T. The human genome in health and disease: a story of four letters. Garland Science; 2019.
2. Bush WS, Moore JH. Chapter 11: genome-wide association studies. PLoS Comput Biol. 2012;8:e1002822.
3. Pritchard JK, Cox NJ. The allelic architecture of human disease genes: common disease–common variant… or not? Hum Mol Genet. 2002;11:2417–23.
4. Visscher PM, Brown MA, McCarthy MI, Yang J. Five years of GWAS discovery. Am J Hum Genet. 2012;90:7–24.
5. Katrib A, Hsu W, Bui A, Xing Y. "Radiotranscriptomics": a synergy of imaging and transcriptomics in clinical assessment. Quant Biol. 2016;4:1–12.
6. Venter JC, Adams MD, Myers EW, et al. The sequence of the human genome. Science. 2001;291:1304–51.
7. Lander ES, Linton LM, Birren B, et al. Initial sequencing and analysis of the human genome. Nature. 2001;409:860–921.
8. Piétu G, Mariage-Samson R, Fayein NA, et al. The Genexpress IMAGE knowledge base of the human brain transcriptome: a prototype integrated resource for functional and computational genomics. Genome Res. 1999;9:195–209.
9. Velculescu VE, Zhang L, Zhou W, Vogelstein J, Basrai MA, Bassett DE Jr, Hieter P, Vogelstein B, Kinzler KW. Characterization of the yeast transcriptome. Cell. 1997;88:243–51.
10. Adams MD, Kelley JM, Gocayne JD, Dubnick M, Polymeropoulos MH, Xiao H, Merril CR, Wu A, Olde B, Moreno RF. Complementary DNA sequencing: expressed sequence tags and human genome project. Science. 1991;252:1651–6.
11. Eisen JA. Badomics words and the power and peril of the ome-meme. Gigascience. 2012;1:6.
12. Dong Z, Chen Y. Transcriptomics: advances and approaches. Sci China Life Sci. 2013;56:960–7.
13. Lowe R, Shirley N, Bleackley M, Dolan S, Shafee T. Transcriptomics technologies. PLoS Comput Biol. 2017;13:e1005457.
14. McGettigan PA. Transcriptomics in the RNA-seq era. Curr Opin Chem Biol. 2013;17:4–11.
15. Wainberg M, Sinnott-Armstrong N, Mancuso N, et al. Opportunities and challenges for transcriptome-wide association studies. Nat Genet. 2019;51:592–9.
16. Sim GK, Kafatos FC, Jones CW, Koehler MD, Efstratiadis A, Maniatis T. Use of a cDNA library for studies on evolution developmental expression of the chorion multigene families. Cell. 1979;18:1303–16.
17. Marra MA, Hillier L, Waterston RH. Expressed sequence tags--ESTablishing bridges between genomes. Trends Genet. 1998;14:4–7.
18. Alwine JC, Kemp DJ, Stark GR. Method for detection of specific RNAs in agarose gels by transfer to diazobenzyloxymethyl-paper and hybridization with DNA probes. Proc Natl Acad Sci U S A. 1977;74:5350–4.
19. Becker-André M, Hahlbrock K. Absolute mRNA quantification using the polymerase chain reaction (PCR). A novel approach by a PCR aided transcript titration assay (PATTY). Nucleic Acids Res. 1989;17:9437–46.
20. The Economist. Life 2.0. 2006. https://www.economist.com/special-report/2006/08/31/life-20. Accessed 15 Feb 2021.
21. Pan Q, Shai O, Lee LJ, Frey BJ, Blencowe BJ. Deep surveying of alternative splicing complexity in the human transcriptome by high-throughput sequencing. Nat Genet. 2008;40:1413–5.
22. Sultan M, Schulz MH, Richard H, et al. A global view of gene activity and alternative splicing by deep sequencing of the human transcriptome. Science. 2008;321:956–60.
23. Lappalainen T, Sammeth M, Friedländer MR, et al. Transcriptome and genome sequencing uncovers functional variation in humans. Nature. 2013;501:506–11.
24. Melé M, Ferreira PG, Reverter F, et al. Human genomics. The human transcriptome across tissues and individuals. Science. 2015;348:660–5.
25. Khademhosseini A, Suh K-Y, Zourob M. Biological microarrays: methods and protocols. Humana Press; 2010.
26. Madden SL, Wang CJ, Landes G. Serial analysis of gene expression: from gene discovery to target identification. Drug Discov Today. 2000;5:415–25.
27. Yamamoto M, Wakatsuki T, Hada A, Ryo A. Use of serial analysis of gene expression (SAGE) technology. J Immunol Methods. 2001;250:45–66.
28. Anisimov SV. Serial Analysis of Gene Expression (SAGE): 13 years of application in research. Curr Pharm Biotechnol. 2008;9:338–50.
29. Weeraratna AT. Serial analysis of gene expression (SAGE): advances, analysis and applications to pigment cell research. Pigment Cell Res. 2003;16:183–9.
30. Datson NA. Scaling down SAGE: from miniSAGE to microSAGE. Curr Pharm Biotechnol. 2008;9:351–61.
31. Lehninger AL, Nelson DL, Cox MM, University Michael M Cox. Lehninger principles of biochemistry. Macmillan; 2005.
32. Strachan T, Read A. Human molecular genetics. Garland Science; 2018.
33. Schena M, Shalon D, Davis RW, Brown PO. Quantitative monitoring of gene expression patterns with a complementary DNA microarray. Science. 1995;270:467–70.
34. Watson JD, Baker TA, Bell SP. Molecular biology of the gene. Benjamin-Cummings Publishing Company; 2014.
35. da Costa JP, Rocha-Santos T, Duarte AC. Analytical tools to assess aging in humans: the rise of geri-omics. Trends Anal Chem. 2016;80:204–12.
36. Rando O. Hybridization to homemade microarrays. Cold Spring Harb Protoc. 2019. https://doi.org/10.1101/pdb.prot096487.
37. Zahurak M, Parmigiani G, Yu W, Scharpf RB, Berman D, Schaeffer E, Shabbeer S, Cope L. Pre-processing Agilent microarray data. BMC Bioinformatics. 2007;8:142.

38. Miller MB, Tang Y-W. Basic concepts of microarrays and potential applications in clinical microbiology. Clin Microbiol Rev. 2009;22:611–33.
39. Hrdlickova R, Toloue M, Tian B. RNA-Seq methods for transcriptome analysis. Wiley Interdiscip Rev RNA. 2017. https://doi.org/10.1002/wrna.1364.
40. Marguerat S, Bähler J. RNA-seq: from technology to biology. Cell Mol Life Sci. 2010;67:569–79.
41. Kukurba KR, Montgomery SB. RNA sequencing and analysis. Cold Spring Harb Protoc. 2015;2015:951–69.
42. Gasperskaja E, Kučinskas V. The most common technologies and tools for functional genome analysis. Acta Med Litu. 2017;24:1–11.
43. Metzker ML. Sequencing technologies - the next generation. Nat Rev Genet. 2010;11:31–46.
44. Conesa A, Madrigal P, Tarazona S, et al. Erratum to: a survey of best practices for RNA-seq data analysis. Genome Biol. 2016;17:181.
45. Neumann HP. Progress in DNA methylation research. Nova Publishers; 2007.
46. Kulkarni A, Anderson AG, Merullo DP, Konopka G. Beyond bulk: a review of single cell transcriptomics methodologies and applications. Curr Opin Biotechnol. 2019;58:129–36.
47. Munsky B, Neuert G, van Oudenaarden A. Using gene expression noise to understand gene regulation. Science. 2012;336:183–7.
48. Crosetto N, Bienko M, van Oudenaarden A. Spatially resolved transcriptomics and beyond. Nat Rev Genet. 2015;16:57–66.
49. Aldridge S, Teichmann SA. Single cell transcriptomics comes of age. Nat Commun. 2020;11:4307.
50. Rozenblatt-Rosen O, Stubbington MJT, Regev A, Teichmann SA. The Human Cell Atlas: from vision to reality. Nature. 2017;550:451–3.
51. Regev A, Teichmann SA, Lander ES, et al. The Human Cell Atlas. Elife. 2017. https://doi.org/10.7554/eLife.27041.
52. Blencowe BJ. Alternative splicing: new insights from global analyses. Cell. 2006;126:37–47.
53. Nilsen TW, Graveley BR. Expansion of the eukaryotic proteome by alternative splicing. Nature. 2010;463:457–63.
54. Wang ET, Sandberg R, Luo S, Khrebtukova I, Zhang L, Mayr C, Kingsmore SF, Schroth GP, Burge CB. Alternative isoform regulation in human tissue transcriptomes. Nature. 2008;456:470–6.
55. Harrington ED, Boue S, Valcarcel J, Reich JG, Bork P. Estimating rates of alternative splicing in mammals and invertebrates. Nat Genet. 2004;36:916–7.
56. Kim E, Magen A, Ast G. Different levels of alternative splicing among eukaryotes. Nucleic Acids Res. 2007;35:125–31.
57. Chen L, Bush SJ, Tovar-Corona JM, Castillo-Morales A, Urrutia AO. Correcting for differential transcript coverage reveals a strong relationship between alternative splicing and organism complexity. Mol Biol Evol. 2014;31:1402–13.
58. Wahl MC, Will CL, Lührmann R. The spliceosome: design principles of a dynamic RNP machine. Cell. 2009;136:701–18.
59. Matera AG, Wang Z. A day in the life of the spliceosome. Nat Rev Mol Cell Biol. 2014;15:108–21.
60. Lee Y, Rio DC. Mechanisms and regulation of alternative pre-mRNA splicing. Annu Rev Biochem. 2015;84:291–323.
61. Fica SM, Tuttle N, Novak T, Li N-S, Lu J, Koodathingal P, Dai Q, Staley JP, Piccirilli JA. RNA catalyzes nuclear pre-mRNA splicing. Nature. 2013;503:229–34.
62. Galej WP, Oubridge C, Newman AJ, Nagai K. Crystal structure of Prp8 reveals active site cavity of the spliceosome. Nature. 2013;493:638–43.
63. Naftelberg S, Schor IE, Ast G, Kornblihtt AR. Regulation of alternative splicing through coupling with transcription and chromatin structure. Annu Rev Biochem. 2015;84:165–98.
64. Cooper TA, Wan L, Dreyfuss G. RNA and disease. Cell. 2009;136:777–93.
65. Padgett RA. New connections between splicing and human disease. Trends Genet. 2012;28:147–54.
66. Suñé-Pou M, Prieto-Sánchez S, Boyero-Corral S, Moreno-Castro C, El Yousfi Y, Suñé-Negre JM, Hernández-Munain C, Suñé C. Targeting splicing in the treatment of human disease. Genes. 2017. https://doi.org/10.3390/genes8030087.
67. Montes M, Sanford BL, Comiskey DF, Chandler DS. RNA splicing and disease: animal models to therapies. Trends Genet. 2019;35:68–87.
68. Pandit S, Zhou Y, Shiue L, Coutinho-Mansfield G, Li H, Qiu J, Huang J, Yeo GW, Ares M Jr, Fu X-D. Genome-wide analysis reveals SR protein cooperation and competition in regulated splicing. Mol Cell. 2013;50:223–35.
69. Änkö M-L. Regulation of gene expression programmes by serine-arginine rich splicing factors. Semin Cell Dev Biol. 2014;32:11–21.
70. Motta-Mena LB, Heyd F, Lynch KW. Context-dependent regulatory mechanism of the splicing factor hnRNP L. Mol Cell. 2010;37:223–34.
71. Krawczak M, Thomas NST, Hundrieser B, Mort M, Wittig M, Hampe J, Cooper DN. Single base-pair substitutions in exon-intron junctions of human genes: nature, distribution, and consequences for mRNA splicing. Hum Mutat. 2007;28:150–8.
72. Stenson PD, Ball EV, Howells K, Phillips AD, Mort M, Cooper DN. The Human Gene Mutation Database: providing a comprehensive central mutation database for molecular diagnostics and personalized genomics. Hum Genomics. 2009;4:69–72.
73. Lim KH, Ferraris L, Filloux ME, Raphael BJ, Fairbrother WG. Using positional distribution to identify splicing elements and predict pre-mRNA processing defects in human genes. Proc Natl Acad Sci U S A. 2011;108:11093–8.
74. Mendell JT, Sharifi NA, Meyers JL, Martinez-Murillo F, Dietz HC. Nonsense surveillance regulates expression of diverse classes of mammalian transcripts and mutes genomic noise. Nat Genet. 2004;36:1073–8.
75. Yepiskoposyan H, Aeschimann F, Nilsson D, Okoniewski M, Mühlemann O. Autoregulation of the nonsense-mediated mRNA decay pathway in human cells. RNA. 2011;17:2108–18.
76. Tani H, Imamachi N, Salam KA, Mizutani R, Ijiri K, Irie T, Yada T, Suzuki Y, Akimitsu N. Identification of hundreds of novel UPF1 target transcripts by direct determination of whole transcriptome stability. RNA Biol. 2012;9:1370–9.
77. Hug N, Longman D, Cáceres JF. Mechanism and regulation of the nonsense-mediated decay pathway. Nucleic Acids Res. 2016;44:1483–95.
78. da Costa PJ, Menezes J, Romão L. The role of alternative splicing coupled to nonsense-mediated mRNA decay in human disease. Int J Biochem Cell Biol. 2017;91:168–75.
79. Pan Q, Saltzman AL, Kim YK, Misquitta C, Shai O, Maquat LE, Frey BJ, Blencowe BJ. Quantitative microarray profiling provides evidence against widespread coupling of alternative splicing with nonsense-mediated mRNA decay to control gene expression. Genes Dev. 2006;20:153–8.
80. Hillman RT, Green RE, Brenner SE. An unappreciated role for RNA surveillance. Genome Biol. 2004;5:R8.
81. Lareau LF, Brooks AN, Soergel DAW, Meng Q, Brenner SE. The coupling of alternative splicing and nonsense-mediated mRNA decay. Adv Exp Med Biol. 2007;623:190–211.
82. Ni JZ, Grate L, Donohue JP, Preston C, Nobida N, O'Brien G, Shiue L, Clark TA, Blume JE, Ares M Jr. Ultraconserved elements are associated with homeostatic control of splicing regulators by alternative splicing and nonsense-mediated decay. Genes Dev. 2007;21:708–18.
83. Peters MJ, Joehanes R, Pilling LC, et al. The transcriptional landscape of age in human peripheral blood. Nat Commun. 2015;6:8570.
84. GTEx Consortium. The Genotype-Tissue Expression (GTEx) project. Nat Genet. 2013;45:580–5.

85. Ju W, Nair V, Smith S, et al. Tissue transcriptome-driven identification of epidermal growth factor as a chronic kidney disease biomarker. Sci Transl Med. 2015;7:316ra193.
86. Park J, Shrestha R, Qiu C, Kondo A, Huang S, Werth M, Li M, Barasch J, Suszták K. Single-cell transcriptomics of the mouse kidney reveals potential cellular targets of kidney disease. Science. 2018;360:758–63.
87. Saben J, Lindsey F, Zhong Y, Thakali K, Badger TM, Andres A, Gomez-Acevedo H, Shankar K. Maternal obesity is associated with a lipotoxic placental environment. Placenta. 2014;35:171–7.
88. Renthal W, Boxer LD, Hrvatin S, Li E, Silberfeld A, Nagy MA, Griffith EC, Vierbuchen T, Greenberg ME. Characterization of human mosaic Rett syndrome brain tissue by single-nucleus RNA sequencing. Nat Neurosci. 2018;21:1670–9.
89. Schmidt M, Hopp L, Arakelyan A, et al. The human blood transcriptome in a large population cohort and its relation to aging and health. Front Big Data. 2020. https://doi.org/10.3389/fdata.2020.548873.
90. Wilson PC, Wu H, Kirita Y, Uchimura K, Ledru N, Rennke HG, Welling PA, Waikar SS, Humphreys BD. The single-cell transcriptomic landscape of early human diabetic nephropathy. Proc Natl Acad Sci U S A. 2019;116:19619–25.
91. Nho K, Nudelman K, Allen M, et al. Genome-wide transcriptome analysis identifies novel dysregulated genes implicated in Alzheimer's pathology. Alzheimers Dement. 2020;16:1213–23.
92. Guo Y, Shang X, Li Z. Identification of cancer subtypes by integrating multiple types of transcriptomics data with deep learning in breast cancer. Neurocomputing. 2019;324:20–30.

Proteomics Principles and Clinical Applications

6

Ixchel Ramírez-Camacho, Gibrán Pedraza-Vázquez, Karla Daniela Rodríguez-Hernández, Elizabeth Sulvaran-Guel, and Nadia Alejandra Rivero-Segura

Abbreviations

BLAST	Basic local alignment search tool
Ca2+	Calcium
CEA	Carcinoembryonic antigen
CSF	Cerebrospinal fluid
CATH	Class Architecture Topology and Homologous superfamily
DNA	Deoxyribonucleic acid
EGF	Epidermal growth factor
GST-ORF	Glutathione-S-transferase open reading frame
HIV	Human immunodeficiency virus
HD	Huntington's disease
PARK7	Parkin 7
PD	Parkinson's disease
PDB	Protein Data Bank
QMEAN	Qualitative Model Energy Analysis
RNA	Ribonucleic acid
SARS-CoV-2	Severe acute respiratory syndrome coronavirus 2
SCOP	Structural Classification of Protein
SOD1	Superoxide dismutase 1
VEGF	Vascular endothelial growth factor

I. Ramírez-Camacho
Dirección de Investigación, Instituto Nacional de Geriatría, Ciudad de México, Mexico

G. Pedraza-Vázquez
Posgrado en Biología Experimental, Departamento de Ciencias de la Salud, Universidad Autónoma Metropolitana Unidad Iztapalapa, Mexico City, Mexico
e-mail: gpv@xanum.uam.mx

K. D. Rodríguez-Hernández
Laboratorio de Estudios sobre Tripanosomiasis, Departamento de Inmunología, Instituto de Investigaciones Biomédicas, Universidad Nacional Autónoma de México, México City, Mexico
e-mail: xandy411@comunidad.unam.mx

E. Sulvaran-Guel
Licenciatura en Ciencias Genómicas UNAM, México, Mexico

N. A. Rivero-Segura (✉)
Dirección de Investigación, Instituto Nacional de Geriatría (INGER), Instituto Nacional de Geriatría, Ciudad de México, Mexico
e-mail: nrivero@inger.gob.mx

Introduction

The proteome refers to the set of proteins present in a cell or organism at any given time. In this context, since the DNA contains the needed information for the creation of proteins, the relation between the genome and the proteome is complex, since one single gene may encode for more than a single protein (alternative splicing). Furthermore, posttranslational modifications and protein cleavage or modifications give rise to several protein isoforms for every single gene. Thus, studying the proteome provides additional information that otherwise we wouldn't notice only by studying the genome or the transcriptome [1]. Moreover, it is important to highlight that since proteome varies among cells and tissues depending on the condition, allowing the discovery of specific biomarkers involved in health and disease [2]. As mentioned before, the phenotype can be better explained by the proteome than by the genome or the transcriptome due to differences in expression levels and to protein modifications [3]. Furthermore, the proteome is more stable and more easily assessed than the metabolome; thus, proteomics studies are a powerful tool preferentially used in studying disease, development, aging, and other conditions.

Proteome

Unlike the genome, the genetic information contained in DNA is the sequencing of the 23 pairs of chromosomes in human and that is internalized in the heart of the cell "the nucleus," "the proteome" is the information of all the pro-

J. C. Gomez-Verjan, N. A. Rivero-Segura (eds.), *Principles of Genetics and Molecular Epidemiology*,
https://doi.org/10.1007/978-3-030-89601-0_6

teins expressed in a cell and how this information can change regarding environmental conditions, genetic, hormonal, or in different stages of human development [4].

Proteomics

Proteomics is the study of the information found in all the proteins expressed in an organism, and the change of these proteins under different situations in which different technologies are used for their identification, expression, and quantification; in addition to these technologies which we will list later, the study of the proteome is supplemented with other transcription and genomic techniques to expose the identity of the proteins of an organism and to know the structure and functions of a particular protein. These technologies are used in different basic and clinical research studies as markers for timely diagnosis. The points that the study of proteomics tries to clarify are identity and expression patterns of proteins, functional pathways of proteins in different diseases, carbonylation, acetylations, and other conformational changes therapeutic white spots mechanisms of pathogens vaccine production [5].

Proteomics Databases and Tools

There are several available protein databases and tools, ranging from sequence databases, alignment tools, domain prediction tools, motif prediction tools, and three-dimensional structure databases and tools, among many others.

Sequence databases provide the amino acid sequence for many proteins and are essential for the creation of other databases and tools [6]. Some of the most important are GenBank, Ref-Seq, UniProt, TrEMBL, and SwissProt [7]. GenBank contains redundant sequence data and allows the alignment of sequences with BLAST [6]. Ref-Seq nonredundant data comes from the GenBank database and includes important sequence annotations, such as domains, CDSs, and variants, among others [6]. UniProt is the most important protein database and is divided into TrEMBL and SwissProt, which contain computationally obtained and manually curated data, respectively [6].

Structure databases contain information obtained in experimentally confirmed protein structure studies [6]. The most important structure database known is Protein Data Bank (PDB) [8]. However, there are other databases available, such as CATH and SCOP [7]. PDB provides three-dimensional data structures along with significant annotations, including helices and sheets, variants in specific sites such as the active site, among other biochemical indicators [8]. Furthermore, there are available tools for the prediction of protein structures, such as SWISS-MODEL, which uses a UniProt identifier for the retrieval of the sequence, the alignment with sequences in the database and the prediction of the protein folding [9]. SWISS-MODEL allows the visualization of homologous protein structures and alignments, including options for amino acid classification by charge, size, entropy, and amino acid groups, among others [9]. Finally, the selected models can be assessed by stoichiometry with a methodology named QMEAN and with Ramachandran plots [9]. Structure data either from a database or from a modeling tool can be downloaded for visualization in other tools, such as Pymol or Jmol [8].

Proteomics Technologies

In 1975, the study of proteomics with the second-dimensional gel electrophoresis technique was used to detect proteins from a mixture of *Escherichia coli* [10]. Twenty-five years later, the biochemical genomics technique and methods for the quantification and analysis of proteins were introduced, which allowed researchers to identify proteins from genes with activities in biochemical metabolic pathways; the GST-ORF technique (opening reading frame fused to glutathione

Table 6.1 The relevant techniques of proteomics

Year of discovery	Technique	Type of technique	Advantages	References
1975	Two-dimensional gel electrophoresis	Conventional and advanced	Separation of subunits	O' Farrell et al. [10]
1999	Biochemical genomics	Advanced	Fast precision	Martzen et al. [11]
2001	Microarrays of Proteins	Advanced	Analyze the complex protein	Zhu H et al. (2001)
2011	MS	Advanced protein mixtures	Analyze of complex	Yates Iii (2011)
2006	ICAT labeling	Quantitative	Isotopes	Shiio and Aebersol (2006)
2006	SILAC	Quantitative	Isotopes amino acids	Ong et al. (2006)
2007	iTAQ	Quantitative	Relative and absolute	Wies et al. (2007)
2000	X-ray crystallography	High throughput	High-throughput 3D structure	Smith MS (2000)
2015	NMR-spectroscopy	High throughput	Understand biological function	Krosveen et al. (2015)

MS Mass spectroscopy, *ICAT* labeling Isotope-codea affinity tag, *SILAC* Stable Isotope labeling with amino acids in cell culture, *iTRAQ* Isobaric tag for relative and absolute quantification

Advanced techniques; Timeline of techniques used in the study of proteomics

S-transferase) is fast and sensitive and is applicable for almost any detectable activity [11]. These and all subsequent techniques in the advancement and evolution of the study of proteomics are listed in Table 6.1.

Proteomics in Health and Disease

Proteomics for the Diagnosis of Neurodegenerative Diseases

Neurodegenerative diseases are a wide and heterogeneous group of diseases that represent a common cause of morbidity and cognitive impairment in the elderly. Unfortunately, the prevalence of such diseases is still growing, and the early diagnosis is challenging since most clinicians are not properly trained to diagnose these conditions. As in most chronic diseases, the opportune diagnosis of neurodegenerative diseases is crucial for a patient's outcome, since this allows an appropriate prescription and management that may impact the prognosis. Thus, in this section we briefly described the state of the art for proteomics applied for the diagnosis of most common neurodegenerative diseases.

Alzheimer's Disease (AD) and Related Dementias Diagnosis

AD is the most common neurodegenerative dementia; in 2017, there were an estimated 46 million people with AD worldwide, and it has been suggested that the prevalence of such disease will continue to grow in the coming years. In the beginning, physicians identified AD as *probable AD*, since they only performed physical examination that revealed mild cognitive impairment or dementia; unfortunately, the definitive AD diagnosis came with the autopsy that showed the presence of amyloid plaques and tau neurofibrillary tangles in the brain [12]. Currently, with the technological advances in neuroimaging (computed tomography, magnetic resonance imaging, or positron tomography), AD diagnosis relies on the evaluation of behavior tests that give information about an individual's memory and cognitive functions; it has been reported that these novel technologies offer sensitive and specific imaging biomarkers for AD diagnosis. Most of these biomarkers reflect glial inflammation, epigenomic alterations, structural and functional brain alteration, and synaptic or cellular degeneration from the early stages of AD [13]. Unfortunately, the main limitation to transfer these technologies to daily medical practice is both the affordability and unavailability of the equipment in most hospitals.

In this context, proteomic biomarkers for AD have gained relevance in the clinical field, since these enable us to understand the etiology and identify potential targets for improving the management of AD patients, leading to the development of personalized medicine [14]. The main biomarkers derived from this *omic* technology are the amyloid plaque, which is composed of Aβ-peptides derived from the APP. Thus, the early-stage AD patients show elevated levels of Aβ40 and Aβ42, and both diminish in the late stage of the disease [12]. Another proteomic study suggests a set of five proteins that differentiates AD from non-AD individuals, which includes α-1-microglobulin, apolipoprotein E (ApoE), brain natriuretic peptide, interleukin-16, and serum glutamic oxaloacetic transaminase; these biomarkers are highly sensitive and (89.36%) and specific (79.1%). However, when these proteins are compared with previous published studies, only ApoE is reproducible among the different cohorts [15]. As well, proteomics are helpful to understand the biological network, pathway, and cell type changes in human tissue; in this sense, a recent study performed in the early 400 CSF samples identifies 3334 proteins that help to differentiate among AD and other dementias, proteins involved in inflammation, sugar metabolism, mitochondrial function, synaptic, RNA-associated proteins, and glial activation [16]. Similarly, another study identifies a list of 26 core proteins linked to neurodegeneration such as PARK7, SOD1, YWHAZ, and YKL-40, reflecting the astrocytic activation, glucose metabolism, and antioxidant defenses in independent multicentric cohorts, with a significantly specificity (87%) and sensitivity for AD individuals. Moreover, a meta-analysis reveals that along the proteomic studies performed in three different cohorts, identify six candidate biomarkers (alpha-2-macroglobulin, pancreatic polypeptide, apolipoprotein A-1, afamin, insulin growth factor binding protein-2, and fibrinogen-gamma-chain) for the early diagnosis of AD [17].

Proteomic Studies in Parkinson's Disease

As mentioned by several authors Parkinson's disease (PD) is a chronic and neurodegenerative disease commonly associated with the aging process. PD is mainly characterized by the loss of dopaminergic neurons in substantia nigra pars compacta region, reduced dopamine level in the striatum, and accumulation of α-synuclein protein aggregates (Lewy bodies). As well, this neurodegenerative disease is featured by motor symptoms such as tremors, rigidity, bradykinesia/akinesia, and postural instability [18, 19].

Currently, PD clinical diagnosis is based on the physical examination; however, similar to AD, the symptoms are visible when 70% of dopaminergic neurons have been lost. In this sense, research is focused on reliable biomarkers that help in the early diagnosis of PD, in an attempt to identify potential therapeutic targets. In this sense, proteomics have contributed to the understanding of the etiology and molecular mechanisms underlying this disease. For instance, several studies performed in brain tissue demonstrate that proteins implicated in Ca^{2+} homeostasis (regucalcin) and transport (kinectin) [20], cell signaling phosphoinositide-dependent protein kinase 1, ERK 1/2, monomeric α-synuclein, and p38 [21], mitochondrial integrity (Prx2, ATP synthase D chain, electron transport chain complexes I and III, and prohibitin) [22, 23], cytoskeleton (profilin, fatty-acid binding protein, and γ-glutamyl hydrolase) [24], aldehyde metabolism (aldehyde dehydrogenase A and cellular retinol-binding protein 1), L-DOPA methylation (S-adenosyl homocysteine hydrolase 1), glial activation (glial fibrillary acidic protein, glial maturation factor-β, galectin 1, and sorcin A) [25], oxidative stress (H-ferritin, SOD1, DJ-1, Prx2, Prx6, and Prx3) [26, 27], energy metabolism (cytochrome b-c1 subunit 2 and ATP synthase subunit D, aldolase A, enolase, and glyceraldehyde dehydrogenase) [27, 28], and ubiquitination pathway (UCHL1) [29] are dysregulated in PD.

Additionally, in the quest of reliable biomarkers derived from less invasive procedures, liquid biopsies (or biofluids) such as cerebrospinal fluid (CSF), tears, blood, and blood-derived components have become relevant for PD diagnosis. For instance, neurexin-1, R-1-acid glycoprotein, β-2-glycoprotein 1, DJ-1, α-synuclein [30], apolipoprotein E (ApoE), autotaxin, SOD1 [31], ceruloplasmin, chromogranin C, and ApoH [32] have been identified in CSF samples from PD patients. Blood samples derived from AD patients show proteomic profiles involved in hemoglobin clearance (haptoglobin-related protein precursor, truncated β-globin) [33, 34], inflammation (PRNP, HSPG2), lipid metabolism (clusterin, complement C1r subcomponent, fibrinogen γ-chain), immunoregulation (immunoglobulin kappa-chain VK-1, Ig-γ-3, chain C region), protein folding (ApoA1, fibrinogen γ-chain), protein aggregation (serum amyloid P component), intracellular transport, cell proliferation immunoregulation, blood clotting (fibrinogen γ-chain, full size inter-α-trypsin inhibitor heavy chain H4), inflammation (transthyretin ApoAq complement factor H) [35], and mitochondria (mitochondrial ATP synthase β-subunit) [36] are dysregulated in PD patients in comparison to healthy controls. Finally, tears are a novel source of potential biomarkers of PD, suggesting that protein involved in immune response, lipid metabolism, and oxidative stress (PRx6, annexin-A-5, glutathione-S-transferase-A, ApoD, ApoA4, Apo AI, lactotransferrin, galectin 3, and profilin 1) [37].

Huntington's Disease

Huntington's disease (HD) is a neurodegenerative disease which is triggered in adulthood, and patients show emotional problems, cannot control their physical movements, have speech disorders and dementia, lose the ability to think, and die within 15–20 years of diagnosis; Huntington's disease is genetically inherited through an autosomal dominant gene located on chromosome 4; it is caused by DNA mutation of generally 37 or more repetition of CAG nucleotides. Proteomic analysis with gel electrophoresis, mass spectroscopy, Western blot, and technology based on chromatography and X-ray crystallography, where it is shown that there are molecular changes that occur in HD, improves the specific treatment [38].

Aggregation of polyglutamine-expanded Huntingtin exon 1 (HttEx1) in Huntington's disease (HD) is characterized by the aggregation of soluble oligomers to late-stage inclusions, until today the nature of the aggregates and how they lead to neuronal dysfunction is not fully understood.

Proteomic analysis of this protein by mass spectrometry (MS) in a murine model was observed that HD has extensive remodeling of the soluble brain proteome, which was correlated with the formation of insoluble aggregates during disease progression. This deep and quantitative analysis shows differences in protein expression levels, sequence characteristics [39], low complexity regions, and spiral domains. In a cell-based model of HD, overexpression of a subset of the sequestered proteins in most cases rescued viability and reduced aggregate size, indicating widespread loss of cellular protein function contributes to mediated toxicity by aggregates of HttEx1 [40].

In addition to the protein aggregation characteristic in HD, it has been discovered that another failure that occurs in HD is the protein misfolding process, which can generate protein oligomers or larger aggregates. This may be due to high temperatures, low pH, oxidative stress, abnormal presence of metal ions, mutations, transcriptional, translational or posttranslational errors, and aging [41, 42].

The toxicity is due to soluble oligomers rather than the large protein inclusions that develop over time; therefore, researchers are currently focusing on developing aggregation inhibitors, although so far this has not been achieved with success [39]

Cancer

The main clinical application of proteomics is the discovery of changes in the proteome in certain diseases for the identification of biomarkers that may lead to more efficient diagnosis and treatments [43]. One example of particular interest is cancer, in which several genes are mutated or rearranged [43]. Cancer cells often show aberrant gene expression, protein localization, and posttranslational modifications, affecting protein function and cell stability [44]. Furthermore, distinct cancer types show different cellular gene expressions and behavior [44]. Therefore, the molecular hallmarks in cancer have a promising future in the understanding of the development of the disease, as well as in prognosis and therapies [44].

Several studies on breast cancer patients' tissues or cells have shown differentially expressed proteins when compared to healthy subjects [45]. Examples include differences in normal epithelial tissues and invasive tissues and between estrogen receptor positive and negative cells [45]. In the latter, ontology term enrichment analysis included focal adhesion and lipid metabolism proteins [45]. Furthermore, studies comparing different subtypes of breast cancer proteomes have been done, with subsequent clustering analysis, achieving the grouping of subtypes according to expression patterns [45]. Some identified differentially expressed proteins include ER, PR, HER2, p53, PIK3CA, and GATA3, among others [45].

On the other hand, proteomic studies on lung cancer, the most prevalent cancer type, have revealed different biomarkers as potential therapeutic targets [46]. The protein carcinoembryonic antigen (CEA) has been found to be overexpressed in lung cancer tissues [46]. Although its diagnostic potential is not high, when used in combination with other proteins, such as CYFRA, it loses lung cancer specificity [46]. Ontology term enrichment analysis has revealed these proteins as components of the cell membrane and the cytoskeleton, respectively [46]. Other differentially expressed proteins useful for treatment include TPA, ProGRP, NSE, and mutations in the epidermal growth factor (EGFR) protein [46].

There are several other studies on different cancer types [47]. Other common cancer types include prostate cancer [47]. Proteins found to be differentially expressed include calgranulin B, radical scavenger enzymes, and GTP-binding proteins [47]. Furthermore, PCOTH overexpressing cells showed elevated phosphorylation of oncoproteins, suggesting an important role in cell growth and its potential target for therapies [47]. Additionally, mass spectrometry revealed chemokines and aberrant isoforms of serum amyloid A protein (the latter in metastasized bone patients) present in serological samples, which functioned for diagnosis with artificial intelligence algorithms, showing promising results [47].

Although proteomics has amazing advantages that have favorable future applications, including its ability to detect protein isoforms, including posttranslational modifications—which would go unnoticed using other technologies, such as transcriptomics—and to detect intracellular localization, it is important to be noticed that clinical proteomics as a still emerging application has several drawbacks [48]. Some of the main problems are the impossibility of protein amplification (in comparison with nucleic acids), limiting studies to the exact protein amount in the cell, and accordingly, the inability to detect important proteins having important roles even under low expression levels [48]. For these reasons, more studies in the field of clinical proteomics must be done, with greater cohorts and larger tissues or cell samples for greater statistical significance, together with other important fields, such as genomics, transcriptomics, and metabolomics [48]. These improvements could make omics technologies scale from laboratories to clinics, with promising results for disease treatments and diagnosis [48].

Mitochondria and Proteomics

Mitochondria is a double membrane organ in which the respiratory complexes responsible for electron transport and oxidative phosphorylation by ATP synthase coexist in the inner mitochondrial membrane; it also contains its own genetic material for the synthesis of some of its high-molecular-weight and highly lipophilic protein complexes. The mitochondrial proteome plays a crucial role in different diseases such as type 1 and 2 diabetes, cardiovascular, and neurodegenerative, among others, in addition to playing a crucial role in cell signaling and different metabolic pathways for the generation of intermediate metabolites. It is known that when there is a condition linked to redox unbalance, the mitochondria are also damaged, specifically complex I and complex III [49].

The human mitochondrial proteome is made up of 1158 genes encoding for proteins; the human mitochondrion 2.0 was updated in which 240 new genes were added to the 918 already existing, and they are compared with the proteome of 14 mouse organs where it was also observed that the proteome consists of 1158 genes encoded for proteins [50].

MitoMiner, developed by Smith and Robinson in 2017, centralizes data on the localization of mitochondrial proteins prioritizing target genes for mitochondrial diseases for research [51]. MitoCarta2.0 and MitoMiner are updated catalogs offering access to all mitochondrial proteins that support and facilitate the investigation of mammalian mitochondrial proteins.

Proteomics of Infectious Diseases

Viral Infections

As we previously described, proteins are responsible for controlling the various signaling pathways of cell function. The field of studying the proteins of an organism has become the main field for the identification and characterization of proteins.

The proteome of a cell, tissue, or organism is influenced by a variety of external and environmental stimuli, including those caused by infectious viral diseases such as HIV, hepatitis C, and currently SARS-COV-2.

HIV The human immunodeficiency virus (HIV) proteome has been extensively explored, and representative genomes of each virus family have been sequenced, Databases maintained by the Los Alamos National Laboratories (www.hiv.lanl.gov) and BioAfrica (www.bioafricrica.net).

The application of proteomics in HIV studies has been achieved to study interactions and measure protein levels quickly and accurately with microarrays, chips, and ELISA tests, where they identified a high expression of vascular endothelial growth factor (VEGF) and the factor of epidermal growth (EGF) in HIV patients who have paradoxically low CD4 + T-cell counts despite a low viral load. If protein experiments are standardized, efficient, and inexpensive, they could pave the way for research on HIV and other diseases [52].

Also, in other experiments analyzing the immune response in patients with antiretroviral treatment (ART), Arnaud et al. data showed a decrease in the immunoaffinity of antibodies vs the original strain of HIV after 4 weeks of ART, which was interpreted as evidence of viral adaptation to patients' immune responses, ART, or both [53].

Hepatitis Several proteomic studies have been analyzed for the hepatitis virus (HV). In particular in hepatitis B virus (HBV), a C-terminal fragment of complement factor C3 and an apolipoprotein A1 (ApoA1) isoform are known to be impaired in this disease [54]. The proteins haptoglobin, transthyretin, antitrypsin, topoisomerase II, ApoA1, and ApoA4, 1-a are overexpressed in patients with chronic hepatitis B [55]. Using the 2D gel technique with serum samples, the amyloid P component was detected in patients with healthy and chronic HBV, but not in patients with HBV-HCC [56].

SARS-CoV-2 In 2019, a highly infectious disease was discovered in Wuhan in the Chinese People's Republic that affects human health; this SAR-Cov 2 virus causes coronavirus disease COVID-19. Although the underlying mechanisms are unknown, patients can develop acute respiratory distress syndrome (ARDS) or even die suddenly in a short period. This sudden change implies a "two-stage" pattern of disease progression.

On March 11, 2020, the World Health Organization declared the coronavirus disease COVID-19 a pandemic, until in March 2021 a little over 117,644,021 cases and 2,612,360 deaths had already been reported (WHO, 2021).

Proteomics studies for COVID-19 have already been carried out, such as the one carried out by Chinese researchers in 2020 with urine samples from 37 patients in which they applied a quantitative proteomic approach of independent data acquisition (DIA), based on mass spectrometry; the analysis consisted of three experimental groups which were (1) healthy, (2) non-COVID-19, and (3) non-COVID-19 pneumonia; 5991 proteins were found in the 37 urine samples, 1986 protein levels changed significantly in the COVID-19 group compared to the other groups.

It can be concluded that patients with COVID-19 in the early stage show changes in the proteomics that reveal immunosuppression, while patients with COVID-19 in late stages show proteomics of immune activation, which provides the basis of molecular biology to understand clinical symptoms and develop strategies to elucidate the stage of the disease that patients will develop. [57]

Bacterial Infections

The study of diseases caused by bacteria has gained relevance in recent years because the number of bacteria with antibiotic resistance has increased alarmingly. Also, diseases that were practically under control are having a rebound in the number of infections per year. Therefore, proteomic approaches to detail how these microorganisms interact with the host and the changes they generate in the host are extremely important [58].

Such interventions can be useful for precision medicine strategies by determining which bacterium is causing the infection, whether it is a variant, and then applying the most appropriate treatment. Also, with antibiotic resistance problems, docking and molecular dynamic techniques make it possible to search or synthesize molecules with antibiotic potential or to reposition drugs in this area [58, 59].

One of the bacterial infections with the greatest impact on public health is *Mycobacterium tuberculosis*. The diagnosis is complex, and this allows it not only to affect the lungs but also to spread to other organs of the body [60]. Besides, the presence of several variants makes treatment complex, so in a recent analysis, they reviewed biomarkers in serum and plasma from 18 studies performed in different parts of the world, using various techniques from ELISA to mass spectrometry [61]. Interestingly, most of the studies show different combinations of cytokines and other proteins that could be characteristic of that regional variant of the bacterium. It is important to continue researching and using proteomic tools to be able to apply them to the clinical setting as soon as possible [58].

Another recurrent infection is caused by *Salmonella* spp. bacteria. In vitro studies in which the proteomic profile of these bacteria has been evaluated during their interaction with human cells found that proteins belonging to metabolic pathways such as glycolysis, pyrimidine degradation, as well as flagellar proteins and glucose transporters are overrepresented and that proteins related to anaerobic respiration, TCA, and chemotaxis are underrepresented [62].

The application of proteomic tools in precision medicine, in this case in the fight against infectious diseases caused by bacteria, is very important to improve diagnostics. This will make it possible to combat the problem of antibiotic resistance by generating new strategies to make existing treatments more efficient and the possibility of generating new ones.

Parasitic Infections

Parasites are organisms responsible for several types of infections and may be classified into three main groups: protozoa, helminths, and ectoparasites. Protozoans are eukaryotic unicellular organisms responsible for infections such as malaria, toxoplasmosis, and trichomoniasis. On the other hand, helminths are worm-like multicellular organisms usually found in the gastrointestinal tracts of humans. Ascariasis, trichuriasis, and hookworm are the most common helminths infections, which present a high prevalence in developing countries [63]. Finally, ectoparasites are organisms that cause skin infections, such as pediculosis and tungiasis [64].

Parasitic infections have several problems: firstly, diagnosis is usually done with classic microscopy technologies [65]. For this reason, diagnosis usually has low specificity and sensitivity and additionally depends on technician abilities, which is time-consuming and costly. Secondly, treatments are commonly unsuccessful, resulting in high mortality and morbidity [66]. Finally, the most commonly used tests are ineffective in current and past infections, making prognosis difficult [65]. Parasitic infections are of special interest in low-income countries, where they are endemic [66]. For these reasons, it is necessary to develop new, more accurate, and affordable diagnosis tools [66].

Proteomic technologies have a promising solution for these problems, since they have the ability to identify specific biomarkers expressed in parasitic infections, facilitating diagnosis and treatments [66]. Furthermore, proteomics may enable the distinction between current and past infections with the detection of pathogen-derived molecules [66]. Several studies analyzing the proteome during parasitic infections have been already done, yielding significant and useful results. For example, in malaria, hypoxanthine phosphoribosyltransferase, phosphoglycerate mutase, lactate dehydrogenase, and fructose-bisphosphate aldolase were found at higher concentrations in patients than in healthy control individuals. Additionally, currently there are rapid malaria diagnosis tests available, which measure histidine-rich protein 2, lactate dehydrogenase, and aldolase levels. Similarly, opisthorchiasis, an infection caused by the helminth *Opisthorchis viverrini*, may cause severe long-term effects, including the development of cholangiocarcinoma. However, most of the patients don't present any symptoms during the early infection, making diagnosis a highly difficult task. Proteomics analysis revealed increases in fibronectin in opisthorchiasis patients [67].

Leishmaniasis, caused by the protozoan parasites of the genus *Leishmania*, benefits from proteomics, since current evidence demonstrates the discovery of potential biomarkers with a wide range of applications such as diagnosis, prognosis, therapeutic targets, monitoring disease progression, treatment follow-up, and identification of vaccine candidates [68].

It is expected that in future years, more studies regarding the proteome in parasitic diseases are developed, thus, allowing the identification of more biomarkers that may be useful both for diagnosis and as therapeutic targets [66]. Furthermore, immunoproteomics, a technique integrating proteomics with immunological responses, may help in antigen discovery for vaccine development. Finally, development and commercialization of new diagnosis devices are required for accurate and the early parasitic infections diagnosis, addressing mortality issues, especially in low-income countries [66].

Aging, Frailty, and Skeletal Muscle Wasting

The advances in medicine and technology have made possible the extension of human longevity, but when it comes to the state of health of older adults there is a negative correlation, seriously affecting their quality of life.

Age-related diseases are one of the obstacles to overcome to improve the health span of older adults and proteomics as a diagnostic tool can contribute to timely detection and treatment. Despite its usefulness, there are not many reports on the status of protein levels throughout life and particularly in aging. However, some studies cover issues related to protein synthesis and its quality control in aging, muscle-related diseases, and frailty.

Protein synthesis and its quality control are biological processes of high importance for the correct functioning of the organism [69, 70], so the changes that these processes undergo throughout life are very relevant. The functionality of the whole system depends on a balance between the synthesis, degradation, and function of each protein, and in aging, changes in this balance are often observed that can trigger pathologies [70].

Some studies suggest that there is a decline in protein synthesis in aging, the ability of cells to fold and degrade proteins [70]. In many cases, this loss of proteostasis in aging leads to the accumulation of damaged proteins and other molecules, which in turn can inhibit cell functionality and thus trigger an aging-associated disease [69, 71].

Undoubtedly, one of the most accepted theories is that during aging there is an accumulation of oxidized proteins [72]. This oxidation modifies the folding of the protein, and this leads to its aggregation, which often interferes with its degradation, making it easier for more oxidized proteins to aggregate, eventually altering and compromising the viability of the cell. The aggregation of proteins, lipids, and other molecules leads to the formation of lipofuscins. It consists of approximately 30%–70% cross-linked proteins and 20%–50% lipids, but carbohydrates were also identified to be a component of lipofuscin [70, 73]. The number of oxidized proteins or the detection of lipofuscins may be useful to detect changes in proteostasis and provide a warning of possible pathology [74].

However, a recent study involving 36 different proteomic analyses identifying proteins that change significantly with age reveals 1128 proteins reported by at least 2 analyses and a set of 32 proteins reported in 5 or more analyses. These 32

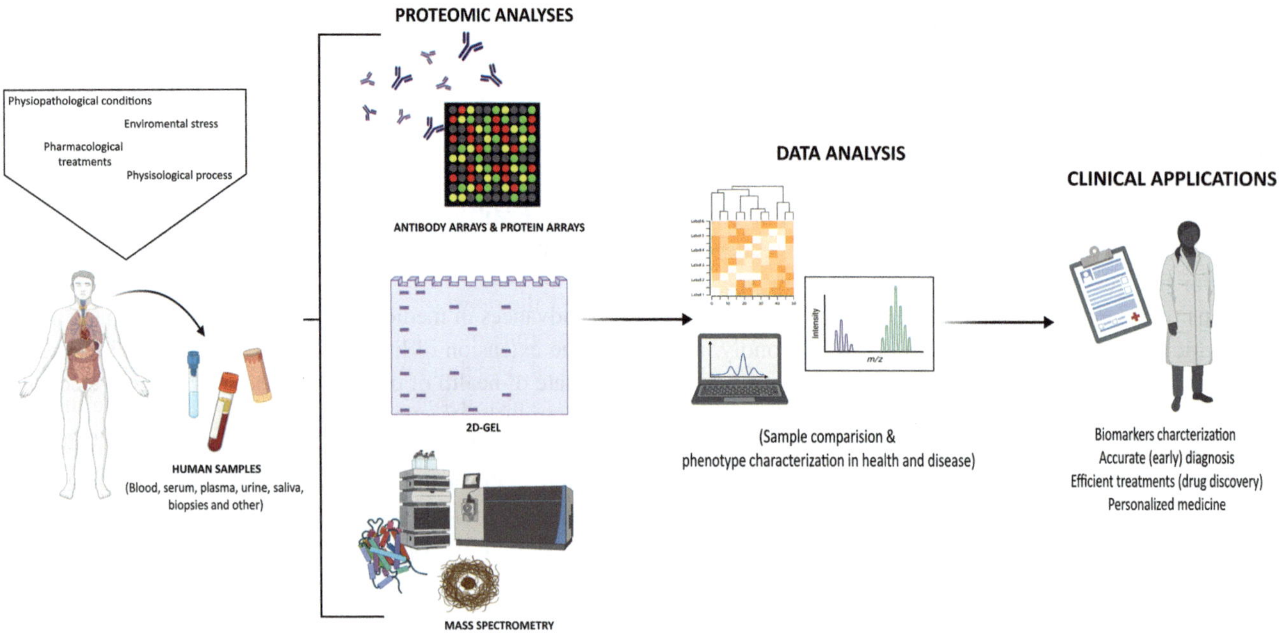

Fig. 6.1 Clinical applications of proteomic analyses. Proteomic studies are large-scale, high-throughput tools that offer valuable information regarding physiological and pathophysiological processes. The current proteomic analyses comprise protein arrays, 2D-electrophoresis gel, and mass spectrometry. Altogether, proteomics have relevant clinical applications such as characterization of biomarkers in health and disease, or during a drug treatment. (Modified from [79])

proteins become more relevant because of their known associations with aging and age-related diseases [75]. Within this list, we find proteins such as EGFR, GDF15, GSTP1, HGF, LAMC1, and VEGF.

As a result of this meta-analysis of the 1128 proteins, they propose a proteomic aging clock based on detectable proteins in plasma, which have been reported to change their levels with aging in at least 3 different studies [75].

One of the major limitations for an older adult has to do with mobility, strength, and control of his or her body. Skeletal muscle plays an important role in this aspect as it is one of the tissues most affected during aging. The loss of this tissue associated with pathologies such as sarcopenia or neuromuscular diseases greatly affects people's quality of life [76].

In this sense, several studies have been carried out in which a proteomic signature has been determined for diseases such as amyotrophic lateral sclerosis in which CRP, NfH/NfL, TDP-43, apoA-I, and clusterin proteins are deregulated. In the case of Duchene muscular dystrophy, the affected proteins are fibronectin, MMP-9, TIMP-1, osteopontin, haptoglobin, myostatin, and dystroglycan. On the other hand, sarcopenia shows important changes in CAF, myostatin, eHSP72, sTnT, C1q, adiponectin, and myokine irisin proteins [77].

Similarly, another study that focused on skeletal muscle proteomics found that in older individuals, ribosomal proteins and proteins associated with energy metabolism (particularly the TCA cycle, mitochondrial respiration, and glycolysis) were underrepresented. On the other hand, proteins associated with innate and adaptive immunity, proteostasis, and alternative splicing processes were overrepresented [78].

Conclusions

Proteome study represents a challenge since it is highly dynamic and interconnected; hence, the advances in this field evolved quickly looking for more sensitive and specific methodologies that depict accurately both physiological and pathophysiological processes in the organism. In this sense, the current technologies based on large-scale, high-throughput proteomics represent a powerful tool for the discovery of potential biomarkers not only for diagnosis or prognosis; it also contributes for the individual follow-up during a drug treatment leading the physicians to make more accurate decisions to obtain the best outcomes (Fig. 6.1). However, there are few limitations such as the lack of harmonized experiment methodologies and data processing that may be solved in the upcoming years in order to move forward proteomics into clinical usage.

References

1. Ahmad Y, Lamond AI. A perspective on proteomics in cell biology. Trends Cell Biol. 2014;24:257–64.
2. Strachan T, Read A. Human Molecular Genetics. Garland Science; 2018.
3. Analytical tools to assess aging in humans: The rise of geri-omics. Trends Analyt Chem. 2016;80:204–12.
4. Reed TT, Sultana R, Butterfield DA. Redox proteomics of Oxidatively modified brain proteins in mild cognitive impairment. Neuroproteomics. 2011;
5. Cheung TK, Lee C-Y, Bayer FP, McCoy A, Kuster B, Rose CM. Defining the carrier proteome limit for single-cell proteomics. Nat Methods. 2021;18:76–83.
6. Wu CH, Chen C. Bioinformatics for comparative proteomics. Humana Press; 2010.
7. Aslam B, Basit M, Nisar MA, Khurshid M, Rasool MH. Proteomics: technologies and their applications. J Chromatogr Sci. 2017;55:182–96.
8. ND L, Lehninger AL, Nelson DL, Cox MM, University Michael M Cox. Lehninger principles of biochemistry. Macmillan; 2005.
9. Waterhouse A, Bertoni M, Bienert S, et al. SWISS-MODEL: homology modelling of protein structures and complexes. Nucleic Acids Res. 2018;46:W296–303.
10. O'Farrell PH. High resolution two-dimensional electrophoresis of proteins. J Biol Chem. 1975;250:4007–21.
11. Martzen MR, McCraith SM, Spinelli SL, Torres FM, Fields S, Grayhack EJ, Phizicky EM. A biochemical genomics approach for identifying genes by the activity of their products. Science. 1999;286:1153–5.
12. Petersen RC. How early can we diagnose Alzheimer disease (and is it sufficient)? The 2017 Wartenberg lecture. Neurology. 2018;91:395–402.
13. Márquez F, Yassa MA. Neuroimaging biomarkers for Alzheimer's disease. Mol Neurodegener. 2019;14:21.
14. Di Resta C, Ferrari M. New molecular approaches to Alzheimer's disease. Clin Biochem. 2019;72:81–6.
15. Guo L-H, Alexopoulos P, Wagenpfeil S, Kurz A, Perneczky R, Alzheimer's Disease Neuroimaging Initiative. Plasma proteomics for the identification of Alzheimer disease. Alzheimer Dis Assoc Disord. 2013;27:337–42.
16. Johnson ECB, Dammer EB, Duong DM, et al. Large-scale proteomic analysis of Alzheimer's disease brain and cerebrospinal fluid reveals early changes in energy metabolism associated with microglia and astrocyte activation. Nat Med. 2020;26:769–80.
17. Rehiman SH, Lim SM, Neoh CF, Majeed ABA, Chin A-V, Tan MP, Kamaruzzaman SB, Ramasamy K. Proteomics as a reliable approach for discovery of blood-based Alzheimer's disease biomarkers: a systematic review and meta-analysis. Ageing Res Rev. 2020;60:101066.
18. Rivero-Segura NA, Guerrero-Cruz AA, Barrera-Vázquez OS. Age-related neurodegenerative diseases: an update. Clinical genetics and genomics of. Aging. 2020:27–41.
19. Dixit A, Mehta R, Singh AK. Proteomics in human Parkinson's disease: present scenario and future directions. Cell Mol Neurobiol. 2019;39:901–15.
20. van Dijk KD, Teunissen CE, Drukarch B, Jimenez CR, Groenewegen HJ, Berendse HW, van de Berg WDJ. Diagnostic cerebrospinal fluid biomarkers for Parkinson's disease: a pathogenetically based approach. Neurobiol Dis. 2010;39:229–41.
21. Lachén-Montes M, González-Morales A, Iloro I, Elortza F, Ferrer I, Gveric D, Fernández-Irigoyen J, Santamaría E. Unveiling the olfactory proteostatic disarrangement in Parkinson's disease by proteome-wide profiling. Neurobiol Aging. 2019;73:123–34.
22. Basso M, Giraudo S, Corpillo D, Bergamasco B, Lopiano L, Fasano M. Proteome analysis of human substantia nigra in Parkinson's disease. Proteomics. 2004;4:3943–52.
23. Dutta D, Ali N, Banerjee E, Singh R, Naskar A, Paidi RK, Mohanakumar KP. Low levels of Prohibitin in substantia Nigra makes dopaminergic neurons vulnerable in Parkinson's disease. Mol Neurobiol. 2018;55:804–21.
24. Licker V, Kövari E, Hochstrasser DF, Burkhard PR. Proteomics in human Parkinson's disease research. J Proteome. 2009;73:10–29.
25. Werner CJ, Heyny-von Haussen R, Mall G, Wolf S. Proteome analysis of human substantia nigra in Parkinson's disease. Proteome Sci. 2008;6:8.
26. Choi J, Rees HD, Weintraub ST, Levey AI, Chin L-S, Li L. Oxidative modifications and aggregation of Cu,Zn-superoxide dismutase associated with Alzheimer and Parkinson diseases. J Biol Chem. 2005;280:11648–55.
27. Licker V, Côte M, Lobrinus JA, Rodrigo N, Kövari E, Hochstrasser DF, Turck N, Sanchez J-C, Burkhard PR. Proteomic profiling of the substantia nigra demonstrates CNDP2 overexpression in Parkinson's disease. J Proteome. 2012;75:4656–67.
28. Gómez A, Ferrer I. Increased oxidation of certain glycolysis and energy metabolism enzymes in the frontal cortex in Lewy body diseases. J Neurosci Res. 2009;87:1002–13.
29. Choi J, Levey AI, Weintraub ST, Rees HD, Gearing M, Chin L-S, Li L. Oxidative modifications and Down-regulation of ubiquitin carboxyl-terminal hydrolase L1 associated with idiopathic Parkinson's and Alzheimer's diseases. J Biol Chem. 2004;279:13256–64.
30. Hong Z, Shi M, Chung KA, et al. DJ-1 and α-synuclein in human cerebrospinal fluid as biomarkers of Parkinson's disease. Brain. 2010;133:713–26.
31. Guo J, Sun Z, Xiao S, Liu D, Jin G, Wang E, Zhou J, Zhou J. Proteomic analysis of the cerebrospinal fluid of Parkinson's disease patients. Cell Res. 2009;19:1401–3.
32. Abdi F, Quinn JF, Jankovic J, et al. Detection of biomarkers with a multiplex quantitative proteomic platform in cerebrospinal fluid of patients with neurodegenerative disorders. J Alzheimers Dis. 2006;9:293–348.
33. Sinha A, Patel S, Singh MP, Shukla R. Blood proteome profiling in case controls and Parkinson's disease patients in Indian population. Clin Chim Acta. 2007;380:232–4.
34. Pan C, Zhou Y, Dator R, et al. Targeted discovery and validation of plasma biomarkers of Parkinson's disease. J Proteome Res. 2014;13:4535–45.
35. Alberio T, Pippione AC, Comi C, Olgiati S, Cecconi D, Zibetti M, Lopiano L, Fasano M. Dopaminergic therapies modulate the T-CELL proteome of patients with Parkinson's disease. IUBMB Life. 2012;64:846–52.
36. Mila S, Albo AG, Corpillo D, Giraudo S, Zibetti M, Bucci EM, Lopiano L, Fasano M. Lymphocyte proteomics of Parkinson's disease patients reveals cytoskeletal protein dysregulation and oxidative stress. Biomark Med. 2009;3:117–28.
37. Boerger M, Funke S, Leha A, Roser A-E, Wuestemann A-K, Maass F, Bähr M, Grus F, Lingor P. Proteomic analysis of tear fluid reveals disease-specific patterns in patients with Parkinson's disease - a pilot study. Parkinsonism Relat Disord. 2019;63:3–9.
38. Kumar S, Singh P, Sharma S, Ali J, Baboota S, Pottoo FH. Proteomic analysis of Huntington's disease. Curr Protein Pept Sci. 2020;21:1218–22.
39. Shacham T, Sharma N, Lederkremer GZ. Protein Misfolding and ER stress in Huntington's disease. Front Mol Biosci. 2019;6:20.
40. Hosp F, Gutiérrez-Ángel S, Schaefer MH, Cox J, Meissner F, Hipp MS, Hartl F-U, Klein R, Dudanova I, Mann M. Spatiotemporal proteomic profiling of Huntington's disease inclusions reveals widespread loss of protein function. Cell Rep. 2017;21:2291–303.

41. Sekijima Y, Wiseman RL, Matteson J, Hammarström P, Miller SR, Sawkar AR, Balch WE, Kelly JW. The biological and chemical basis for tissue-selective amyloid disease. Cell. 2005;121:73–85.
42. Chu CT, Plowey ED, Wang Y, Patel V, Jordan-Sciutto KL. Location, location, location. J Neuropathol Exp Neurol. 2007;66:873–83.
43. Altelaar AFM, Munoz J, Heck AJR. Next-generation proteomics: towards an integrative view of proteome dynamics. Nat Rev Genet. 2013;14:35–48.
44. Hao Y, Ye M, Chen X, Zhao H, Hasim A, Guo X. Discovery and validation of FBLN1 and ANT3 as potential biomarkers for early detection of cervical cancer. Cancer Cell Int. 2021;21:125.
45. Mardamshina M, Geiger T. Next-generation proteomics and its application to clinical breast cancer research. Am J Pathol. 2017;187:2175–84.
46. Cho J-Y, Sung H-J. Proteomic approaches in lung cancer biomarker development. Expert Rev Proteomics. 2009;6:27–42.
47. Reymond MA, Schlegel W. Proteomics in cancer. Adv Clin Chem. 2007;44:103–42.
48. Macklin A, Khan S, Kislinger T. Recent advances in mass spectrometry based clinical proteomics: applications to cancer research. Clin Proteomics. 2020;17:17.
49. Palmfeldt J, Bross P. Proteomics of human mitochondria. Mitochondrion. 2017;33:2–14.
50. Calvo SE, Clauser KR, Mootha VK. MitoCarta2.0: an updated inventory of mammalian mitochondrial proteins. Nucleic Acids Res. 2016;44:D1251–7.
51. Smith AC, Robinson AJ. MitoMiner v3.1, an update on the mitochondrial proteomics database. Nucleic Acids Res. 2016;44:D1258–61.
52. Arnaud M-C, Gazarian T, Palacios Rodriguez Y, Gazarian K, Sakanyan V. Array assessment of phage-displayed peptide mimics of human immunodeficiency virus type 1 gp41 immunodominant epitope: binding to antibodies of infected individuals. Proteomics. 2004;4:1959–64.
53. List EO, Berryman DE, Bower B, Sackmann-Sala L, Gosney E, Ding J, Okada S, Kopchick JJ. The use of proteomics to study infectious diseases. Infect Disord Drug Targets. 2008;8:31–45.
54. Steel LF, Shumpert D, Trotter M, Seeholzer SH, Evans AA, London WT, Dwek R, Block TM. A strategy for the comparative analysis of serum proteomes for the discovery of biomarkers for hepatocellular carcinoma. Proteomics. 2003;3:601–9.
55. Tan X-F, Wu S-S, Li S-P, Chen Z, Chen F. Alpha-1 antitrypsin is a potential biomarker for hepatitis B. Virol J. 2011;8:274.
56. Sheraz M, Cheng J, Tang L, Chang J, Guo J-T. Cellular DNA topoisomerases are required for the synthesis of hepatitis B virus covalently closed circular DNA. J Virol. 2019; https://doi.org/10.1128/JVI.02230-18.
57. Tian W, Zhang N, Jin R, et al. Immune suppression in the early stage of COVID-19 disease. Nat Commun. 2020;11:1–8.
58. Saleh S, Staes A, Deborggraeve S, Gevaert K. Targeted proteomics for studying pathogenic bacteria. Proteomics. 2019;19:e1800435.
59. Peng B, Li H, Peng X. Proteomics approach to understand bacterial antibiotic resistance strategies. Expert Rev Proteomics. 2019;16:829–39.
60. Flores-Villalva S, Rogríguez-Hernández E, Rubio-Venegas Y, Cantó-Alarcón JG, Milián-Suazo F. What can proteomics tell us about tuberculosis? J Microbiol Biotechnol. 2015;25:1181–94.
61. Bisht D, Sharma D, Sharma D, Singh R, Gupta VK. Recent insights into mycobacterium tuberculosis through proteomics and implications for the clinic. Expert Rev Proteomics. 2019;16:443–56.
62. Zhang B, Liu B, Zhou Y, Zhang X, Zou Q, Liu X. Contributions of mass spectrometry-based proteomics to understanding -host interactions. Pathogens. 2020; https://doi.org/10.3390/pathogens9070581.
63. Hotez PJ, Brindley PJ, Bethony JM, King CH, Pearce EJ, Jacobson J. Helminth infections: the great neglected tropical diseases. J Clin Invest. 2008;118:1311–21.
64. Heukelbach J, Feldmeier H. Ectoparasites--the underestimated realm. Lancet. 2004;363:889–91.
65. Ndao M. Diagnosis of parasitic diseases: old and new approaches. Interdiscip Perspect Infect Dis. 2009;2009:278246.
66. Sánchez-Ovejero C, Benito-Lopez F, Díez P, Casulli A, Siles-Lucas M, Fuentes M, Manzano-Román R. Sensing parasites: proteomic and advanced bio-detection alternatives. J Proteome. 2016;136:145–56.
67. Barrett J, Jefferies JR, Brophy PM. Parasite proteomics. Parasitol Today. 2000;16:400–3.
68. Capelli-Peixoto J, Mule SN, Tano FT, Palmisano G, Stolf BS. Proteomics and Leishmaniasis: potential clinical applications. Proteomics Clin Appl. 2019;13:e1800136.
69. Klaips CL, Jayaraj GG, Hartl FU. Pathways of cellular proteostasis in aging and disease. J Cell Biol. 2018;217:51–63.
70. Anisimova AS, Alexandrov AI, Makarova NE, Gladyshev VN, Dmitriev SE. Protein synthesis and quality control in aging. Aging. 2018;10:4269–88.
71. Hartl FU. Protein Misfolding diseases. Annu Rev Biochem. 2017;86:21–6.
72. Harman D. Aging: a theory based on free radical and radiation chemistry. J Gerontol. 1956;11:298–300.
73. Reeg S, Grune T. Protein oxidation in aging: does it play a role in aging progression? Antioxid Redox Signal. 2015;23:239–55.
74. Korovila I, Hugo M, Castro JP, Weber D, Höhn A, Grune T, Jung T. Proteostasis, oxidative stress and aging. Redox Biol. 2017;13:550–67.
75. Systematic review and analysis of human proteomics aging studies unveils a novel proteomic aging clock and identifies key processes that change with age. Ageing Res Rev. 2020;60:101070.
76. Cruz-Jentoft AJ, Bahat G, Bauer J, et al. Sarcopenia: revised European consensus on definition and diagnosis. Age Ageing. 2019;48:16–31.
77. Murphy S, Zweyer M, Mundegar RR, Swandulla D, Ohlendieck K. Proteomic serum biomarkers for neuromuscular diseases. Expert Rev Proteomics. 2018;15:277–91.
78. Ubaida-Mohien C, Lyashkov A, Gonzalez-Freire M, et al. Discovery proteomics in aging human skeletal muscle finds change in spliceosome, immunity, proteostasis and mitochondria. 2019; https://doi.org/10.7554/eLife.49874.
79. Whittaker K, Burgess R, Jones V, Yang Y, Zhou W, Luo S, Wilson J, Huang R-P. Quantitative proteomic analyses in blood: a window to human health and disease. J Leukoc Biol. 2019;106:759–75.

Metabolomics: From Scientific Research to the Clinical Diagnosis

7

E. A. Estrella-Parra, A. M. Espinosa-González, A. M. García-Bores, E. Nolasco-Ontiveros, J. C. Rivera-Cabrera, C. T. Hernández-Delgado, I. Peñalosa-Castro, and J. G. Avila-Acevedo

Abbreviations

^{1}H NMR	Nuclear magnetic resonance of proton
AADC	L-amino-acid decarboxylase
AD	Alzheimer's disease
ADPKD	Autosomal dominant polycystic kidney disease
APO	Apolipoproteins
CE-MS	Capillary electrophoresis-mass spectrometry
CEs	Cholesteryl ester
CG/MS	Gas chromatography with mass spectrometry
DAD	Diode-array detector
DHEA-S	Serum dehydroepiandrosterone sulfate
EIA	Immunoassay
GC × GC-ToFMS	Two-dimensional gas chromatography-time-of-flight mass spectrometry
GC/Q-TOF	Gas chromatography coupled to quadrupole time-of-flight mass spectrometry
GDM	Gestational *diabetes mellitus*
HCC	Hepatocellular carcinoma
HETCOR	Heteronuclear simple chemical-shift correlation spectroscopy
HMBC	Heteronuclear multiple bond correlation
HPLC	High-performance liquid chromatography
HSQC	Heteronuclear single-quantum correlation of ^{1}H-^{13}C
HS-SPME/GC-qMS	Headspace solid-phase microextraction gas chromatography-mass spectrometry with quadrupole mass spectrometry
LA-REIMS	Laser-assisted rapid evaporative ionization mass spectrometry
LC/MS	Liquid chromatography with mass spectrometry
LC-ESI-MS/MS	Liquid chromatography electrospray ionization tandem mass spectrometry
LC-HRMS	Liquid chromatography with ultrahigh-resolution mass spectrometry
LPCs	Lysophosphatidylcholine
MCI	Mild cognitive impairment
MS	Mass spectrometry
NIR	Near-infrared spectroscopy
NMR	Nuclear magnetic resonance
Orbitrap	Orbiting trap
PCR	Polymerase chain reaction
PCs	Phosphatidylcholine
PEs	Lysophosphatidylethanolamine
PPGLs	Pheochromocytomas and paragangliomas
RAMAN	RAMAN spectroscopy
SPME	Solid-phase microextraction
TD-GC × GC-FID/Qms	Two-dimensional gas chromatography with dual flame ionization and quadrupole mass spectrometric detection
TGs	Triacylglycerol

E. A. Estrella-Parra · A. M. Espinosa-González · A. M. García-Bores
E. Nolasco-Ontiveros · I. Peñalosa-Castro · J. G. Avila-Acevedo (✉)
Laboratorio de Fitoquímica, Unidad de Biología y Prototipos (UBIPRO), FES-Iztacala, Universidad Nacional Autónoma de México, Estado de México, Mexico
e-mail: estreparr@iztacala.unam.mx; ericknolasco@iztacalaunam.mx

J. C. Rivera-Cabrera
Laboratorio de Cromatografía de Líquidos, Departamento de Farmacología, Escuela Médico Militar, Mexico City, Mexico

C. T. Hernández-Delgado
Laboratorio de Bioactividad de Productos Naturales, Unidad de Biología y Prototipos (UBIPRO), FES-Iztacala, Universidad Nacional Autónoma de México, Estado de México, Mexico
e-mail: tzasna@unam.mx

J. C. Gomez-Verjan, N. A. Rivero-Segura (eds.), *Principles of Genetics and Molecular Epidemiology*,
https://doi.org/10.1007/978-3-030-89601-0_7

UHPLC/MS/MS	Ultra high-performance liquid chromatography with tandem mass spectrometry
USE	Ultrasound elastography

Introduction

Medicine is undergoing a revolution that will transform the practice of healthcare in virtually every way [1]. Personalized medicine aims to assess medical risks and monitor, diagnose, and treat patients according to their specific genetic composition and molecular phenotype [2]. The "omics" data, referring to large-scale datasets in the biological and molecular field [3], contributes to the integration of quantitative data within order to system biology approach [4, 5] through computer technology [6]. Even more the "connectomics" big data era allows the development of high-performance platforms needed for the exploration of the biological mechanism [7] characterizing the molecular underpinning of human disease [5] and allowing the medical treatments to be individualized [8]. Metabolomics is considered as "totalomic" or "panomic," because it is part of a technological tool in molecular biology [9] being defined as the study of small molecules called metabolites [4]. The applications of metabolomics support to uncover relevant pathophysiological mechanisms and to identify biomarkers of risk and progression in obesity and diabetes, as well as in digestive [10] and other diseases. This personalized medicine/therapy is based on tailored biology attributes as genetics, proteomics, and metabolomics [11], with mass spectrometry as the most useful technique in biomarker discovery toward clinical applications and improvement in clinical diagnostics [12].

Biomarkers are a key part of precision of personalized or individualized medicine [10]. Metabolomics encompasses the diverse metabolic activity of cells [13], which allows directing a more supportive diagnosis as well as a personalized treatment [14]. Inclusive, metabolomic analysis has made it possible to determine the impact of traditional Chinese medicine in human pathologies such as osteoporosis [15] and polycystic ovary [16].

The focus of this chapter is to analyze the performance of metabolomics in the diagnosis of various diseases, as well as its implementation in the clinical therapy.

The "Omics Science" in Human Health

The "omics technologies" are now applied in all aspects of knowledge of life. The various subdisciplines include genomics (DNA), transcriptomics (mRNA), proteomics (proteins), and metabolomics (metabolites), among a host of other emerging areas [17]. Metabolomics is the "systematic study of the unique chemical fingerprints that specific cellular processes leave behind" [18]. Specifically, it focuses on the study of the profiles of small metabolites that play a key role in understanding the phenotype of an organism and the changes it undergoes due to various factors such as perinatal asphyxia, the action of drugs, changes in diet, and the physiological impacts of the environment [19]. Moreover, metabolomics is a rapidly evolving field that aims to identify and quantify the concentration changes of all the metabolites in a given bio-fluid, or tissue extract, from a patient [20]. Using metabolomic markers, diagnostic accuracy may in the near future outperform mammography and ultrasound and set new standards for breast cancer screening and diagnosis [21]. In recent years, metabolomics, also called "metabonomics," has been successfully applied in the field of cancer research, in which prognostic markers that can distinguish indolent from aggressive prostate cancer could have substantial benefit in patients [22], providing insights into the dynamics of cellular response to ionizing radiation [4]. Even, this metabolic approach could also provide new insights into the pathophysiology of airway dysfunction, suggesting novel pathways for drug discovery [23].

Furthermore, metabolomics and proteomics have allowed the characterization of the proteins and metabolites of COVID-19, as well as the dysregulation of multiple apolipoproteins APO-A1, APO-A2, APO-H, APO-L1, APO-D, and APO-M [24]. Another disease that can be predicted via metabolic profiles is polycystic ovary syndrome, which presented biomarkers [25]. Tuberculosis meningitis presents 20 metabolites in contrast to healthy controls [26]. Moreover, metabolomic and lipidomic analyses have been used for the profiling of neurodegenerative processes [27] and pathophysiological conditions as gestational *diabetes mellitus* [28]. Nowadays, the Duke researchers continue to study the bacteria and metabolomics of the babies, with the goal to improve survival and reduce illness in this population [29]. Recently, the potential use of an automated metabolomic robotic platform, employing the principle of laser-assisted rapid evaporative ionization mass spectrometry (LA-REIMS) in cervical cancer screening 130 women, improves the accuracy and efficiency of the current national screening program [30]. The evidence states that "omics" will irrevocably change the practice of medicine [31]. In this manner, the "omics sciences" arrived with the best understanding of diseases, being complementary to the traditional clinic.

Toward Personalized Medicine

Predictive Technologies in Human Diseases

The explosion of new technologies and knowledge, particularly in the field of "omics," calls for further efforts to implement these new and promising diagnostic tools in clinical practice [31]. Chemical analytical methods, such as mass spectrometry and nuclear magnetic resonance (NMR), are noninvasive methods that have been widely useful in the diagnosis of several diseases [Table 7.1].

A biomarker is a biochemical entity used to measure the progress of a disease or the effects of treatment on clinical. In medicine, the term refers to a protein/metabolite measured in blood, whose concentration reflects the presence or severity of a disease state [64]. In addition, exhaled biomarkers "denominated as breathomics" can be validated in the diagnosis, monitoring, and treatment of patients in respiratory diseases, contributing to the development of personalized medicine [65]. Inclusive, "pharmacometabolomics" focuses on the analysis of the pre-dose bio-fluid metabolite profile, which could reflect the complex interactions among physiopathological conditions [66].

Table 7.1 Methodology, instrumentation, and human diseases diagnosed by metabolomics

	Technology used			Analysis/post-analysis	
Pre-analysis	MS	NMR	Other		
Organism/methods	Instrumentation			Results/conclusion	Author
35 children with allergic asthma: breath samples	HS-SPME /GCMS	–	–	44 volatiles: MS rapid and noninvasive diagnostic tool	[32]
54 patients with ADPKD: urine	–	^{1}H, HSQC	–	Identification of ADPKD patients	[33]
21 gout patients: urine and serum	–	–	HPLC-DAD	Multiple biomarkers can provide an overall pattern to predict the disease	[34]
35 persons with hepatic steatosis: plasma	UHPLC/MS/MS. GC-MS	–	–	437 metabolites. Marked changes in bile salts and in biochemicals related to glutathione	[35]
135 children with asthma: breath	–	^{1}H: 600 MHz	–	Difference of metabolites between individuals with asthma and healthy	[23]
365 patients with PPGLs: plasma	–	–	PCR	O-methylated metabolite of dopamine is the biomarker in patients with metastasis	[36]
32 children with asthma: breath	GC × GC-ToFMS	–	–	134 metabolites regarding diagnostic, prognostic, and treatment follow-up	[37]
42 patients with breast cancer: serum	–	^{1}H: 600.29 MHz	–	12 metabolites as part of metabolic syndrome associated with a poor response in breast cancer	[38]
10 women: blood plasma	–	^{1}H: 499.97 MHz	–	Changes across menstrual stages, helping predictive fertility analysis	[39]
21,788 newborns: dried blood spot	LC-MS	–	–	Differences in metabolites by gestational age, birth weight, gender, and season	[40]
15 subjects during the graft recovery process of kidney transplantation: urine	–	^{1}H: 600 MHz	–	NMR methods allow monitoring of kidney graft recovery patients who are not progressing within the normal range.	[41]
12 children hospitalized by spontaneous micturition: urine	GC-MS	–	–	GC-MS method aiming to support pediatric clinics and assist in diagnostics.	[42]
452 participants (majority obese): serum	UPLC-QTOF-MS	–	–	Free fatty acids levels with metabolic phenotypes among several groups of obese participants	[43]
6 newborns with perinatal asphyxia: urine	–	^{1}H:600MHz-	–	Increase: lactate, threonine, 3-OH isovalerate, glucose, and aspartate. Decrease: acetate, formate, urea, aconitate, creatinine, dimethylamine, dimethylglycine,and betaine	[19]
127,987 newborns were screened for AADC deficiency: plasma	LC-MS/MS	–	RT-PCR	Newborn screening of AADC deficiency was achieved with a 100% positive predictive rate	[44]
136 elderly person >55 years: plasma	LC/MS-MS	^{1}H: 600 MHz	–	Deficiency of ergothioneine predisposing individuals to neurodegenerative diseases	[45]
24 pregnant women with GDM: plasma	GC-MS	–	–	2-Hydroxybutyrate and 3-hydroxybutyrate to predict the onset of diabetic complications in women with GDM	[46]
40 patients (miscarriage): maternal blood samples and urine	UPLC-MS	–	–	Urine metabolites as a noninvasive screening tool for the risk stratification of women presenting with threatened miscarriage	[47]
252 persons having prevalent hypertension: plasma sample	LC/MS-MS	–	–	Elevated F2-isoprostane levels do not increase the risk of hypertension	[48]

(continued)

Table 7.1 (continued)

Pre-analysis	Technology used MS	NMR	Other	Analysis/post-analysis	
Organism/methods	Instrumentation			Results/conclusion	Author
113 adult outpatients with cirrhosis: Plasma sample	UPLC-MS/MS	–	–	Ascorbate and aldarate metabolism, methylation, and cellular glucuronidation as metabolomic signature	[49]
15 first-episode drug-naïve major depressive disorder patients: plasma sample	LC-MS	–	–	Biomarkers in plasma lipid species such as LPCs, PCs, PEs, CEs, and TGs are correlated with several depressive sub-symptoms	[50]
13 mother carrying fetus: posterior urethral valves, urinary sample of fetus	CE-MS	–	–	The potential of cumulative different omics traits in biomarker research	[51]
46 patients with diabetic nephropathy: serum	GC/MS	–	–	Serum citric acid level is potentially a biomarker that could assist in the diagnosis of diabetic nephropathy	[52]
217 pregnant women with GDM: serum. Metformin vs insulin	–	^{1}H	–	Compared to insulin, metformin caused an increase in alanine, isoleucine, and lactate concentrations	[28]
9125 patient samples (breast cancer)	HRMS	–	RT-PCR	Metabolic heterogeneity within and across cancer types	[53]
60 patients with Parkinson's disease: plasma	GC–MS	–	–	Integrating blood metabolomics enhances the diagnostic discrimination power	[54]
33 patients with diabetes type 2: urine	–	^{1}H: 600 MHz	–	Metabolites in diabetic subjects: urinary creatine, glutamic acid, and 5-hydroxyindoleacetic acid	[5]
VOC's exhaled breath of patients	TD-GC × GC-FID/qMS	–	–	Analysis of breath VOCs by GC × GC is clinically viable with low storage cost.	[55]
660 patients (fertility status): seminal plasma	LC-MS	–	–	Phthalates may affect the semen quality by causing disorders of seminal plasma	[56]
46 volunteers with dengue: breath sample	SPME-GC/Q-TOF	–	–	Six dengue breath biomarkers. Rapid and easy diagnosis of dengue disease	[57]
658 volunteers (metabolic syndrome): plasma	LC-ESI-MS/MS	–	–	Numerous lipid species that were associated with metabolic risk factors cross-sectionally	[58]
100 volunteers (idiopathic cervical dystonia): plasma	LC-HRMS	–	–	289 metabolite biomarkers	[15]
1515 volunteers with tuberculosis: urine sample	–	–	EIA	7 biomarkers showed potential as tuberculosis diagnostic	[59]
104 patients with HCC: blood sample	GC-MS	–	PCR	Good performance in predicting early HCC in patients who had tumor size <2 cm	[60]
2969 women (DHEA-S): blood sample	LC-MS/MS	–	EIA	Low DHEA-S is associated with impaired lung function, predicted airflow limitation	[61]
31 patients: Saliva (Sjögren's syndrome)	–	–	USE, RAMAN	Molecular composition of saliva yielded an overall accuracy of 81%	[62]
39 patients (breast cancer): blood sample	HPLC-TQ/M	–	–	Metabolomic profiling proposed biomarkers in discovery of early breast cancer patients	[21]
46 patients with COVID-19: serum	UPLC-MS/MS	–	–	105 proteins were expressed in COVID-19 patients but not in the non-COVID-19 patients	[24]
100 patients: plasma (idiopathic cervical dystonia	LC-HRMS	–	–	Provide potential novel insights into the biology of cervical dystonia	[63]

Thereby, numerous success stories have been widely reported. The urinary proteomics by NMR spectroscopic fingerprinting was applied in patients with autosomal dominant polycystic kidney disease (ADPKD) [33] and hepatic steatosis [35]. Furthermore, metabolomics helped to detect the methoxytyramine in metastatic pheochromocytomas and paragangliomas (PPGLs) [36], as well as in breast cancer [38, 53], in the process of kidney transplantation, [41] and in Parkinson diagnosis [54].

In the same way, metabolic profile allows the amount of acylcarnitines in the plasma to vary depending on the type of depressive disorder in patients [67]. NMR of proton (^{1}H NMR) allows fast and reliable detection of a large number of metabolites. In this way, it was detected in a cohort of 180 serum samples that 30% were related to mild cognitive impairment (MCI) and increased risk of Alzheimer's disease (AD), showing the coexistence of inflammation, metabolic syndrome, as well as elevated glycoproteins [68]. In addition, metabolomics was employed to screen and identify novel biomarkers of gout based on human serum and urine samples [34] in asthma [23, 69] and arthritis diseases [70]. Moreover, metabolic syndrome is commonly observed in metastatic breast cancer and associated with a poor response to chemotherapy, suggesting the importance of metabolome correlation with personalized therapy [38]. Furthermore, in cystic fibrosis, metabolomic profiling of regulatory lipid allows for the creation of a unique set of biomarkers for further characterization and biologic impact in lung function [71]. Inclusive, breath metabolic profile allows the diagnosis of dengue disease and achieved a 100% rate [57].

Thereby, not only in adults these technologies have been applied; some examples are applied in the following: metabolomic assay to find biomarker molecules in children with allergic asthma [32, 37] and in children with spontaneous micturition [42], the presence of phthalate compounds in the urinary metabolome from children with cystic fibrosis [72], as well as an increase in fat and body weight, altering the endocrine system [73]. Metabolomics is a tool that has been used in studies in newborn with perinatal asphyxia [43] as well as in deficiency of L-amino-acid decarboxylase [44], even in fetus diagnosis [51]. On the other hand, it has been used in the diagnosis in human fertility issues due to the accumulation of phthalates in human semen [56]. Thus, metabolomic profile of metaphase I and II oocytes was obtained by near-infrared spectroscopy (NIR), which allows for the prediction of embryo viability [74].

Otherwise, metabolomic techniques permit detection of gestational *diabetes mellitus* in pregnant women [46] [28], inclusive in women with risk of threatened miscarriage [47]. In persons with *diabetes mellitus*, biomarkers have been detected to diagnose nephropathy [52] as well as in metabolic syndrome [58]. In addition, metabolomics is a useful tool in patients with depressive disorder [50], Sjögren's syndrome [62], in women with impaired lung function [61], idiopathic cervical dystonia [15], as well as in patients with tuberculosis disease [59]. Finally, genomic sciences have allowed the study and analysis of the SARS-CoV-2 virus present in infected people [24].

The application of "omics technologies" in various human diseases has been widely reported, with great success and without presenting an invasive sampling technique toward the patient.

How to Implement Clinical Analysis Through Metabolomics

Instrumentation in the Metabolomic Clinic

The complexities of translating basic discoveries into clinical trials and studies, followed by the implementation of results in medical practice, must be considered a fundamental part of the curriculum for laboratory medicine residents [31]. The medical system, by contrast, is holistic and utilizes all types of biological information – DNA, RNA, protein, metabolites, small molecules, interactions, cells, tissues, organs, individuals, social networks, and external environmental signals – integrating them, so as to lead to predictive and actionable models for health and disease [1]. In recent years, there has been a tremendous effort to develop biomarkers for prognosis and prediction of clinical response to a given treatment by studying the difference between the normal tissue and tumors as well as the differences in the tumors across a cohort of patients [75], being the metabolomic evaluation crucial to assisted reproduction technology [76]. For example, hepatocellular carcinoma is the sixth most widespread tumor and the third leading cause of cancer-related death worldwide [60], being the cancer treatment lies in a personalization of medicine, where each patient's treatment regime is tailored to the genetic diversity of their tumors [77]. Today, the combination of these techniques (NMR, LC/MS, and CG/MS) is desirable in order to detect, identify, and quantify hundreds of thousands of metabolites in a given automatism and useful in biomarker discovery toward clinical applications [12, 19].

Therefore, mass spectrometry (MS) has been a routine technique in the diagnosis of various diseases, being coupled to various other techniques. Thus, in the HS-SPME/GC technique, the facility to acquire the biological sample was reported, which is noninvasive [32]. Moreover, the analysis of exhaled air composition could be especially useful because metabolomic changes occur in the human body at the incipient phase of a disease, which are transmitted to the alveolar exhaled air via the lungs [57].

Also, the liquid chromatography technique (LC) has been reported widely, with different experimental conditions, as well as the mass analyzer, which is more powerful and robust, allowing the metabolic identification with greater assertiveness. Generally, the biological samples obtained from most of the patients were kept frozen at −80 °C, until their subsequent spectrometric analysis. Before the introduction of the sample into the mass analyzer, the samples first pass through a chromatographic column (reversed phase) with very different characteristics. Thus, high-resolution mass spectrometry (HRMS) in an Orbitrap mass spectrometer has been used to analyze the seminal plasma metabolites [56] as well as the therapeutic drug monitoring and quantification of antidepressants [78]. In addition, quadrupole mass spectrometer has been employed successfully for the detection of phthalate in urine samples [72] and oxylipin molecules, which act as biomarkers for monitoring renal function in the post-renal transplantation period [79], as well as to monitor the amino acid composition in bone to correlate possible risks in bone fractures in older adults [80]. Meanwhile, triple-quadrupole mass analyzer has been employed to the detect metabolites related to cardiovascular disease [81], to identify biomarkers in pulmonary hypertension [82], and to monitor the tryptophan in inflammatory bowel diseases [83]. Besides, a double quadrupole has been employed to identify the acylcarnitine in Alzheimer's disease [84]. Other mass spectrometers – such as quadrupole time-of-flight (qTOF) – permit analysis of the glycosylation profiles as prognostic markers in patients with granulomatosis and polyangiitis [85], as well as monitoring the metabolic transition from pregnancy to postpartum in gestational *diabetes mellitus* [86] and the identification of biomarkers in tuberculosis diseases [87]. Even MS imaging presents versatility in clinical applications such as biomarker diagnostics of different diseases [12].

Meanwhile, NMR allows for the identification of tumor metabolism in hepatocellular carcinoma using 600 MHz ^{1}H NMR [88]. By means of ^{1}H NMR of 500 MHz, the metabolic profile in the cerebrospinal fluid of children with tuberculous meningitis was characterized [26]. Even, using ^{1}H NMR 600 MHz, it allows the evaluation of hemodialysis efficiency in patients affected with end-stage renal disease [89] as well as to characterize serum samples of mild cognitive impairment in Alzheimer's disease [68]. Using ^{1}H NMR at 600 MHz, it also allows us to understand the pathogenesis of osteoporosis [15].

Thus, technological improvement in spectroscopic and spectrometric equipment allows clinical analysis, something that was not possible until a few years ago.

Mass Spectral Library and Bioinformatics

Posteriorly to the acquisition of metabolome data, it is necessary to identify and quantify the metabolites in whole metabolome [Fig. 7.1]. The "big data" bring the capacity to improve the clinical observations and measurements to corroborate

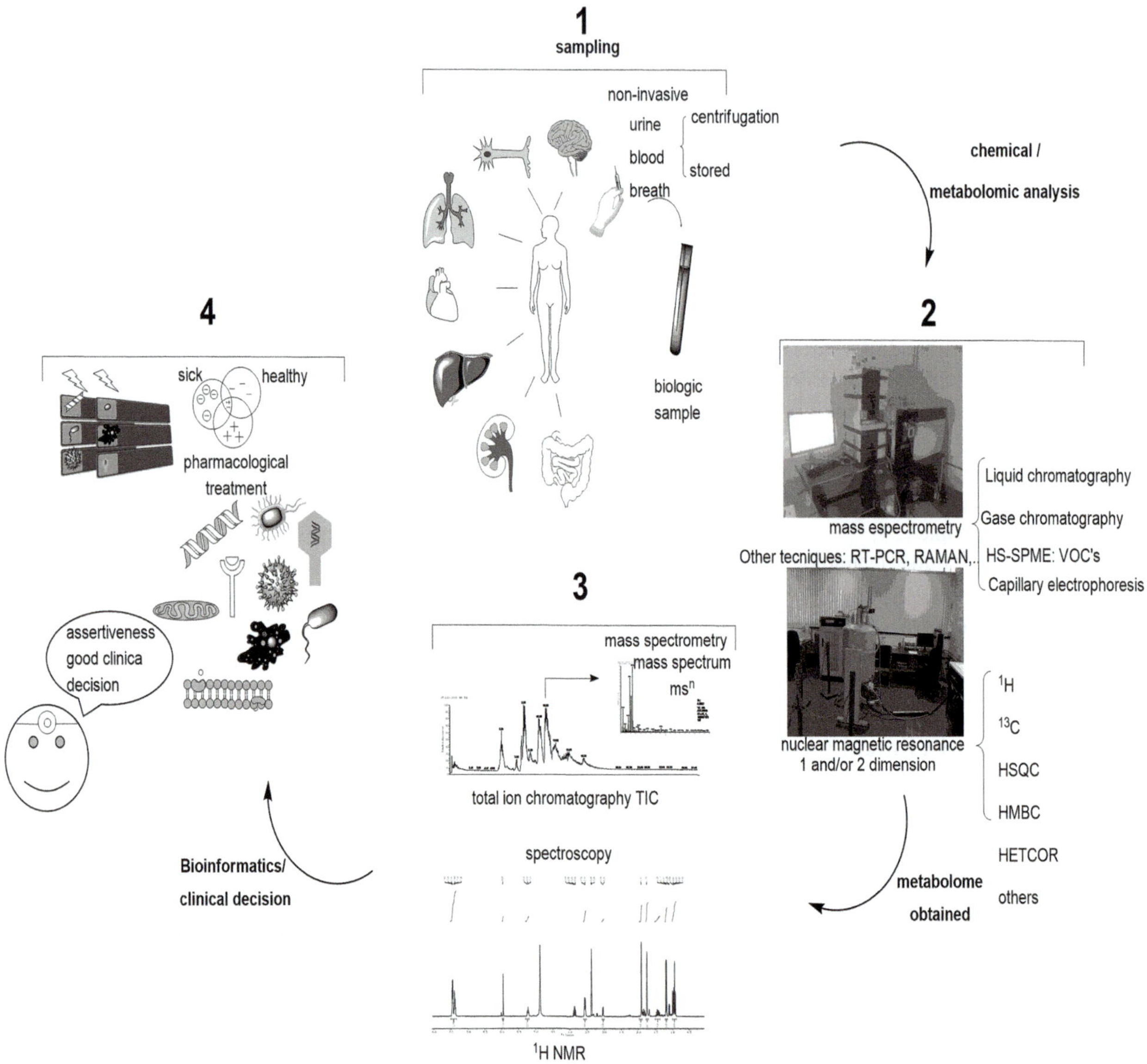

Fig. 7.1 General strategy of metabolomic analysis. *1*. Sampling: non-invasive sample (urine, blood, plasma, and breath) and samples ultra-frozen until use (−80 °C). *2*. Metabolomic technique: LC-MS, HS-SPME, GC-MS, CE, and NMR. *3*. Metabolome obtained: mass spectra and/or resonance spectra. *4*. Bioinformatics: consultation of biological libraries and clinical decision

patient's outcomes [3]. The generation of biological information has led to the creation of databases such as BioBankWarden, which can be used to store and retrieve specific information from different clinical fields linked to biomaterials collected from patients, providing the functionalities required to support translational research in the field of cancer [90]. The Georgetown Database of Cancer (G-DOC) is a web platform that enables basic and clinical research by integrating patient characteristics and clinical outcome data from more than 2500 breast cancer patients and 800 gastrointestinal cancer patients, which includes bioinformatics and biology tools for analysis and visualization of "omics" types [20].

For example, LC-MS raw data were imported into Progenesis QI data software and compared with the Human Metabolome Database (HMDB) and LIPID MAPS database [91]. Using the MATLAB script, the metabolites were extracted by using unique mass channels and retention indices in the mass spectral library at the Swedish Metabolomics Centre [86]. Meanwhile, in NMR experiments, the spectra were exported to MATLAB R2010 and compared in a combined analysis of Nigerian and Egyptian data [88] as well as pure compounds in the spectral libraries [26, 68]. In addition, metabolites were identified by databases such as HMDB and SDBS, MetaboAnalyst platform, KEGG, and SMPDB [15].

Therefore, the generation of biological libraries of different diseases – due to use of the omics' technologies – will allow a more precise diagnosis in a context of globalization in health.

Conclusions

In clinical diagnosis, it is essential to obtain with greater precision a diagnosis as well as a treatment that restores the health of the patient, whatever the disease. For this – besides the traditional clinical diagnosis and evaluation – the use of new technologies such as "omics science" is necessary, allowing greater precision in the diagnosis and follow-up of the treatment of any disease. For this purpose, analytical chemistry tools such as NMR and MS that were previously only used in scientific matters and in industry have been applied in medical matters such as the monitoring and diagnosis of various diseases such as SARS-COV-2. This is why it is crucial to understand, comprise, and apply the omics sciences in medical therapy for an assertive and unambiguous diagnosis. Moreover, the globalization of medicine allows the creation of clinical libraries based on the "omics technologies," which can be consulted by medical staff in "real time" in any particular case, being a reference for a precise diagnosis.

References

1. Hood L, Flores MA. A personal view on systems medicine and the emergence of proactive P4 medicine: predictive, preventive, personalized and participatory. N Biotechnol. 2012;29(6):613–24. PubMed PMID: 22450380
2. Chen R, Mias GI, Li-Pook-Than J, Jiang L, Lam HY, Chen R, et al. Personal omics profiling reveals dynamic molecular and medical phenotypes. Cell. 2012;148(6):1293–307. PubMed PMID: 22424236; PMCID: PMC3341616.
3. Coleman AL. How big data informs us about cataract surgery: the LXXII Edward Jackson memorial lecture. Am J Ophthalmol. 2015;160(6):1091–1103. e3. PubMed PMID: 26432566
4. Bentzen SM. From cellular to high-throughput predictive assays in radiation oncology: challenges and opportunities. Semin Radiat Oncol. 2008;18(2):75–88. PubMed PMID: 18314062
5. Keshavan MS, Clementz BA, Pearlson GD, Sweeney JA, Tamminga CA. Reimagining psychoses: an agnostic approach to diagnosis. Schizophr Res. 2013;146(1–3):10–6. PubMed PMID: 23498153
6. Rhea M, Bettles C. Future changes driving dietetics workforce supply and demand: future scan 2012-2022. J Acad Nutr Diet. 2012;112(3 Suppl):S10–24. PubMed PMID: 22709858
7. Xia M, He Y. Functional connectomics from a "big data" perspective. Neuroimage. 2017;160:152–67. PubMed PMID: 28232122
8. Liu L, Teague WG, Erzurum S, Fitzpatrick A, Mantri S, Dweik RA, et al. Determinants of exhaled breath condensate pH in a large population with asthma. Chest. 2011;139(2):328–36. PubMed PMID: 20966042; PMCID: PMC3032364.
9. Chetty IJ, Martel MK, Jaffray DA, Benedict SH, Hahn SM, Berbeco R, et al. Technology for innovation in radiation oncology. Int J Radiat Oncol Biol Phys. 2015;93(3):485–92. PubMed PMID: 26460989; PMCID: PMC4610140.
10. Carethers JM, Braun J, Sands BE. Genetics, genetic testing, and biomarkers of digestive diseases. Gastroenterology. 2015;149(5):1131–3. PubMed PMID: 26327133; PMCID: PMC4589521.
11. Colvin M, Sweitzer NK, Albert NM, Krishnamani R, Rich MW, Stough WG, et al. Heart failure in non-caucasians, women, and older adults: a white paper on special populations from the Heart Failure Society of America guideline committee. J Card Fail. 2015;21(8):674–93. PubMed PMID: 26051012
12. Ye H, Gemperline E, Li L. A vision for better health: mass spectrometry imaging for clinical diagnostics. Clin Chim Acta. 2013;420:11–22. PubMed PMID: 23078851; PMCID: PMC3574966.
13. van Belkum A, Tassios PT, Dijkshoorn L, Haeggman S, Cookson B, Fry NK, et al. Guidelines for the validation and application of typing methods for use in bacterial epidemiology. Clin Microbiol Infect. 2007;13(Suppl 3):1–46. PubMed PMID: 17716294
14. Bissonnette F, Masip MJ, Kadoch IJ, Librach C, Sampalis J, Yuzpe A. Individualized ovarian stimulation for in vitro fertilization: a multicenter, open label, exploratory study with a mixed protocol of follitropin delta and highly purified human menopausal gonadotropin. Fertil Steril. 2020;30:S0015-0282(20)32400-6. PubMed PMID: 33267959
15. Liu S, Yuan X, Ma C, Zhao J, Xiong Z. ^{1}H-NMR-based urinary metabolomic analysis for the preventive effects of gushudan on glucocorticoid-induced osteoporosis rats. Anal Biochem. 2020;610:113992. PubMed PMID: 33075315.

16. Song J, Xiang S, Yang Y, Sun Z. Assessment of follicular fluid metabolomics of polycystic ovary syndrome in kidney yang deficiency syndrome. Eur J Integr Med. 2019;30:100944.
17. Gallagher WM, Tweats D, Koenig J. Omic profiling for drug safety assessment: current trends and public-private partnerships. Drug Discov Today. 2009;14(7–8):337–42. PubMed PMID: 19340928
18. Warren RB, Griffiths CE. Future therapeutic directions for the treatment of psoriasis. Actas Dermosifiliogr. 2009;100(Suppl 2):28–31. PubMed PMID: 20096159
19. Longini M, Giglio S, Perrone S, Vivi A, Tassini M, Fanos V, et al. Proton nuclear magnetic resonance spectroscopy of urine samples in preterm asphyctic newborn: a metabolomic approach. Clin Chim Acta. 2015;444:250–6. PubMed PMID: 25727514
20. Madhavan S, Gusev Y, Harris M, Tanenbaum DM, Gauba R, Bhuvaneshwar K, et al. G-DOC: a systems medicine platform for personalized oncology. Neoplasia. 2011;13(9):771–83. PubMed PMID: 21969811; PMCID: PMC3182270.
21. Kohl M, Megger DA, Trippler M, Meckel H, Ahrens M, Bracht T, et al. A practical data processing workflow for multi-OMICS projects. Biochim Biophys Acta. 2014;1844(1PtA):52–62. PubMed. PMID: 23501674.
22. Teahan O, Bevan CL, Waxman J, Keun HC. Metabolic signatures of malignant progression in prostate epithelial cells. Int J Biochem Cell Biol. 2011;43(7):1002–9. PubMed PMID: 20633696
23. Saude EJ, Skappak CD, Regush S, Cook K, Ben-Zvi A, Becker A, et al. Metabolomic profiling of asthma: diagnostic utility of urine nuclear magnetic resonance spectroscopy. J Allergy Clin Immunol. 2011;127(3):757–764. e1–6. PubMed PMID: 21377043
24. Shen B, Yi X, Sun Y, Bi X, Du J, Zhang C, et al. Proteomic and metabolomic characterization of COVID-19 patient sera. Cell. 2020;182(1):59–72.e15. PubMed PMID: 32492406; PMCID: PMC7254001.
25. Zhou W, Hong Y, Yin A, Liu S, Chen M, Lv X, et al. Non-invasive urinary metabolomics reveals metabolic profiling of polycystic ovary syndrome and its subtypes. J Pharm Biomed Anal. 2020;185:113–262. PubMed PMID: 32222648.
26. van Zyl CW, Loots DT, Solomons R, van Reenen M, Mason S. Metabolic characterization of tuberculous meningitis in a south African pediatric population using ^{1}H NMR metabolomics. J Infect. 2020;81(5):743–52. PubMed PMID: 32712206
27. Carta MG, Moro MF, Lorefice L, Trincas G, Cocco E, Del Giudice E, et al. The risk of bipolar disorders in multiple sclerosis. J Affect Disord. 2014;155:255–60. PubMed PMID: 24295600
28. Huhtala MS, Tertti K, Pellonperä O, Rönnemaa T. Amino acid profile in women with gestational diabetes mellitus treated with metformin or insulin. Diabetes Res Clin Pract. 2018;146:8–17. PubMed PMID: 30227169
29. Duff E. International news-april 2017. Midwifery. 2017;47:A1–5. https://doi.org/10.1016/S0266-6138(17)30177-8.
30. Paraskevaidi M, Cameron SJS, Whelan E, Bowden S, Tzafetas M, Mitra A, et al. Laser-assisted rapid evaporative ionisation mass spectrometry (LA-REIMS) as a metabolomics platform in cervical cancer screening. EBioMedicine. 2020;60:103017. PubMed PMID: 32980699; PMCID: PMC7522750.
31. Plebani M. The changing scenario in laboratory medicine and the role of laboratory professionals in translational medicine. Clin Chim Acta. 2008;393(1):23–6. PubMed PMID: 18423398
32. Caldeira M, Barros AS, Bilelo MJ, Parada A, Câmara JS, Rocha SM. Profiling allergic asthma volatile metabolic patterns using a headspace-solid phase microextraction/gas chromatography based methodology. J Chromatogr A. 2011;1218(24):3771–80. PubMed PMID: 21546028
33. Gronwald W, Klein MS, Zeltner R, Schulze BD, Reinhold SW, Deutschmann M, et al. Detection of autosomal dominant polycystic kidney disease by NMR spectroscopic fingerprinting of urine. Kidney Int. 2011;79(11):1244–53. PubMed PMID: 21389975
34. Liu Y, Sun X, Di D, Quan J, Zhang J, Yang X. A metabolic profiling analysis of symptomatic gout in human serum and urine using high-performance liquid chromatography-diode array detector technique. Clin Chim Acta. 2011;412(23–24):2132–40. https://doi.org/10.1016/j.cca.2011.07.031. Epub 2011 Aug 16. PMID: 21867696.
35. Kalhan SC, Guo L, Edmison J, Dasarathy S, McCullough AJ, Hanson RW, et al. Plasma metabolomic profile in nonalcoholic fatty liver disease. Metabolism. 2011;60(3):404–13. PubMed PMID: 20423748; PMCID: PMC2950914.
36. Eisenhofer G, Lenders JW, Siegert G, Bornstein SR, Friberg P, Milosevic D, et al. Plasma methoxytyramine: a novel biomarker of metastatic pheochromocytoma and paraganglioma in relation to established risk factors of tumour size, location and SDHB mutation status. Eur J Cancer. 2012;48(11):1739–49. PubMed PMID: 22036874; PMCID: PMC3372624.
37. Caldeira M, Perestrelo R, Barros AS, Bilelo MJ, Morête A, Câmara JS, et al. Allergic asthma exhaled breath metabolome: a challenge for comprehensive two-dimensional gas chromatography. J Chromatogr A. 2012;1254:87–97.
38. Stebbing J, Sharma A, North B, Athersuch TJ, Zebrowski A, Pchejetski D, et al. A metabolic phenotyping approach to understanding relationships between metabolic syndrome and breast tumour responses to chemotherapy. Ann Oncol. 2012;23(4):860–6. PubMed PMID: 21821546
39. McRae C, Baskind NE, Orsi NM, Sharma V, Fisher J. Metabolic profiling of follicular fluid and plasma from natural cycle in vitro fertilization patients--a pilot study. Fertil Steril. 2012;98(6):1449–1457.e6. PubMed PMID: 22921074
40. Ryckman KK, Berberich SL, Shchelochkov OA, Cook DE, Murray JC. Clinical and environmental influences on metabolic biomarkers collected for newborn screening. Clin Biochem. 2013;46(1–2):133–8. PubMed PMID: 23010448; PMCID: PMC3534803.
41. Calderisi M, Vivi A, Mlynarz P, Tassin M, Banasik M, Dawiskiba T, Carmellini M. Using metabolomics to monitor kidney transplantation patients by means of clustering to spot anomalous patient behavior. Transplant Proc. 2013;45(4):1511–5. PubMed PMID: 23726608
42. Christou C, Gika HG, Raikos N, Theodoridis G. GC-MS analysis of organic acids in human urine in clinical settings: a study of derivatization and other analytical parameters. J Chromatogr B Analyt Technol Biomed Life Sci. 2014;964:195–201. PubMed PMID: 24480519
43. Ni Y, Zhao L, Yu H, Ma X, Bao Y, Rajani C, Loo LW, et al. Circulating unsaturated fatty acids delineate the metabolic status of obese individuals. EBioMedicine. 2015;2(10):1513–22. PubMed PMID: 26629547; PMCID: PMC4634820.
44. Chien YH, Chen PW, Lee NC, Hsieh WS, Chiu PC, Hwu WL, et al. 3-O-methyldopa levels in newborns: result of newborn screening for aromatic l-amino-acid decarboxylase deficiency. Mol Genet Metab. 2016;118(4):259–63. PubMed PMID: 27216367
45. Cheah IK, Feng L, Tang RMY, Lim KHC, Halliwell B. Ergothioneine levels in an elderly population decrease with age and incidence of cognitive decline; a risk factor for neurodegeneration? Biochem Biophys Res Commun. 2016;478(1):162–7. PubMed PMID: 27444382
46. Dudzik D, Zorawski M, Skotnicki M, Zarzycki W, García A, Angulo S, et al. GC-MS based gestational diabetes mellitus longitudinal study: identification of 2-and 3-hydroxybutyrate as potential prognostic biomarkers. J Pharm Biomed Anal. 2017;144:90–8. PubMed PMID: 28314466

47. Ku CW, Tan ZW, Lim MK, Tam ZY, Lin CH, Ng SP, et al. Spontaneous miscarriage in first trimester pregnancy is associated with altered urinary metabolite profile. BBA Clin. 2017;8:48–55. PubMed PMID: 28879096; PMCID: PMC5574812.
48. Melton CD, Luo R, Wong BJ, Spasojevic I, Wagenknecht LE, D'Agostino RB Jr, et al. Urinary F2-isoprostanes and the risk of hypertension. Ann Epidemol. 2017;27(6):391–6. PubMed PMID: 28558917; PMCID: PMC7147630
49. Mindikoglu AL, Opekun AR, Putluri N, Devaraj S, Sheikh-Hamad D, Vierling JM, et al. Unique metabolomic signature associated with hepatorenal dysfunction and mortality in cirrhosis. Transl Res. 2018;195:25–47. PubMed PMID: 29291380; PMCID: PMC6037419
50. Kuwano N, Kato TA, Setoyama D, Sato-Kasai M, Shimokawa N, Hayakawa K, Ohgidani M, et al. Tryptophan-kynurenine and lipid related metabolites as blood biomarkers for first-episode drug-naïve patients with major depressive disorder: an exploratory pilot case-control study. J Affect Disord. 2018;231:74–82. PubMed PMID: 29454180
51. Buffin-Meyer B, Klein J, Breuil B, Muller F, Moulos P, Groussolles M, et al. Combination of the fetal urinary metabolome and peptidome for the prediction of postnatal renal outcome in fetuses with PUV. J Proteomics. 2018;184:1–9. PubMed PMID: 29929039
52. Gao H, Yu X, Sun R, Yang N, He J, Tao M, et al. Quantitative GC-MS assay of citric acid from humans and db/db mice blood serum to assist the diagnosis of diabetic nephropathy. J Chromatogr B Analyt Technol Biomed Life Sci. 2018;1077-1078:28–34. PubMed PMID: 29413574
53. Peng X, Chen Z, Farshidfar F, Xu X, Lorenzi PL, Wang Y, et al. Molecular characterization and clinical relevance of metabolic expression subtypes in human cancers. Cell Rep. 2018;23(1):255–269 e4. PubMed PMID: 29617665; PMCID: PMC5916795
54. Glaab E, Trezzi JP, Greuel A, Jäger C, Hodak Z, Drzezga A, et al. Integrative analysis of blood metabolomics and PET brain neuroimaging data for Parkinson's disease. Neurobiol Dis. 2019;124:555–62. PubMed PMID: 30639291
55. Wilde MJ, Cordell RL, Salman D, Zhao B, Ibrahim W, Bryant L, et al. Breath analysis by two-dimensional gas chromatography with dual flame ionisation and mass spectrometric detection - Method optimisation and integration within a large-scale clinical study. J Chromatogr A. 2019;1594:160–72. PubMed PMID: 30755317; PMCID: PMC6491496.
56. Wang YX, Wu Y, Chen HG, Duan P, Wang L, Shen HQ, et al. Seminal plasma metabolome in relation to semen quality and urinary phthalate metabolites among Chinese adult men. Environ Int. 2019;129:354–63. PubMed PMID: 31150977
57. Welearegay TG, Durán-Acevedo C, Jaimes-Mogollón AL, Pugliese G, Ionescu F, Perez-Ortiz OM, et al. Exhaled air analysis as a potential fast method for early diagnosis of dengue disease. Sens Actuators B Chem. 2020;310:127859.
58. Yin X, Willinger CM, Keefe J, Liu J, Fernández-Ortiz A, Ibáñez B, et al. Lipidomic profiling identifies signatures of metabolic risk. EBioMedicine. 2020;51:102520. PubMed PMID: 31877415; PMCID: PMC6938899.
59. Eribo OA, Leqheka MS, Malherbe ST, McAnda S, Stanley K, van der Spuy GD, et al. Host urine immunological biomarkers as potential candidates for the diagnosis of tuberculosis. Int J Infect Dis. 2020;99:473–81. PubMed PMID: 32800854
60. Omran MM, Farid K, Omar MA, Emran TM, El-Taweel FM, Tabll AA. A combination of α-fetoprotein, midkine, thioredoxin and a metabolite for predicting hepatocellular carcinoma. Ann Hepatol. 2020;19(2):179–85. PubMed PMID: 31648804
61. Pesce G, Triebner K, van der Plaat DA, Courbon D, Hustad S, Sigsgaard T, et al. Low serum DHEA-S is associated with impaired lung function in women. EClinicalMedicine. 2020;23:100389. PubMed PMID: 32529179; PMCID: PMC7280766.
62. Moisoiu V, Badarinza M, Stefancu A, Iancu SD, Serban O, Leopold N, et al. Combining surface-enhanced Raman scattering (SERS) of saliva and two-dimensional shear wave elastography (2D-SWE) of the parotid glands in the diagnosis of Sjögren's syndrome. Spectrochim Acta A Mol Biomol Spectrosc. 2020;235:118267. PubMed PMID: 32276224
63. Liu C, Scorr L, Kilic-Berkmen G, Cotton A, Factor SA, Freeman A, et al. A metabolomic study of cervical dystonia. Parkinsonism Relat Disord. 2021;82:98–103. PubMed PMID: 33271463
64. Mobasheri A, Cassidy JP. Biomarkers in veterinary medicine: towards targeted, individualized therapies for companion animals. Vet J. 2010;1:1–3. PubMed PMID: 20541693
65. van der Schee MP, Paff T, Brinkman P, van Aalderen WMC, Haarman EG, Sterk PJ. Breathomics in lung disease. Chest. 2015;147(1):224–31. PubMed PMID: 25560860
66. He C, Liu Y, Wang Y, Tang J, Tan Z, Li X, et al. ^{1}H NMR based pharmacometabolomics analysis of metabolic phenotype on predicting metabolism characteristics of losartan in healthy volunteers. J Chromatogr B Analyt Technol Biomed Life Sci. 2018;1095:15–23. PubMed PMID: 30041085.
67. Ahmed AT, Mahmoudian DS, Bhattacharyya S, Arnold M, Liu D, Neavin D, et al. Acylcarnitine metabolomic profiles inform clinically-defined major depressive phenotypes. J Affect Disord. 2020;264:90–7. PubMed PMID: 32056779; PMCID: PMC7024064.
68. Tukiainen T, Tynkkynen T, Mäkinen VP, Jylänki P, Kangas A, Hokkanen J, et al. A multi-metabolite analysis of serum by ^{1}H NMR spectroscopy: early systemic signs of Alzheimer's disease. Biochem Biophys Res Commun. 2008;375(3):356–61. PubMed PMID: 18700135
69. Adamko DJ, Sykes BD, Rowe BH. The metabolomics of asthma: novel diagnostic potential. Chest. 2012;141(5):1295–302. PubMed PMID: 22553262
70. Giera M, Ioan-Facsinay A, Toes R, Gao F, Dalli J, Deelder AM, et al. Lipid and lipid mediator profiling of human synovial fluid in rheumatoid arthritis patients by means of LC-MS/MS. Biochim Biophys Acta. 2012;1821(11):1415–24. PubMed PMID: 22841830; PMCID: PMC3433634.
71. Yang J, Eiserich JP, Cross CE, Morrissey BM, Hammock BD. Metabolomic profiling of regulatory lipid mediators in sputum from adult cystic fibrosis patients. Free Radic Biol Med. 2012;53(1):160–71. PubMed PMID: 22580336; PMCID: PMC3412514.
72. Keller BO, Davidson AG, Innis SM. Phthalate metabolites in urine of CF patients are associated with use of enteric-coated pancreatic enzymes. Environ Toxicol Pharmacol. 2009;27(3):424–7. PubMed PMID: 21783974
73. Chang CH, Chen CF, Tsai YA, Wang SL, Huang PC, Chen BH, et al. The sex-specific association of phthalate exposure with DNA methylation and characteristics of body fat in children. Sci Total Environ. 2020;737:139833. PubMed PMID: 32526583
74. Nagy ZP, Jones-Colon S, Roos P, Botros L, Greco E, Dasig J, et al. Metabolomic assessment of oocyte viability. Reprod Biomed Online. 2009;18(2):219–25. PubMed PMID: 19192342
75. Chung CH, Wong S, Ang KK, Hammond EH, Dicker AP, Harari PM, et al. Strategic plans to promote head and neck cancer translational research within the radiation therapy oncology group: a report from the translational research program. Int J Radiat Oncol Biol Phys. 2007;69(2Suppl):S67–78. PubMed PMID: 17848300; PMCID: PMC2064008.
76. Marhuenda-Egea FC, Martínez-Sabater E, Gonsálvez-Alvarez R, Lledó B, Ten J, Bernabeu RA. Crucial step in assisted reproduction technology: human embryo selection using metabolo-

mic evaluation. Fertil Steril. 2010;2:772–4. PubMed PMID: 19962138

77. McMillan EA, Ryu MJ, Diep CH, Mendiratta S, Clemenceau JR, Vaden RM, et al. Chemistry-first approach for nomination of personalized treatment in lung cancer. Cell. 2018;173(4):864–878. e29. PubMed PMID: 29681454; PMCID: PMC5935540.
78. Lindner J, Vogeser M, Sorg K, Grimm SH. A semi-automated, isotope-dilution high-resolution mass spectrometry assay for therapeutic drug monitoring of antidepressants. Clin Mass Spectrom. 2019;14B:89–98.
79. Medina S, De Las H-GI, Casas-Pina T, Bultel-Poncé V, Galano JM, Durand T, et al. Urinary oxylipin signature as biomarkers to monitor the allograft function during the first six months post-renal transplantation. Free Radic Biol Med. 2020;146:340–9. PubMed PMID: 31734358
80. Su Y, Elshorbagy A, Turner C, Refsum H, Chan R, Kwok T. Circulating amino acids are associated with bone mineral density decline and ten-year major osteoporotic fracture risk in older community-dwelling adults. Bone. 2019;129:115082. PubMed PMID: 31622772; PMCID: PMC6925590.
81. Hampel R, Breitner S, Kraus WE, Hauser E, Shah S, Ward-Caviness CK, et al. Short-term effects of air temperature on plasma metabolite concentrations in patients undergoing cardiac catheterization. Environ Res. 2016;151:224–32. PubMed PMID: 27500855
82. Lewis GD, Ngo D, Hemnes AR, Farrell L, Domos C, Pappagianopoulos PP, et al. Metabolic profiling of right ventricular-pulmonary vascular function reveals circulating biomarkers of pulmonary hypertension. J Am Coll Cardiol. 2016;67(2):174–89. PubMed PMID: 26791065; PMCID: PMC4962613.
83. Nikolaus S, Schulte B, Al-Massad N, Thieme F, Schulte DM, Bethge J, et al. Increased tryptophan metabolism is associated with activity of inflammatory bowel diseases. Gastroenterology. 2017;153(6):1504–1516. e2. PubMed PMID: 28827067
84. Ciavardelli D, Piras F, Consalvo A, Rossi C, Zucchelli M, Di Ilio C, et al. Medium-chain plasma acylcarnitines, ketone levels, cognition, and gray matter volumes in healthy elderly, mildly cognitively impaired, or Alzheimer's disease subjects. Neurobiol Aging. 2016;43:1–12. PubMed PMID: 27255810
85. Kemna MJ, Plomp R, van Paassen P, Koeleman CAM, Jansen BC, Damoiseaux JGMC. Galactosylation and sialylation levels of IgG predict relapse in patients with PR3-ANCA associated vasculitis. EBioMedicine. 2017;17:108–18. PubMed PMID: 28169190; PMCID: PMC5360573.
86. Chorell E, Hall UA, Gustavsson C, Berntorp K, Puhkala J, Luoto R, et al. Pregnancy to postpartum transition of serum metabolites in women with gestational diabetes. Metabolism. 2017;72:27–36. PubMed PMID: 28641781
87. Isa F, Collins S, Lee MH, Decome D, Dorvil N, Joseph P, et al. Mass spectrometric identification of urinary biomarkers of pulmonary tuberculosis. EBioMedicine. 2018;31:157–65. PubMed PMID: 29752217; PMCID: PMC6013777.
88. Shariff MIF, Kim JU, Ladep NG, Gomaa AI, Crossey MME, Okeke E, et al. The plasma and serum metabotyping of hepatocellular carcinoma in a Nigerian and Egyptian cohort using proton nuclear magnetic resonance spectroscopy. J Clin Exp Hepatol. 2017;7(2):83–92. PubMed PMID: 28663670; PMCID: PMC5478965.
89. Kromke M, Palomino-Schätzlein M, Mayer H, Pfeffer S, Pineda-Lucena A, Luy B, et al. Profiling human blood serum metabolites by nuclear magnetic resonance spectroscopy: a comprehensive tool for the evaluation of hemodialysis efficiency. Transl Res. 2016;171:71–82. PubMed PMID: 26924041
90. Ferreti Y, Miyoshi NSB, Silva WA Jr, Felipe JC. BioBankWarden: a web-based system to support translational cancer research by managing clinical and biomaterial data. Comput Biol Med. 2017;84:254–61. PubMed PMID: 25959800
91. Wang D, Cheng SL, Fei Q, Gu H, Raftery D, Cao B, et al. Metabolic profiling identifies phospholipids as potential serum biomarkers for schizophrenia. Psychiatry Res. 2019;272:18–29. PubMed PMID: 30579177

Microscopy Principles in the Diagnosis of Epidemic Diseases

8

Nadia Alejandra Rivero-Segura,
Sandra Lizbeth Morales-Rosales,
and Ruth Rincón-Heredia

Abbreviations

CLSM	Confocal laser scanning microscopy
CMV	Cytomegalovirus
CSF	Cerebrospinal fluid
EM	Electron microscopy
FM	Fluorescence microscopy
HE	Hematoxylin and eosin
HPV	Human papillomavirus
HSV	Herpes simplex virus
LSCM	Laser scanning confocal microscopy
RCM	Reflectance confocal microscopy
SSCM	Slit scanning confocal microscopy
TEM	Transmission electron microscopy
VZV	Varicella-zoster virus

Introduction

Epidemiological diseases and pandemics are not a new issue; they have had great importance in the history of humanity. Globalization, the constant exploitation of natural resources, the great dependence we have on livestock, and the ease with which we can travel today, has allowed humans to be in contact with various infectious agents [1, 2].

Pandemics are caused by specific etiological agents; the conditions of human-animal proximity and environmental changes promoted the appearance of zoonotic diseases. Some very clear examples are the agents of measles, smallpox, tuberculosis, and many other pandemic diseases that evolved from diseases that only had affected domestic animals and are now capable of infecting humans [3].

The biomedical and clinical sciences have developed various technologies not only to discover the etiological agents of diseases that affect humans, but they have also been expected to understand their infection mechanisms to make an accurate and timely diagnosis but also to find possible cures for these diseases [4].

One of the most used tools for the discovery of infectious agents that cause epidemiological diseases is the microscope, which has accompanied doctors and researchers for more than three centuries [5].

Brief History of the Development of the Microscope

The cell is the morphological and functional unit of every living being, and it is the smallest element that can be considered alive. A typical animal cell measures between 10 and 20 μm in diameter, which is about one fifth of the smallest particle observable by the human eye. Because cells are small and complex, it is difficult to see their structure, discover their molecular structure, and even more difficult to know how their components work.

The tools that scientists and physicians have at their disposal determine how much we can learn about cells. The introduction of new techniques frequently results in an advance in the knowledge of cell biology, medicine, pathology, and the diagnosis of diseases. To understand how cells work, one needs to know the identity and structure of their molecular components and how they interact. One of the technological advances that have allowed the development

N. A. Rivero-Segura
Dirección de Investigación, Instituto Nacional de Geriatría (INGER), Instituto Nacional de Geriatría, Ciudad de México, Mexico
e-mail: nrivero@inger.gob.mx

S. L. Morales-Rosales
Posgrado en Biología Experimental, Universidad Autónoma Metropolitana, Unidad Iztapalapa, Mexico City, Mexico

R. Rincón-Heredia (✉)
Microscopy Core Unit, Instituto de Fisiología Celular, Universidad Nacional Autónoma de México, Mexico City, Mexico
e-mail: rrincon@ifc.unam.mx

of knowledge of biology, physiology, and pathology is the microscope [6].

The compound microscope was invented by Zacharias Janssen in 1590; however, it was not until 1665 that it took on real importance, when Robert Hook published his work *Micrographia* in which diagrams of images obtained in optical microscopy appeared for the first time. It is also Hook who, when observing a slice of cork, notes that its structure is composed of small cavities organized like cells, minting the term "cell." Even though Hook observed dead cells, his discovery started the cell theory movement.

On the other hand, in the mid-seventeenth century, Antonie Van Leeuwenhoek made improvements to the simple microscope using more powerful lenses. In this way they recorded the first observations of muscle fibers, sperm, red blood cells, protozoa, and bacteria using simple microscopes of their own manufacture.

Only after the first light microscopes became available in the early part of the nineteenth century did Schwann and Schleiden suggest that cells are the basic units of life, formally proposing the cell theory and formally initiating the cell biology.

Currently microscopy depends as much on specimen preparation techniques as on the development of microscopy itself [7].

Photonic Microscope

The Use of Microscopy in the Life Sciences

As it was demonstrated a long time ago, life is not limited to what we can observe with the naked eye, and for this modern science has several simple optical microscopes for general or already very specialized use. These microscopes can be differentiated mainly in factors such as the wavelength with which the sample is illuminated, the physical alteration of the light that reaches or emanates from the sample, and the specific analytical processes that can be applied to the final image.

With these technologies it has been possible to cover a wide variety of needs within life sciences, since these technologies have provided tools to study the structure and functions of molecules, cells, and organisms. It is used in forensic science to study biomolecules at crime scenes, in biotechnology and pharmaceutical industry for product quality control, and in clinical medicine as a useful tool for detection, diagnosis, interventionist guidance, monitoring of response to treatment, and treatment of the disease, since they have provided abundant biochemical and structural information in biological samples, which has increased sensitivity and specificity for the detection and localization of diseases, which has resulted in a practical, safe (minimally invasive), and relatively affordable technique.

Generalities

Generally, we can describe the existence of two types of microscopes that use light as a source of energy to form enlarged and detailed images of objects that the human eye could never observe:

1. *Simple photon microscope or magnifying glass*
2. *Compound photonic microscope*

Currently, there are various types of compound photonic microscopes, ranging from general use or to microscopes that are highly specialized, and their differences mainly settle in factors such as the wavelength with which the sample is illuminated, the physical alteration of the light that reaches the sample or emanates from it, and the specific analytical processes that can be applied to the final image. These types of microscopes have a wide range of utility ranging from teaching, research, and clinical diagnostic laboratories, among others [Fig. 8.1].

Components of the Photon Microscope

The compound photon microscope is mainly made up of three types of components:

1. *Mechanical components*: It is basically the structure of the microscope, which serves to support, move, and hold the optical and lighting systems as well as the samples or objects to be observed.
2. *Optical components*: This system is formed by the objectives, the eyepieces, the condenser, and the prisms.
3. *Lighting components*: Any instrument that provides light energy to the microscope is considered within these components, while there are natural and artificial sources of light energy.

Types of Photon Microscopes

Bright-Field Microscope

The microscope used by most students, researchers, and physicians is the bright-field microscope which comes directly from the microscopes that were used in the nineteenth century and that inaugurated the first great era of microscopic research. In this microscope, light is passed through the sample, or it can be reflected by said sample; however, for the image to be clearly visible, it is necessary to stain the sample of interest. The staining allows the cellular and tissue components of the structure to be contrasted by specific colorants that absorb and transmit certain wavelengths of the visible spectrum. The rest of the microscopic field will be

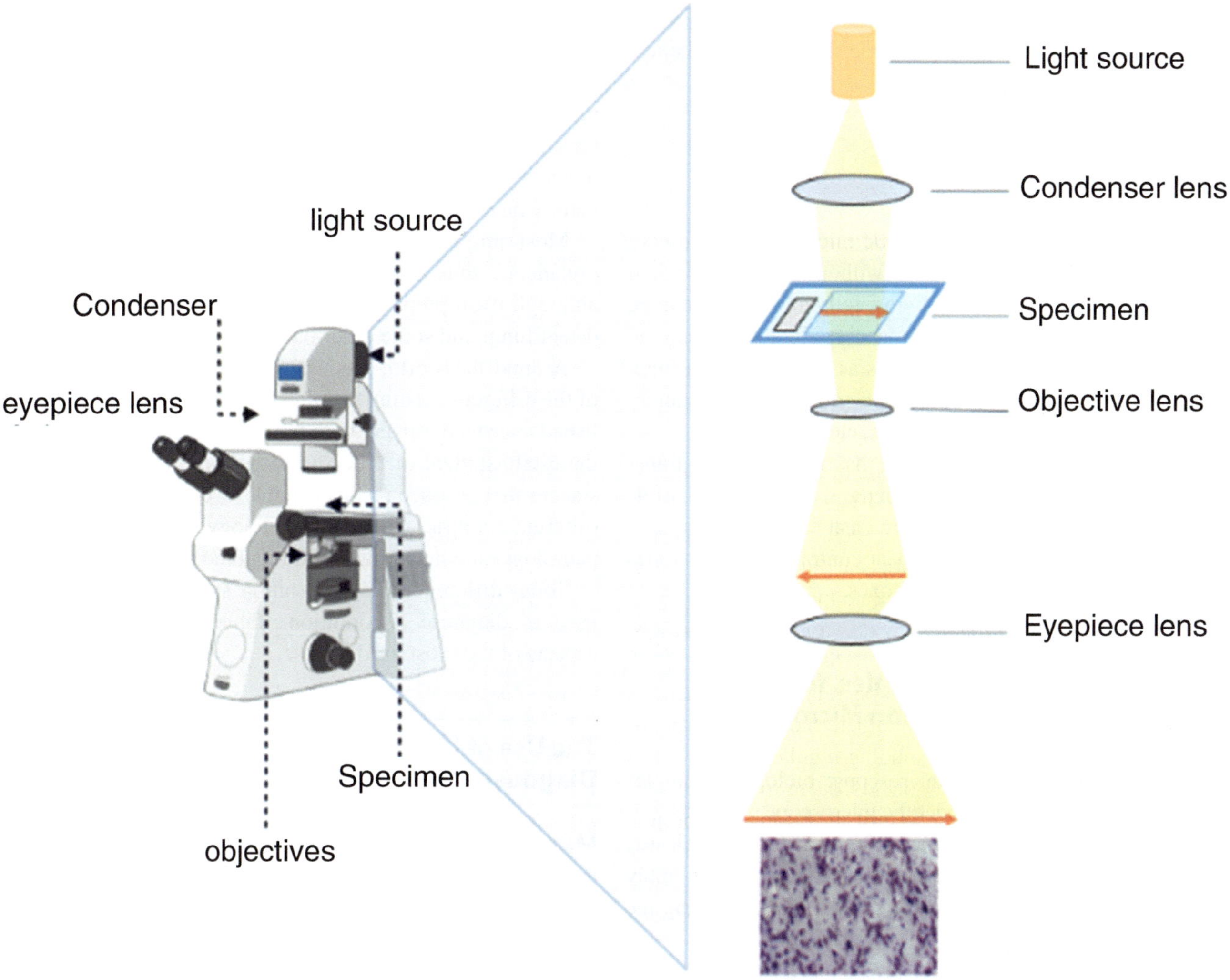

Fig. 8.1 *Diagram of light microscope*. Schematic representation of an inverted microscope, where the main components of the instrument are observed (left side). Internal diagram of the light path (summarized): The light source represented by the yellow cylinder generates the light rays (light yellow) that pass through the converging lenses of the condenser allowing the light to follow a parallel path; below the condenser is the diaphragm (not schematized), in charge of regulating the incident light cone. Rays of light pass through the sample and reach the objective lens. This is where the first real magnified image of the sample is formed (this magnification depends on the magnification of the objective). Then the image passes through the eyepieces generating a virtual image, which is ten times larger than the real image. The image at the bottom of the schematic was obtained at Olympus IX71 by María Fernanda Ramírez, IFC-UNAM (right side)

clear or transparent, since the light rays pass directly to the objective; on the other hand, if samples are examined without staining, the image will only show little contrasted details, and they can even be seen transparent.

However, it is important to mention that the samples to be observed must be thin with a thickness of 5 μm to 50 μm, to be able to take full advantage of the resolving power and sharper contours.

Dark-Field Microscope

It is called this way because the image is formed by a series of bright structures on a dark background; these structures refract the light rays toward the objective, and to observe this it is necessary that the microscope is equipped with a special condenser that illuminates the sample with a lot of intensity in an oblique way, making the field look dark and on it the sample stands out, which reflects part of the light and appears in a bright way.

The dark-field microscope has been used to observe a large number of samples, such as seeing living cells (protozoa, bacteria, desquamated cells, etc.); it should be mentioned that in this type of samples, it is not necessary to stain them or use substances. On the other hand, since the oblique light rays form a halo of bright luminous light around the samples, it is possible to see particles smaller than 0.2 micrometers. It has been observed that by using this microscope, the spirochete bacterium *Treponema pallidum*, which is the cause of syphilis, can be easily distinguished, since the shape and helical movement of this bacterium can be easily observed.

Phase Contrast Microscope

It is the most widely used photonic microscope to observe transparent objects or structures without staining, since it facilitates the observation of living cells in which it can be differentiated and thus be able to analyze in a more detailed way the morphological components as well as certain functions that they may develop (phagocytosis, mitosis, amoeboid, ciliary or flagellar movements, etc.).

The phase-contrast microscope has the ability to transform the small refractive indices in the samples of each of its components into different light intensities, offering images where the object's structures appear contrasted in dark tones or bright tones and intermediate tones.

Staining of Biological Samples, for Use in Different Types of Photon Microscopy

During the emergence of microscopy, biological samples were used to be observed under the microscope without staining; however, the Dutchman Antonie van Leeuwenhoek was probably the first scientist to use a biological dye to apply color to microscopic objects, immensely revolutionizing the way of observing tiny structures more clearly visible than they were unstained, which revolutionized biological science. On the other hand, the use of dyes for use in microscopy did not become common until the middle of the nineteenth century. Today stains have become an indispensable tool in biology and medicine, where they are used in various disciplines, mainly histology, cytology, and microbiology [8].

Most samples are treated with stains that color the microorganisms, thus highlighting them from the background, although fresh preparations without staining can be used to detect fungi and some other pathogens.

Around the world, methods and technical specifications of the different existing staining techniques have been established and standardized, with the sole objective of promoting the establishment of procedures that produce coloring substances that produce results capable of being reproducible in different countries in areas of cytology, bacteriology, histopathology, and hematology, among others [Table 8.1] [9].

Today different biological sample staining procedures are used to diagnose a condition. Table 8.1 describes some aspects of the most used stains.

The Use of the Photon Microscope in Clinical Diagnosis

Microscopy currently plays a crucial role in both research and diagnostic aspects; this generally involves the use of optical microscopes for the analysis of microbiological,

Table 8.1 Histological techniques commonly used for pathological or bacteriological diagnosis

Staining	Description and characteristics	Staining color	Objective of staining
Hematoxylin (basic)	As hematoxylin is cationic or basic, it stains acidic structures (basophils), such as cell nuclei	Blue and purple tones	In histopathology it is used to give a morphological diagnosis in tissues [9, 10]
Eosin (acidic)	Eosin stains basic components (acidophiles), due to their anionic or acidic nature, like the cytoplasm	Shades of pink	
Gram staining	Initially, all bacteria absorb the crystal violet dye; however, with the use of solvent, the lipid layer of gram-negative organisms dissolves, losing the primary staining. The basic fuchsin stain is then used to give the discolored gram-negative bacteria a pinkish color for easier identification	Gram-positive bacteria that appear purple Gram-negative bacteria that appear pink and red	Microbiology, bacterial cell morphology, so as to be able to make a first approximation to bacterial differentiation [11]
Ziehl-Neelsen stain	It is used for the identification of acid-fast bacillus (AFB), which will be stained for the incorporation of basic fuchsin, which requires heat and is resistant for the discoloration process that uses an alcohol-acid combination and then a counterstain with methylene blue	AFB appear red	Used in the diagnosis of *Mycobacterium tuberculosis* [12]
Kinyoun cold stain	Is a modified Ziehl-Neelsen staining. The incorporation of fuchsin is allowed by phenol as a chemical mordant and heat is not required	AFB appear red	Used in the diagnosis of *Mycobacterium tuberculosis* [12]
Giemsa (Romanowsky-Giemsa effect "RGE")	Gives a differential coloring. Contains a mixture of methylene blue, azure A and azure B as basic dyes, and eosin Y as an acidic dye	Chromatin in purple, basophil cytoplasm in blue, neutrophil granules in purple, eosinophil granules in red-brown, basophil granules in purple-black, erythrocytes in pink-red	Applied in bacteriology, cytogenetics, cytology, hematology, histopathology, and parasitology [13, 14]

cytological, and pathological samples, among others. The introduction of the microscope has had an influential effect in the field of medical diagnosis, since in some cases it has turned out to be a simple, fast, and low-cost test, and in addition to that it can be performed close to the patient. Microscopy diagnosis only requires the microscope, the sample, and, for some techniques, a staining process. Microscopy has been the primary diagnostic tool for decades, and although other techniques may exist, the results often still require confirmation by microscopy. Microscopy can turn out to be very sensitive and specific in some cases and ideal for systematic detections, which has helped to reinforce the diagnoses of certain pathologies, which in turn contribute and guide a better treatment [15]. It is worth mentioning that to obtain good results, a good technique is essential when obtaining the sample to be analyzed, in addition to the fact that the interpretation of the microscopic image is a skill that requires training, which implies a good practical knowledge of the microscope.

Microscopy as a Diagnostic Method

As mentioned above, stains are vastly practical for detecting most human pathogens, but there are limitations in specificity, so other additional diagnostic techniques are not ruled out. The sensitivity of special stains to detect pathogens depends on a number of factors such as:

1. Number of microorganisms present at the site
2. Technical factors associated with obtaining samples
3. Observer capacity
4. Magnification used
5. Availability of specific confirmatory tests

Therefore, through appropriate morphological diagnosis, one can greatly contribute to the timely diagnosis of infectious diseases.

The Use of the Photon Microscope in Histopathology for the Diagnosis of Diseases

For more than a century, conventional optical microscopy has been the basic tool for the evaluation of tissues, cells, and microorganisms, so it has played a fundamental role in pathological diagnosis; for this type of studies it is necessary to obtain a biopsy which is a diagnostic procedure that consists of the extraction of a total or partial sample of tissue or a smear which consists of the extension of a sample of body fluid to be examined under a microscope and thus be able to study the composition, structure, and characteristics of the tissue extracted through the use of stains with dyes or more specific through antibodies (histochemical), in order to highlight particular characteristics that are being sought [16].

Through histopathology, it seeks to understand the normal structure and function of the different tissues, it can be very useful to make a diagnosis and to determine the severity and progression of a disease, and it has been used initially and is still used today to diagnose diseases, infectious, degenerative, or neoplastic in humans or animals. These qualitative diagnoses are based on a sum of observable changes in the morphology of the analyzed tissue. The cognition of these changes is based on the recognition of patterns by the observer and the comparison of these patterns with the known physiological variation in the morphology of the tissues in the respective species [17].

Histological examination of tissues can help to diagnose the disease, because each condition produces a characteristic set of changes in their structure. Although diseases are very diverse, the body's responses are more limited and are classified into specific categories.

Inflammation is a characteristic process of most infectious diseases, and it can also be associated with neoplasia, dysplasia, autoimmune diseases, allergic conditions, and idiopathic disorders. It is important to be able to differentiate between inflammatory conditions caused by infectious agents from those with non-infectious etiologies; now the probability that the inflammatory response is due to an infection will be determined, and the next step is to find out which etiological agents are the possible causes of infection, taking into account that each type of pathogen tends to provoke a particular response [18]. Some of the most common pathological processes are described below, with examples of their analysis under the microscope.

Diagnosis of Viral Infections

During their replication in the cells they infect, it has been observed that viruses are capable of inducing morphological changes, such as the formation of inclusion bodies in the host cell nucleus, the cytoplasm, or both, the formation of multinucleated giant cells, the presence of a perinuclear halo around the infected cell, lymphocytic infiltration, or even cell necrosis, among others [Table 8.2] [19].

Diagnosis of Bacterial Infections

Accurate and rapid identification and characterization of pathogens are essential for the proper treatment of infections, which represent a growing problem. The gold standard in the detection of bacterial infections has been based on the growth of pathogens in cell culture, followed by evaluation

Table 8.2 The most recognized viral infections due to histopathological changes

	Microorganism	Typical histopathologic features	Histochemical stains
Viruses	HSV, VZV	Nuclear molded multinucleated cells, where the nuclei are clean "Glassy"	HE, Papanicolaou, Giemsa, or Wright stains
	CMV	Inflammation with endothelial damage is usually observed	HE, Papanicolaou stains
	HPV	Koilocytosis	HE, Papanicolaou stains
	Adenovirus	Smudge cells present	HE stain

Table 8.3 Classic bacterial stains, for the diagnosis of diseases of clinical relevance

	Microorganism	Typical histopathologic features	Histochemical stains
Bacteria	*Helicobacter pylori*	Neutrophilic and/or chronic inflammation, bacteria are often visible in HE-stained sections	Giemsa and Warthin-Starry are the most used stains
	Bartonella	Groups of bacilli can be found within the formation of granulomas	Warthin-Starry or comparable silver stain
	Legionella pneumophila	Neutrophilic; bacillary forms are not discernible on HE or tissue gram stain	Warthin-Starry or comparable silver stain
	Mycobacteria	Associated with necrotizing and non-necrotizing granulomas, as well as acute inflammation	Gram-positive, beaded, non-branching bacilli; Ziehl-Neelsen or auramine-rhodamine stain

by biochemical methods designed to identify strains and species of microorganisms; many of these methods are performed through stains that help to speculate or highlight the presence of bacteria through the help of microscopy [Table 8.3]. Bacterial culture is cost-effective and generally results in a diagnosis with good specificity [20].

Diagnosis of Parasitic Infections

Parasitic infections are truly devastating and prevalent in the world, causing millions of morbidities and mortality annually. Quick and accurate diagnosis is of utmost importance in the fight against parasitic infections. However, the morphological identification of the life cycle stages of parasites is the main way of diagnosing parasitic diseases.

The main method for the diagnosis of possible infection by a parasite is to examine biological samples in search of stages of the parasite's life cycle, based on morphology, through the use of a microscope on wet samples of sputum, urine, vaginal smears, aspirates, duodenal, sigmoidoscopic material, abscesses, and tissue biopsies [21]. In Table 8.4 are some examples of parasitic infections and the use of the photonic microscope as a diagnostic tool.

The comparatively low cost of these methods has perpetuated their sustained use, especially in economically disadvantaged regions of the world, where the cost of molecular and immunological kits remains prohibitive.

Use of the Microscope in Examination of Urine Sediment

Microscopic examination of centrifuged urinary sediment by an experienced nephrologist is an important tool in diagnosing and monitoring a number of conditions that affect the kidneys. Sometimes, although not routinely, it is necessary to use staining techniques in the sediment that allow the identification of particular elements, eosinophils, bacteria, and fats, or differentiate some elements from others: cubic cells of leukocytes, red cells of yeast, and amorphous salts of bacteria, among others [Table 8.5] [32].

Fluorescence Microscopy

Fluorescence microscopy (FM) requires that objects of interest have fluorescence. Fluorescence is the emission of light that occurs within a few nanoseconds after the absorption of light that is typical of a short wavelength. The difference between excitation and emission wavelengths is known as the "Stokes's shift"; it is the property of converting the invisible of ultraviolet (UV) light into longer wavelength, visible radiation [33].

Fluorophores

The molecules that are used in this type of microscopy are called fluorophores. The outermost electron orbitals in the fluorophore molecule determine its efficacy as a fluorescent compound and the absorption and emission wavelengths. When fluorescent compounds absorb the energy of light (photons) in their ground state, alterations occur in the electronic, vibrational, and rotational states of the molecule. The absorbed energy sometimes moves an electron into a different orbital, which is on average the furthest from the nucleus. This transition to an excited state occurs in femtoseconds. Usually, the excitation process also establishes molecular vibrations in which the internuclear distances vary in time. All that stored energy is eventually lost. Vibrational relax-

Table 8.4 Diagnostic microscopy for the detection of protozoan and helminth infections

	Pathogen	Microscopy detection of parasite	Test used for diagnosis
Detection of blood-borne protozoa infection	*Leishmania* species	(Amastigote) in aspirates from spleen, bone marrow, or lymph nodes [22]	Direct examination, saline
	Toxoplasma gondii	Blood or CSF as well as parasite detection from stained tissues [23]	
	Trypanosoma brucei	(Trypomastigote) in the blood (first stage) or CSF (second stage)[24]	
	Trypanosoma cruzi	(Trypomastigote) in blood smears [25]	
	Plasmodium species	Blood smears [26]	
	Babesia microti	Examination of blood smears [27]	
Detection of intestinal protozoan infections	*Cryptosporidium parvum* and *C. hominis*	Procedure to detect oocysts from stool [28]	Modified acid-fast staining
	Giardia lamblia	Detect cysts from stool [15]	Trichrome or iron hematoxylin staining Sedimentation/concentration techniques followed by microscopy
	Entamoeba histolytica	Staining of stool samples [29]	Microscopy almost obsolete because *E. histolytica* cysts and trophozoites are morphologically identical to those of *Entamoeba dispar*
Detection of helminth Infections	*Schistosoma* species	Stool for intestinal schistosomiasis, detection of eggs in urine for urinary schistosomiasis [15]	Kato-Katz technique
	Soil-transmitted helminths	Detection of eggs in stool [15]	Fresh and after Giemsa stain, in direct peripheral blood preparations
	Lymphatic filariasis (*Brugia malayi* and *Wuchereria bancrofti*)	Examination of concentrated blood smears [30]	Giemsa or Wright stains (taken during the nocturnal activity period) for microfilariae
	Taenia solium	Detection of eggs in stool [15]	Kato-Katz technique or direct examination
	Onchocerca volvulus	Skin snips placed to detect larva and examine surgically removed nodules for adult worms [31]	Direct examination, saline

Table 8.5 Example of stains usable in microscopic examination of urine sediment

	Structures	Staining	Equipment
Study of sediment in urine	Lipid granules and fatty oval bodies	Sudán III Oil Red	Bright field, phase contrast, and polarization at 40x and 10x Ó or 100x staining
	Bacteria, yeasts	Gram	
	Epithelial cells and leukocytes	Alcian blue-Pyronine Toluidine blue Methylene blue	
	Eosinophils	Hansel	
	Hemosiderin	Prussian blue	
	Starch and starch granules	Lugol's iodine	

ation and fluorescence emission are the main ways that a fluorophore returns to its low-energy state [33].

Excitation Spectrum

When a fluorophore absorbs light, all the energy possessed by a photon is transferred to the fluorophore. This energy is inversely proportional to the wavelength of the photon. If the energy of the absorbed photon is greater than that necessary for the transition from the ground state to the lowest energy level, the molecule also undergoes a change in vibration, rotation, or motion even within a larger electron orbital. It is necessary to consider that just as a photon of the appropriate energy causes this transition, it is also possible that several photons add their energy to bring a molecule to its excited state.

The Fluorescence Microscope

The preferred approach in modern fluorescence microscopy is epi-illumination. In this configuration, the microscope objective not only has the known function of imaging and magnifying the sample but also serves as the condenser that illuminates it. The advantage of this approach over transmission, or diascopic, microscopy is that in fluorescence microscopy (in which the excitation light comes through the condenser and the emission is picked up by the objective), only a small percentage of the path of the excitation light that is reflected off the sample has to be blocked in the path of the return light in the epi-illumination mode. The main technical hurdle with this approach is that the excitement light and emission fluorescence overlapping the light path require a special type of beam splitter, a dichroic mirror, to separate the excitation light from the emission light. The dichroic mirror is designed to be used in 45° light paths. In ordinary fluorescence microscopes, the dichroic mirror reflects the short wavelength light from the light source and

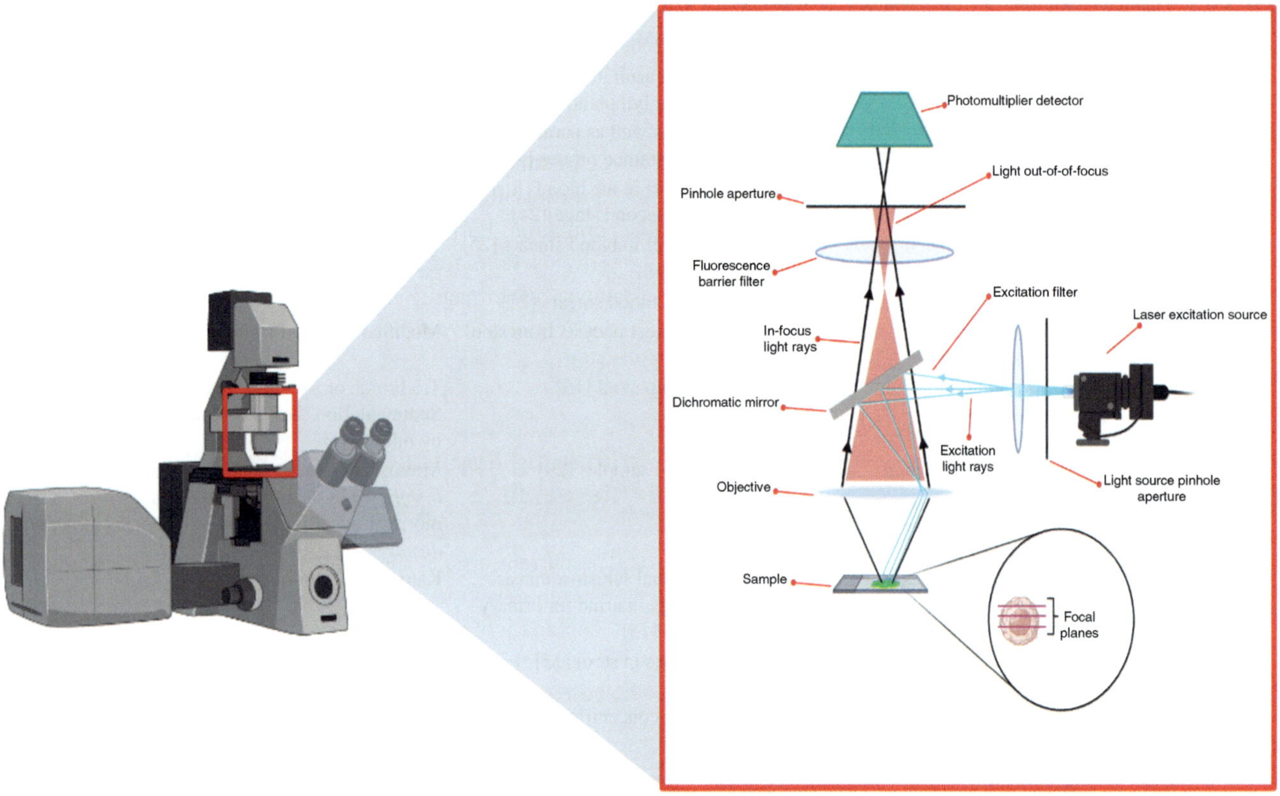

Fig. 8.2 *Diagram of confocal microscope working principle*. Confocal microscopy allowed the analysis of fluorescent labeled samples

transmits longer wavelengths of the emitted fluorescence. Each dichroic is designed to have a transition from reflection to transmission that resides between the excitation and emission peaks for which each fluorophore is designed. Dichroics are rarely used without two additional filters: the excitation filter, which preselects the excitation wavelength, and the sweep filter, which only allows longer wavelength light to pass behind the detector.

Confocal Microscopy

In 1957 the first confocal microscope was patented by Marvin Minsky; however, it was not until 1987 that they became commercially available for biological systems. Confocal microscopy has made it possible to obtain a better quality of biological images, increasing their resolution and processing.

Recall that resolution is defined as the ability to distinguish the distance between two points and is determined by Abbe's point spread function.

In conventional microscopy, a condenser lens is used to illuminate an extended area and the volume of the specimen simultaneously and uniformly. In thick samples, this results in blurry and unfocused areas from volumes above and below the plane of focus. This out-of-focus light can reduce contrast and resolution and significantly prevents the interpretation of detail in images obtained under the microscope. In an effort to minimize these problems, thin or ultrathin sections of tissue have been used in conventional microscopy to allow a clear view of the structure of the specimen to be obtained.

In confocal microscopes, the illumination is sequential and is concentrated in a small volume of the sample, where the more distant regions of the focal plane receive less illumination, which reduces both the out-of-focus area and the blurring of the image. Also, the detection and lighting systems are focused on the same volume element of the sample. Therefore, the illumination, the sample, and the detector are all centered in the same volume and are therefore confocal. By adding a carefully aligned aperture at a focal point in the optical path, further reduction of out-of-focus areas is achieved [34].

Architecture of the Confocal Microscope

A light source, usually a laser, is reflected by a dichroic mirror or beam splitter and brought to a focus point on the objective lens at the level of the "plane of focus." Fluorescence is emitted by the sample from the point of focus and passes back through the objective lens, through the dichroic mirror and the confocal aperture (pinhole) for the detector. However, fluorescence is also emitted in various planes above and below the plane of focus, but thanks to the confocal aperture, light that is outside the field of focus is prevented from passing into the detector. So, image quality is better than in other kinds of fluorescence imaging [Fig. 8.2].

Confocal Microscopy in Molecular Epidemiology: Clinical Applications and Further Directions

Confocal microscopy is widely used in research laboratories, and currently its application has successfully penetrated the clinical field, since this tool provides high-resolution images with the minimum damage to the sample. According to the available literature, confocal microscopy is useful in the diagnosis of both communicable and noncommunicable diseases and leads to the early detection and management of a disease, improving patient's outcomes [35].

On the other hand, there are two confocal systems currently available for clinical applications: the reflectance confocal microscopy (RCM), based on the natural differences in refractive indices of subcellular structures within a tissue, and the fluorescence confocal microscopy, based on the use of fluorochromes, increasing the cell-to-soma contrast [36]. In this context, both confocal systems have benefits in the field of ophthalmology (monitoring the retina and the cornea pathologies), dermatology (diagnosis of neoplastic and inflammatory skin diseases), and oncology (histopathologic diagnosis and for the excision margins in surgery). Hence, in the following paragraphs we aim to depict briefly the most outstanding examples of the clinical application of confocal microscopy.

Confocal Microscopy in Ophthalmology

As mentioned above ophthalmology is one of the most benefited areas from confocal microscopy, since this represents a noninvasive tool for the diagnosis of the eye's surface disorders, including allergies, ulcers, corneal erosions, allergies, keratitis, infectious diseases, corneal dystrophies, and corneal opacity; as well, corneal confocal microscopy is widely used to take decisions in cataract surgery or to analyze the feasibility of donor corneas for keratoplasty. Additionally, confocal microscopy helps to analyze cornea thickness to assess changes in the dynamics and structure or to monitor the status of graft after a transplant [37, 38].

Currently, the RCM is the most employed microscopy in ophthalmology since it brings accurate information for the diagnosis and requires minimal contact between the cornea and the objective lenses of the microscope. In this sense, there are two kinds of confocal microscopes available for eye scanning [38, 39], the slit scanning confocal microscope (SSCM), that uses a fixed laser beam and the preparation is scanned by a motorized stage on the microscope, and the laser scanning confocal microscope (LSCM) which performs the scanning by moving the laser beam through galvanometric mirrors which allows modifying the laser beam incidence through the eye [40].

A major number of studies have been performed in the cornea; thus, we will focus on describing the main dystrophies and diseases that are currently diagnosed in this structure by confocal microscopy. Cornea is the tissue which allows the light to pass from the outside to inside the eye; this also provides protection to the iris and the crystalline lens and protects the eye from infections and other risks from the environment. The cornea conformation comprises five layers [Fig. 8.3].

Confocal microscopy is a noninvasive tool which allows the analysis of all the layers that the cornea comprises, even in cornea with decreased transparency. Hence the current knowledge in this field has help in the diagnosis of the most common corneal dystrophies such as dystrophy of the basal layer, characterized by the split of the intraepithelial basal membrane from the normal epithelial cells, leading to a reduplication of basement membrane; Meesmann corneal dystrophy, identified by hyporeflective areas and punctual reflective points; and Lisch corneal epithelial dystrophy featured by rounded dark injuries with reflective areas in the center. Other types of corneal dystrophy are the Bowman's layer that is currently diagnosed by confocal microscopy; Reis-Bücklers corneal dystrophy, characterized by well-defined deposits in Bowman's membrane and epithelium; Thiel-Behnke corneal dystrophy; and diagnosis of dry eye.

Meesmann corneal dystrophy is a rare condition that has been attributed to mutations in the keratin 3 gene or keratin 12 gene, on the chromosome 12 or 17, respectively. Clinically, such a condition is characterized by cystic changes in the corneal epithelium, which may be identified by histological assessment, electron microscopy, or confocal laser scanning microscope. Such assessment reveals both clumped keratin into the intraepithelial cysts that migrate to the corneal surface and irregularities in cell arrangement including granular deposits and thickened basement membrane; however, the confocal microscopy represents a noninvasive method that confirms the diagnosis accurately over the other two approaches [40] .

As well, laser *in vivo* confocal microscopy aids in the diagnosis of infectious diseases such as keratitis caused by parasites, bacteria, virus, or fungus. Particularly, keratoconjunctivitis caused by adenoviruses leads to the development of nummular lesion in the keratocytes, which conduct to the reduction of visual acuity and increased glare sensitivity, due to the accumulation of immune cells such as lymphocytes, histiocytes, and fibroblasts. The diagnosis of viral keratoconjunctivitis by confocal microscopy derives from the identification of hyperreflective punctate structures in the epithelium; additionally, the appearance of corpuscular changes with dendritic extension has been reported that may correspond to the Langerhans's cells, which migrate to the central cornea in response to traumatic, chemical, or inflammatory stimuli. On the other hand, keratitis may be elicited by *Acanthamoeba*;

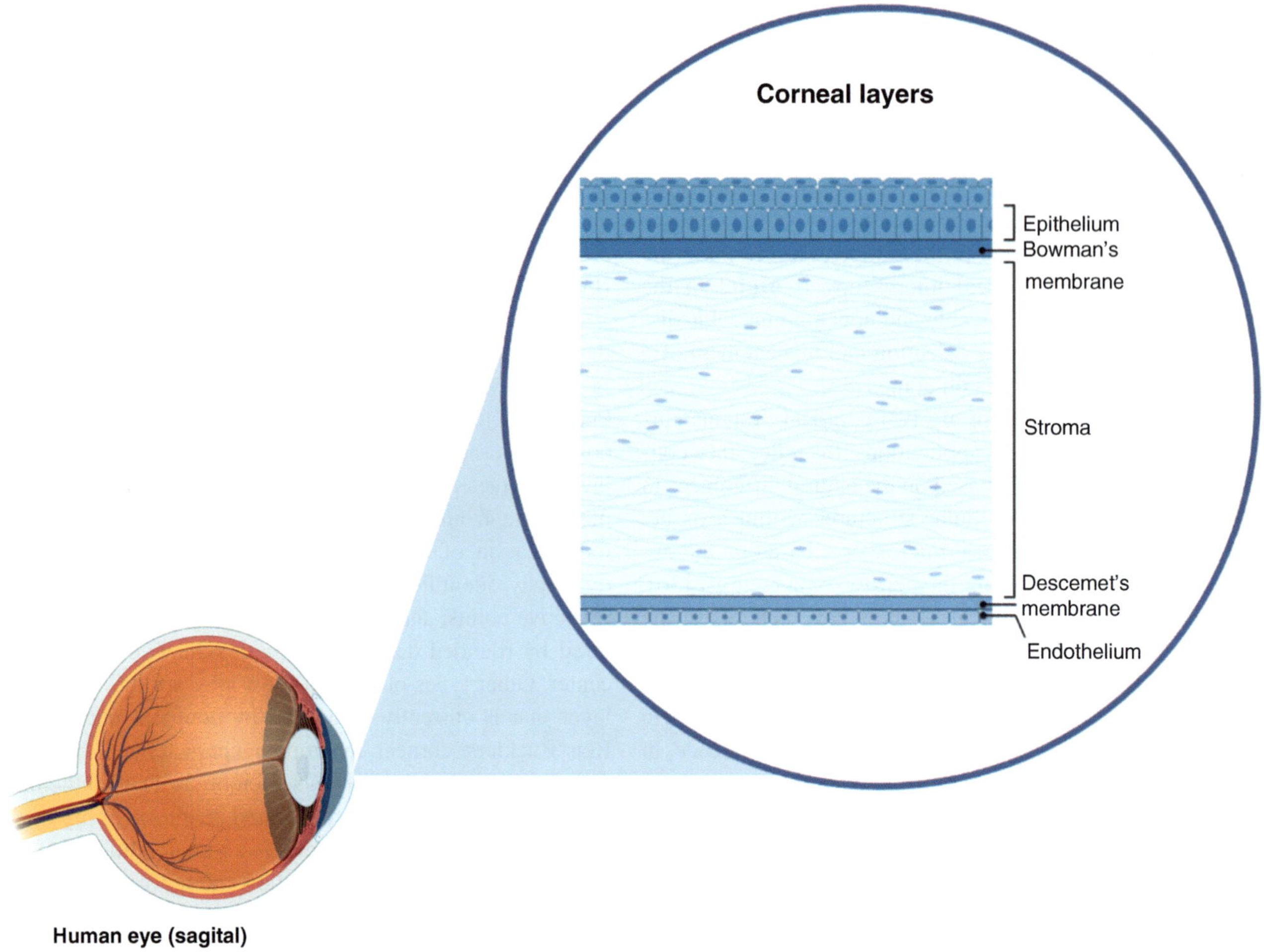

Fig. 8.3 *Cornea comprises five layers.* From the upper to the lower layer, the layers are the epithelium, Bowman's membrane, stroma, Descemet's membrane, and the endothelium. The main function of the cornea is to give protection to the inner eye components from infections or external threats

unfortunately such infection is diagnosed at a late stage, causing blindness. Hence, the role of confocal microscopy is quite useful since *Acanthamoeba* cysts are easily identified in the cornea. However, despite the high-resolution images that confocal microscopy offers in a short period of time, this tool does not substitute the traditional laboratory analysis (culture or smears), yet. As well, fungal keratitis (keratomycosis), induced by *Aspergillus* spp. and *Fusarium* spp., is currently diagnosed by confocal microscopy so efficiently since this brings a rapid *in vivo* visualization of the fungal filaments in the cornea, enabling the opportune therapeutic management of the infection.

As well another use of corneal confocal microscopy regards on both the quantification of the corneal C-fiber pathology and the identification of the severity of peripheral neuropathy in diabetic individuals [41, 42]. As well, in recent years it has been developed in novel noninvasive protocols for neuropathy, particularly in Fabry disease; such is characterized by the accumulation of glycosphingolipids (globotriaosylceramide) in blood vessels, renal epithelia, myocardium, skin, dorsal root ganglion cells, and cornea. Since Fabry disease induces progressive renal failure, cardiac hypertrophy, arrhythmias, and cerebral infarction, the life span of the individuals is approximately 50 years. So, the accurate diagnosis and progression of the disease is crucial for patients. Hence, Fabry disease is characterized by a small nerve fiber damage which is difficult to quantify by conventional neurophysiology and quantitative sensory testing; corneal confocal microscopy provides a novel noninvasive tool to quantify nerve damage in patients with Fabry disease [43].

As described above confocal microscopy has shown great results; particularly the laser in vivo confocal microscopy has demonstrated that in fact such tool has been used in this sense, and confocal microscopy has been suggested as a useful noninvasive tool for in vivo imaging of the cornea, leading to assess, monitor, and diagnose several infections and corneal disease accurately. For instance, according to Bhutani *et al.* confocal microscopy may be helpful to detect and manage conditions such as dystrophy, ecstasy, graft rejection, or endothelial decomposition. Also, confocal microscopy may contribute to assessing tissue repair following surgery or assessing corneal nerve fiber damage.

In fact, the high-resolution imaging of the living cornea is comparable with the histochemical methods traditionally used for such tasks [44].

Moreover, confocal microscopy has represented a useful tool for systemic disease such as metabolic and neurological, since this represents good sensitivity and specificity for identifying potential risk of neuropathy and foot ulceration.

Confocal Microscopy in Dermatology

Dermatology is another field where confocal microscopy use has permeated successfully, mainly because this approach is low invasive and offers accuracy during the early diagnosis of skin cancers or in the evaluation of benign skin diseases [45]. Currently, the confocal laser scanning microscopy (CLSM) facilitates imaging acquisition *in vivo* with the minimum invasiveness, leading to avoiding cosmetic scars in sensitive areas.

In this sense, CLSM working principle bases on differences in reflectivity induced by the endogenous skin chromophores (i.e., melanin) found in keratinocytes and melanocytes and contrasts with other skin structures such as keratin, collagen, and hemoglobin [46].

Particularly, clinical applications of CLSM have been used in the early diagnosis of skin cancer, leading to an opportune management of the disease and reducing both the mortality and morbidity of the patients, concomitantly. Among the skin cancers diagnosed by CLMS, melanoma and the nonmelanoma skin cancers stand out (basal cell carcinoma, actinic keratosis, and squamous cell carcinoma) [47, 48]. As well, *in vivo* CLSM is widely used for the diagnosis of contact dermatitis, since the conventional analysis is low accurate and yields significantly false-positive test results, offering a sensitivity/specificity higher as compared to the traditional evaluation of patch test [49]. As well, both benign non-melanocytic neoplasia (solar lentigo, seborrheic keratosis, lichen planus-like keratosis, sebaceous hyperplasia, clear cell acanthoma, lymphangioma) and benign melanocytic neoplasms (benign nevus, dysplastic nevus, Spitz nevus, blue nevus) are currently diagnosed by confocal microscopy with excellent results, since this tool enabled feasible results and an early diagnosis, thus avoiding pain [50]

Besides diagnosis confocal microscopy has become a useful tool for monitoring skin cancer therapies based on topical chemotherapy, photodynamic therapy, cryosurgery, or radiation therapy [45], suggesting a wide range of clinical applications such tool represent.

Moreover, confocal microscopy has also been used for the diagnosis of the most common pigmentary skin disorder, melasma, which is characterized by abnormal melanin deposits. In this sense, several studies report that the CLSM facilitates the diagnosis, follow-up, and the prognosis following the therapeutic interventions [51, 52].

Rare Diseases and Confocal Microscopy

Primary ciliary dyskinesia (PCD) appears in one in 15,000 individuals in the population; such disease is characterized by defective ciliary beating and reduced mucociliary clearance. The conventional diagnosis consists in the ciliary function assessment using high-speed microscopy and electron microscopy. However, such devices and experimented technicians are not easily available elsewhere. Hence, to make the PCD diagnosis more achievable, several research groups have focused on developing strategies based on the use of fluorescent antibodies to target ciliary proteins often altered in this disease. In this sense, in two cohorts (35 individuals diagnosed with PDC and 368 individuals suspected of PCD), confocal microscopy showed immunofluorescence antibodies that target altered proteins such as DNAL1, DNAH5, and RSPH4A, which are involved in ultrastructural defects including outer dynein arm defect and central complex/transposition defect [53].

Electron Microscopy

Several centuries after the development of the light microscope (sixteenth century), revolutionary studies for innovation in the field of microscopy (twentieth century) began. The history of electron microscopy was born thanks to the contributions of Louis de Broglie with his novel proposals that small particles, such as electrons, could behave as waves [54]; and Hans Busch, who developed electromagnetic lenses [55]. Busch suggested that the electromagnetic field generated in the short coil of a cathode ray tube could be used to direct electron beams in the same way that light passes through a convex optical lens. His work published in the academic journal *Archiv für Elektrotechnik* was read by Ernst Ruska, who together with Max Knoll in 1931 developed the first *T*ransmission *E*lectron *M*icroscope (TEM) prototype [56].

Although the idea of an electron microscope (EM) was revolutionary and innovative, no company wanted to risk commercializing it, despite having the ability to magnify a specimen 12,000 times; there was no commercial appeal. This completely changed when the young physician Helmut Ruska (younger brother of 1986 Nobel Laureate Ernst Ruska for the invention of EM) began to investigate the medical/

biological applications of the instrument and convinced his mentor, Dr. Richard Siebeck (director of the medical clinic of the Hospital Charité in Berlin), to make a review of the electron microscope. Siebeck wrote that the electron microscope would help him in his investigations into the causality of diseases and infectious agents that are not possible to solve with light microscopy; such would be the case with many viruses [57][58]. And as we will review later, this was the case.

Wave Particle Duality

Based on quantum mechanics, both electrons and light can behave as waves or particles; in both cases its speed and wavelength can be variable [54, 59]. Thinking particularly in the case of electron microscopes, it is essential to understand that the wavelength of an electron is a function of its speed. To decrease the wavelength of an electron, it is necessary to increase the acceleration voltage, and because of this decrease in the wavelength, the resolving power of the instrument increases.

Despite the differences in the source of "illumination" between optical and electron microscopes, in both cases the best resolution is sought, a good depth of focus and well-contrasted images. That is the heart of microscopy.

Unlike light microscopy, electron microscopes work with a beam of electrons. However, it is necessary to consider that these last instruments require that the specimen to be observed be subjected to a vacuum chamber, which will be radiated with the beam. For that reason, the preparation of a sample that will be observed in the TEM must be carefully carried out.

Components of the TEM

The TEM design hasn't changed much since the Ruska model in 1931. However, improvements have been implemented in the instrument correcting spherical and chromatic aberrations and astigmatism in electromagnetic lenses, improving the resolution, which was 100 nm in the first models and is now less than 0.1 nm (1 Angstrom, Å).

The principal components of the TEM are the electron gun, the column that contains diverse electromagnetic lenses and which work in a vacuum environment, the apertures, the sample holder, and the camera.

At the top of the TEM, the electron gun is located, and the electron beam travels its way along the column through the high vacuum. The column is made up of a series of electromagnetic lenses along the optical axis of the microscope. At the top of the column, there are at least two condenser lenses, the first (upper), allowing regulation of the size of the spot of the electron beam. Below this, the next condenser lens controls intensity of the beam; the function of both condenser lenses is to control the shape of the electron beam. Below is the objective lens; most TEMs are designed with a double objective lens composed of two electromagnetic fields, positioned so that they flank the sample; the top one allows additional control over the electron beam, while the bottom one magnifies the image about 50 times. Next are the projector lenses, fluorescent screen, and camera [Fig. 8.4].

The next parameters should be considered in each component of the TEM:

- *Electron gun*. The stability (variation of the electron current), brightness (the current density per unit solid angle), and coherence (electrons that are in the same phase and have the same wavelength from the origin) are determiners of the performance of the source of electrons.
- *Electromagnetic lenses*. Can present three types of aberrations: spherical aberration, chromatic aberration, and astigmatism. As a result of this aberration the image is obtained with a reduced quality. For this reason, the correct adjustment of the condenser and objective lenses is necessary. The quality of the electron beam is determined by condenser lenses, while the objective lenses can magnify the aberrations several thousand times [60].
- *Camera*. The TEMs have a coupled CCD camera, which translates the photons into a scintillation layer, in this way the electrons are registered indirectly, the thicker the scintillation layer the better sensitivity will be. The limitation of CCD cameras is that they are slow and require long exposure times to obtain a good quality image, although they remain limited for protein applications. However, the development of DDD (direct detection device) cameras has brought about an improvement in resolution, since DDDs are in the ability to detect electrons directly and consequently blur is avoided and the speed of capture is improved [61].

Sample Preparation

So far it is clear that the samples that you want to observe at TEM cannot be alive, since they will be subjected to high vacuum (which inevitably dehydrates the specimen) and to the radiation of the electron beam [Table 8.6]. Therefore, it is necessary to preserve them properly, since by having such a high resolution, the TEM is able to detect if the membranes of a cell were preserved correctly in order to make an accurate diagnosis in the case of pathological samples. The method chosen to prepare the sample must maintain the cellular state of the sample as it was in the native state; it must allow adequate contrast to be carried out for its observation, and it needs to have an achievable resolution. In the case of

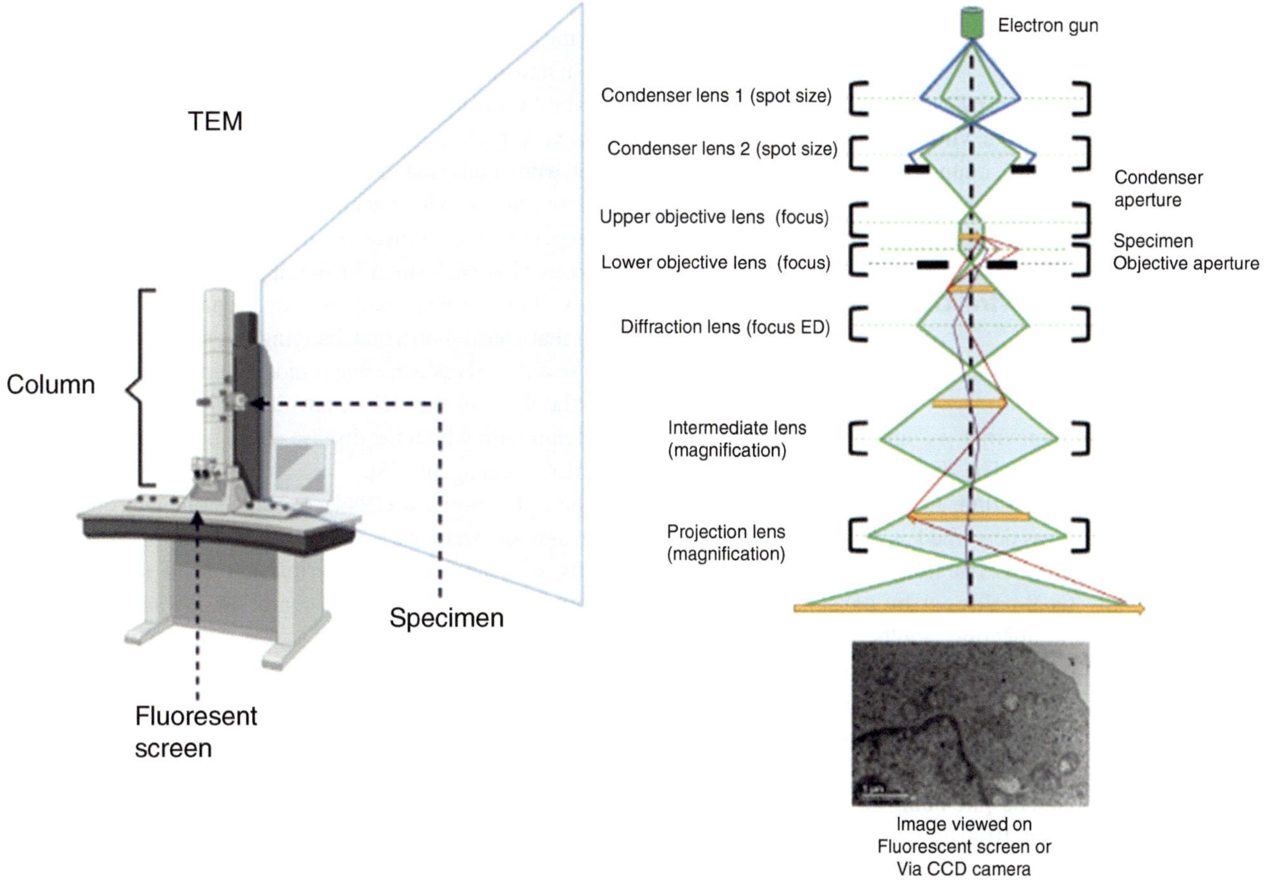

Fig. 8.4 *Diagram of a transmission electron microscope.* Schematic representation of a transmission electron microscope where the electron source is observed and the high vacuum column where the electromagnetic lenses and the fluorescent screen are housed (left side). Internal diagram of the TEM: The electron gun is represented by the green cylinder, the green lines represent the electron beam, on the sides of the beam path the different lenses are outlined with black brackets along with the name of the lenses and the function they perform in the TEM, and the thick black bars positioned vertically at the height of the condenser lens and the objective lens are the apertures. The different paths that scattered electrons can take are represented in red and purple. The yellow arrow is the specimen and its intermediate images (right side) (Modified from Franken et al. [61]). The TEM image at the bottom of the schematic was obtained at a JEOL-JEM-1200 by Rodolfo Díaz Paredes, IFC-UNAM

Table 8.6 Analysis techniques that can be performed in the TEM

Analysis method	Information provided
Electron diffraction	Crystalline structure
Bright field	Morphology, size, and distribution
Dark field	Morphology in relation to crystal structure
High-resolution or network images	Crystalline structure
Electron energy loss	Electronic composition and structure
X-ray energy	Elemental chemical composition

biological samples, staining is the best option to post fix, preserve, and observe the sample; additionally it protects the sample from dehydration and electron beam radiation and improves contrast.

The standard technique used in most laboratories that work with biological samples performs a chemical fixation with aldehydes (generally a mixture of buffered paraformaldehyde and glutaraldehyde), followed by a post fixation (regularly with osmium tetroxide), then dehydration of the sample and its inclusion in resins (generally epoxy). A very important intermediate step must be added to the preparation of samples and observation at TEM: the realization of semi-fine and ultrafine cuts using an ultramicrotome which will be mounted on the grids (with different mesh) and which will undergo different techniques such as negative staining or immunostaining with gold [62, 63].

However, even though TEM allows us to observe cell ultrastructure, it has the disadvantage that standard protocols have a long processing time, which can last between 3 and 5 days. It is important to mention that at present, there are rapid processing protocols, with which the sample preparation can be carried out in 6 h, which allows TEM to be a useful tool in the diagnosis of diseases [64].

The Role of the Electron Microscopy in Virus Diagnostic

The use of electron microscopy as a diagnostic tool has been a fundamental component of viral diagnosis [65]. From its invention to the present day, countless viruses have been discovered, allowing not only their correct classification but also the diagnosis of the diseases they cause.

The Role of the TEM in Viral Gastrointestinal Diseases

The Norovovirus

Norwalk virus or Norovirus is an RNA virus of the *Caliciviridae* family [66]. The disease caused by this virus was described for the first time in 1929 and called "winter vomiting disease"; it was described as the causative agent of nonbacterial gastroenteritis. The transmission of the virus is carried out mainly fecal-orally, or by consuming contaminated food or water, affecting adults and children [67]. Patients with this disease have a characteristic feature: vomiting and diarrhea; and it has a mean duration of illness of 12 to 60 hours and negative stool cultures (routine studies).

Despite attempts to characterize the etiological agent *in vitro*, it was not found despite being the second most common disease found in a 10-year family study [68]. In an attempt to find the etiologic agent, thought to be of viral origin, an immune electron microscopy protocol was adapted that had already proven useful in other virus findings. They used filtered stool from a patient who had developed viral gastroenteritis and inoculated ten volunteers with this filtrate, six of whom developed gastroenteritis. Then, inactivated serum from inoculated convalescent patients was used as a specific source of antibodies. Stool-serum mixtures were made and incubated for 1 hour. After processing the samples with several centrifugations, the mixtures were placed on 400 mesh copper grates covered with Formvar/carbon; observations were made at 400, 000x magnification. The results of the observation to the EM showed the particle aggregates of 27 nm in diameter, which were directly associated as the etiological agent of viral gastroenteritis [69].

Viral gastroenteritis caused by the Norwalk virus has seen outbreaks around the world, many of which have been diagnosed, in just 2 days, using the solid-phase immune electron microscopy technique [70–73]. This common disease is not diagnosed correctly due to its similarity to other gastrointestinal disorders, and it only attracts attention when outbreaks occur in populations, so the calculation of the impact it has is only an estimate [74].

The Rotavirus

Rotavirus is a double-stranded RNA virus of the *Reoviridae* family. The disease it causes is viral gastroenteritis in infants (0–5 years). An attempt was made to characterize the disease after the appearance of six separate epidemics of acute diarrhea in newborns in three hospitals in Baltimore-Washington air over 2 years (1941–1942). The last four epidemics (1942) had a very high mortality. During the activity of the cases, fluids were collected from sick infants (feces, blood, and nasal washing), which were injected into different animal species, but no conclusive results were obtained; stool culture was also performed in the appropriate media, and the results obtained were negative for organisms (known until then) that cause diarrhea. Changing the route of administration, several calves were again inoculated via the nasal route with the feces of the sick infants until the disease was replicated and with which the disease could be described, but not the etiological agent [75].

The following year (1944), transmission electron microscopy studies were carried out, using fecal samples from infants with acute gastroenteritis and fecal samples from newborn calves with the same condition. It was found that in both cases the causative agent was a virus whose both size and shape were indistinguishable between the samples taken from infants and calves. The virus found differed morphologically from the reoviruses and orbiviruses; it was proposed to call these viruses "rotavirus," due to their resemblance to a wheel. Different authors reported virion sizes between 65 and 75 nm; however, mean estimates were 72 nm [76].

The route of infection is oral-fecal; patients with viral gastroenteritis caused by rotavirus develop vomiting, watery diarrhea, and fever. It is a disease that can be fatal due to the high dehydration suffered by patients due to acute diarrhea. Unlike Norovirus, rotaviruses are cultivable viruses, which has allowed the development of rotavirus vaccines, which are currently used in global health programs [76, 77]. Currently it is diagnosed by molecular biology techniques such as ELISA and in some cases also by TEM.

The Role of TEM in Viral Respiratory Diseases

Throughout human history, many diseases evolved from animal pathogens that changed hosts to transform into human infectious agents. As human populations have migrated with increasing ease, these etiological agents have initiated epidemics and pandemics.

The Influenza Virus

From the last century to the present day, we are aware of five pandemics caused by influenza viruses. The first of them was registered in 1918 and the most important during the twentieth century, the so-called Spanish flu caused by the H1N1 virus and which caused the death of more than 50 million people around the world [78]. The first signs of the disease emerged in the military camps of the United States, which sent soldiers to Europe during the First World War. One of the theories of the emergence of the disease postulates that the virus changed host, possibly being the pigs that were used to feed the troops, the original carriers. The military authorities did not stop sending soldiers to the war front, so outbreaks quickly began in different parts of the world and the consequence was a pandemic that lasted 3 years. The virus circulating in 1918 mainly affected young adults. Despite worldwide concern, the etiologic agent remained unknown. Fifteen years later studies were conducted in ferrets, which allowed the virus to be isolated [79]. Molecular characterization was carried out in the 1990s [80], and it was not until the beginning of the twenty-first century that a fragment of the virus sequence was able to reconstruct the virus to evaluate its pathogenicity [Fig. 8.5] [81, 82].

In 1957 another pandemic was registered; this time it appeared in Asia and was caused by the H2N2 influenza virus, which originated from an avian influenza virus, causing more than one million deaths. In 1968, the H3N2 influenza virus claimed the lives of one million people. It originated in Hong Kong and later spread to other parts of the world. This virus is the causative agent of bird flu and swine flu [83, 84].

In 1977, there was a resurgence of the H1N1 virus; this time the age group of those under 25 years of age had mild symptoms. On this occasion, TEM images were made, which allowed for the elucidation of the appearance of the virion, the viral envelope, and the viral nucleus [85]. In 2009, the H1N1 influenza virus began a pandemic resulting in the death of 282,000 people. That same year we learned about the structure of this causal agent with images obtained by TEM [Fig. 8.5] [86].

The Role of Electron Microscopy and the Last Pandemic

Coronavirus epidemics have also been present throughout the history of humanity; however, we are currently facing the first pandemic of the twenty-first century.

The SARS-CoV-2 virus is the causal agent of the disease called COVID-19, which emerged at the end of 2019 in China and very quickly spread throughout the world. The most common symptoms experienced by COVID-19 patients are fever, dry cough, and fatigue; however, some patients also have diarrhea, conjunctivitis, headache, loss of sense of taste and smell, some skin rashes, and/or loss of finger color. It is a highly contagious virus that is transmitted from person to person and that infects the host for 10 to 14 days [87, 88].

Thanks to electron microscopy, we have been able to know the structure of the virus [Fig. 8.6], just a couple of months after the pandemic was declared. In addition to the viral structure, various studies have also been carried out on various organs that are infected by SARS-CoV-2, such as the lungs, brain, and heart, among others. Particularly in the case of this virus, the use of microscopy has made it possible to understand its pathogenesis and the consequences it leaves in the human body; however, it is not yet known exactly what sequelae it will leave in people with the disease [89–94].

Concluding Remarks

As it was demonstrated a long time ago, life is not limited to what we can observe with the naked eye; for more than a century, conventional light microscopy has been the basic tool for the evaluation of tissues, cells, and microorganisms, which is why it has played a fundamental role in pathological diagnosis; it has been tirelessly sought to overcome the limits of optical resolution in search of biological understanding; consequently, innovation in optical microscopy has been parallel to important advances in the understanding of the biological mechanisms. In addition, in many cases it is usually a tool for rapid clinical diagnosis and is a low-cost test.

With these technologies it has been possible to cover a wide variety of needs within the life sciences, since these technologies have provided tools to study the structure and functions of molecules, cells, and organisms; however, to expand the range of possibilities in the obtention of biological information, various types of microscopes have been designed, which allow, among other things, to increase the contrast and sharpness of the image, modifying either the lighting system, or the type of lenses used to form the images, or whatever another element that conforms the structure of the instrument. Based on the above considerations, dark-field, phase-contrast, confocal, and electronic microscopes, among others, have been created, being called microscopes with special applications.

In general terms in the clinical field, with the help of microscopy, it has been sought to understand the normal structure and function of the different biological samples to be studied, since it can be very useful to make a diagnosis and to determine the severity and progression of diseases; it has been used initially and is used today to diagnose infectious, degenerative, or neoplastic diseases. These qualitative diagnoses are based on a sum of observable changes in the morphology of the analyzed tissue. The cognition of these changes is based on the recognition of patterns by the observer and the comparison of these patterns with the known physiological variation in the morphology of the tissues in the respective species.

Additionally, it is important to mention that microscopy is about solving structures that are not observable with the naked eye. For this the approaches are varied; although light microscopy has a resolution of 200 nm, this is not a limitation, as new protocols are being developed for rapid sample preparation for electron microscopy. Many authors mention that the golden age of electron microscopy occurred during

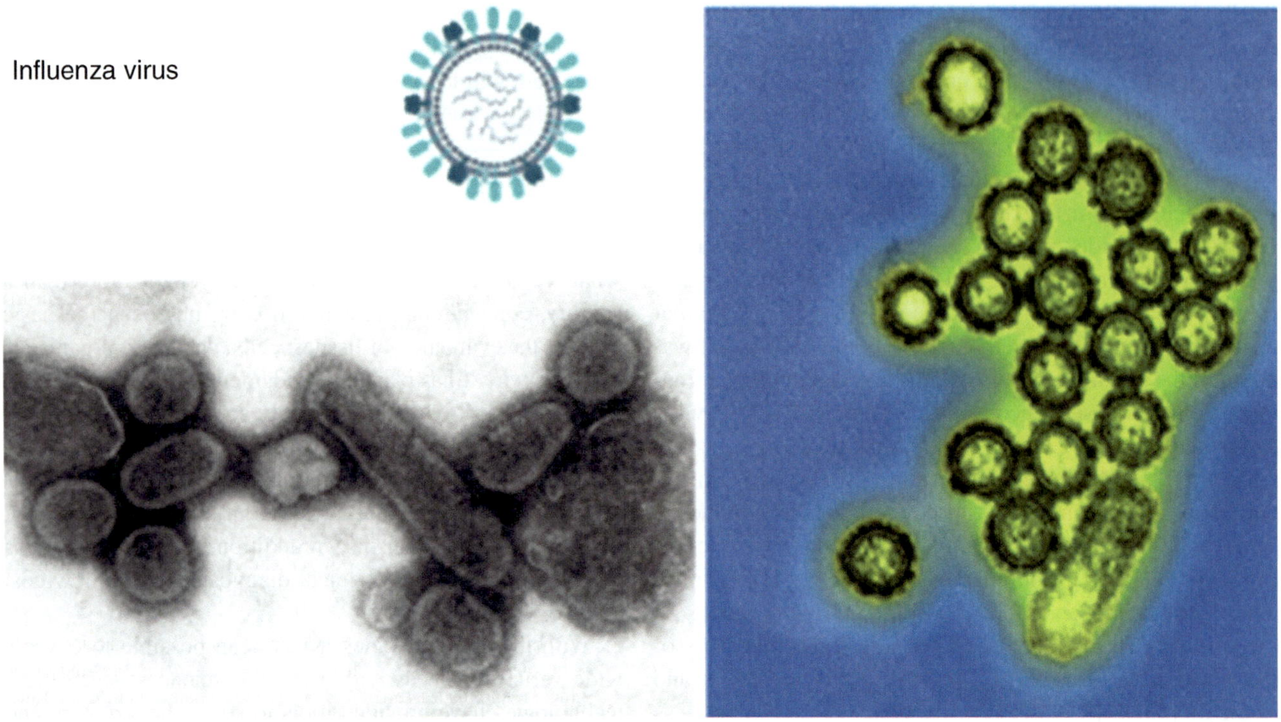

Fig. 8.5 *Micrograph of reconstructed influenza virus (Spanish flu, 1918)* (**a**), *versus H1N1 influenza virus from the 2009 pandemic* (**b**), *observed by TEM.* Photo credit: Cynthia Goldsmith Content Providers(s): CDC/ Dr. Terrence Tumpey/Cynthia Goldsmith – This media comes from the Centers for Disease Control and Prevention's Public Health Image Library (PHIL), with identification number #8243 (A), and digitally colorized transmission electron microscopic (TEM), H1N1 influenza virus particles. Contributed by the Public Health Image Library (PHIL) [86]

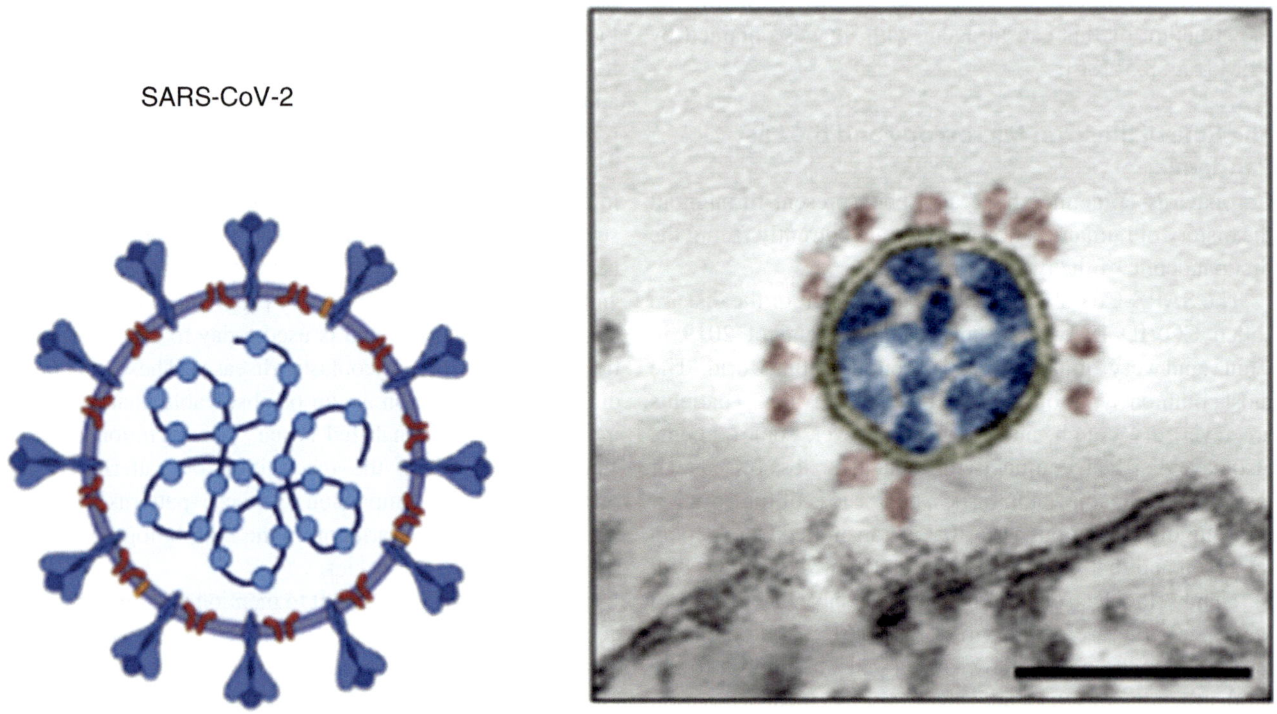

Fig. 8.6 SARs-CoV-2. Two-dimensional electron micrograph of a SARS-CoV-2 virus particle (scale bar, 100 nm). (With permission from Michael Laue [90])

the decades of the 1950s and 1960s. However, the challenge of COVID-19 has once again put this tool on the rise, indispensable in the study of viruses and cellular ultrastructure, allowing accurate diagnoses to be made.

Acknowledgments The authors would like to thank Rodolfo Paredes Díaz, María Fernanda Ramírez, and Dr. Michael Laue for allowing us to use their images to illustrate this chapter. We also thank the Public Health Image Library for keeping your images free. And finally, we want to thank to Dr. Augusto César Poot-Hernández for his technical assistance in the images performed. All images were created using the BioRender.com online program.

References

1. Sommerfeld J. Plagues and peoples revisited. Basic and strategic research for infectious disease control at the interface in the life, health and social sciences. EMBO Rep 4 Spec No:S32–4. 2003.
2. Dobson AP, Carper ER. Infectious diseases and human population history: throughout history the establishment of disease has been a side effect of the growth of civilization. Bioscience. 1996;46:115–26.
3. Diamond JM. Guns, germs and steel: a short history of everybody for the last 13,000 years. Random House; 1998.
4. Morens DM, Daszak P, Markel H, Taubenberger JK. Pandemic COVID-19 joins History's pandemic legion. MBio. 2020. https://doi.org/10.1128/mBio.00812-20
5. Medical microscopy. A guide to the use of the microscope in medical practice. JAMA. 1892; XIX:734.
6. Antony PPM, Trefois C, Stojanovic A, Baumuratov AS, Kozak K. Light microscopy applications in systems biology: opportunities and challenges. Cell Commun Signal. 2013;11:1–19.
7. Website. Clara Sue Ball. (1966). The early history of the compound microscope. Bios, 37(2), 51-60. Retrieved 29 April 2021, from http://www.jstor.org/stable/4606667. Accessed 29 Apr 2021.
8. Schulte EK. Standardization of biological dyes and stains: pitfalls and possibilities. Histochemistry. 1991;95:319–28.
9. Alturkistani HA, Tashkandi FM, Mohammedsaleh ZM. Histological stains: a literature review and case study. Glob J Health Sci. 2015;8:72–9.
10. Wick MR. The hematoxylin and eosin stain in anatomic pathology-an often-neglected focus of quality assurance in the laboratory. Semin Diagn Pathol. 2019;36:303–11.
11. Tripathi N, Sapra A (2021) Gram staining. StatPearls.
12. Kurup R, Chester K. Comparative evaluation of ziehl neelsen staining and knowledge, attitudes and practices of laboratory personnel in relation to ziehl nielsen. West Indian Med J. 2014;63:34–9.
13. Wittekind DH. On the nature of Romanowsky--Giemsa staining and its significance for cytochemistry and histochemistry: an overall view. Histochem J. 1983;15:1029–47.
14. Stockert JC, Blázquez-Castro A, Horobin RW. Identifying different types of chromatin using Giemsa staining. Methods Mol Biol. 2014;1094:25–38.
15. Ricciardi A, Ndao M. Diagnosis of parasitic infections: what's going on? J Biomol Screen. 2015;20:6–21.
16. Saco A, Ramírez J, Rakislova N, Mira A, Ordi J. Validation of whole-slide imaging for histolopathogical diagnosis: current state. Pathobiology. 2016;83:89–98.
17. Klopfleisch R. Multiparametric and semiquantitative scoring systems for the evaluation of mouse model histopathology--a systematic review. BMC Vet Res. 2013;9:123.
18. Gupta E, Bhalla P, Khurana N, Singh T. Histopathology for the diagnosis of infectious diseases. Indian J Med Microbiol. 2009;27:100–6.
19. Procop GW, Wilson M. Infectious disease pathology. Clin Infect Dis. 2001;32:1589–601.
20. Rentschler S, Kaiser L, Deigner H-P. Emerging options for the diagnosis of bacterial infections and the characterization of antimicrobial resistance. Int J Mol Sci. 2021. https://doi.org/10.3390/ijms22010456
21. Momčilović S, Cantacessi C, Arsić-Arsenijević V, Otranto D, Tasić-Otašević S. Rapid diagnosis of parasitic diseases: current scenario and future needs. Clin Microbiol Infect. 2019;25:290–309.
22. Srividya G, Kulshrestha A, Singh R, Salotra P. Diagnosis of visceral leishmaniasis: developments over the last decade. Parasitol Res. 2012;110:1065–78.
23. Robert-Gangneux F, Darde M-L. Epidemiology of and diagnostic strategies for toxoplasmosis. Clin Microbiol Rev. 2012;25:264–96.
24. Yansouni CP, Bottieau E, Lutumba P, et al. Rapid diagnostic tests for neurological infections in Central Africa. Lancet Infect Dis. 2013;13:546–58.
25. Lescure F-X, Le Loup G, Freilij H, Develoux M, Paris L, Brutus L, Pialoux G. Chagas disease: changes in knowledge and management. Lancet Infect Dis. 2010;10:556–70.
26. Wilson ML. Malaria rapid diagnostic tests. Clin Infect Dis. 2012;54:1637–41.
27. Hunfeld K, Hildebrandt A, Gray J. Babesiosis: recent insights into an ancient disease. Int J Parasitol. 2008;38:1219–37.
28. Weber R, Bryan RT, Bishop HS, Wahlquist SP, Sullivan JJ, Juranek DD. Threshold of detection of cryptosporidium oocysts in human stool specimens: evidence for low sensitivity of current diagnostic methods. J Clin Microbiol. 1991;29:1323–7.
29. Gonin P, Trudel L. Detection and differentiation of entamoeba histolytica and entamoeba dispar isolates in clinical samples by PCR and enzyme-linked immunosorbent assay. J Clin Microbiol. 2003;41:237–41.
30. Dietrich CF, Chaubal N, Hoerauf A, Kling K, Piontek MS, Steffgen L, Mand S, Dong Y. Review of dancing parasites in lymphatic filariasis. Ultrasound Int Open. 2019;5:E65–74.
31. Demeler J, Schein E, von Samson-Himmelstjerna G. Advances in laboratory diagnosis of parasitic infections of sheep. Vet Parasitol. 2012;189:52–64.
32. Cavanaugh C, Perazella MA. Urine sediment examination in the diagnosis and management of kidney disease: core curriculum 2019. Am J Kidney Dis. 2019;73:258–72.
33. Lichtman JW, Conchello J-A. Fluorescence microscopy. Nat Methods. 2005;2:910–9.
34. Conchello J-A, Lichtman JW. Optical sectioning microscopy. Nat Methods. 2005;2:920–31.
35. Bhutani J, Chakinala RC, Bhutani S, Sachdeva S. Endocrine and metabolic disease: confocal microscopy as a diagnostic aid. Indian J Endocrinol Metab. 2015;19:171–3.
36. Ragazzi M, Piana S, Longo C, Castagnetti F, Foroni M, Ferrari G, Gardini G, Pellacani G. Fluorescence confocal microscopy for pathologists. Mod Pathol. 2014;27:460–71.
37. Cavanagh HD, Petroll WM, Alizadeh H, He YG, McCulley JP, Jester JV. Clinical and diagnostic use of in vivo confocal microscopy in patients with corneal disease. Ophthalmology. 1993;100:1444–54.
38. You JY, Botelho PJ. Corneal in vivo confocal microscopy: clinical applications. R I Med J. 2016;99:30–3.
39. Papanas N, Ziegler D. Corneal confocal microscopy: a new technique for early detection of diabetic neuropathy. Curr Diab Rep. 2013;13:488–99.
40. Guthoff RF, Stave J. In vivo micromorphology of the cornea: confocal microscopy principles and clinical applications Essent Ophthalmol. 2006; 173–208.

41. Kallinikos P, Berhanu M, O'Donnell C, Boulton AJM, Efron N, Malik RA. Corneal nerve tortuosity in diabetic patients with neuropathy. Invest Ophthalmol Vis Sci. 2004;45:418–22.
42. Malik RA, Kallinikos P, Abbott CA, van Schie CHM, Morgan P, Efron N, Boulton AJM. Corneal confocal microscopy: a non-invasive surrogate of nerve fibre damage and repair in diabetic patients. Diabetologia. 2003;46:683–8.
43. Tavakoli M, Marshall A, Thompson L, Kenny M, Waldek S, Efron N, Malik RA. Corneal confocal microscopy: a novel noninvasive means to diagnose neuropathy in patients with Fabry disease. Muscle Nerve. 2009;40:976–84.
44. Tavakoli M, Hossain P, Malik RA. Clinical applications of corneal confocal microscopy. Clin Ophthalmol. 2008;2:435–45.
45. Que SKT, Grant-Kels JM, Rabinovitz HS, Oliviero M, Scope A. Application of handheld confocal microscopy for skin cancer diagnosis: advantages and limitations compared with the wide-probe confocal. Dermatol Clin. 2016;34:469–75.
46. Ulrich M, Lange-Asschenfeldt S. In vivo confocal microscopy in dermatology: from research to clinical application. J Biomed Opt. 2013;18:061212.
47. Gerger A, Koller S, Kern T, Massone C, Steiger K, Richtig E, Kerl H, Smolle J. Diagnostic applicability of in vivo confocal laser scanning microscopy in melanocytic skin tumors. J Invest Dermatol. 2005;124:493–8.
48. González S, Tannous Z. Real-time, in vivo confocal reflectance microscopy of basal cell carcinoma. J Am Acad Dermatol. 2002;47:869–74.
49. González S, Gilaberte-Calzada Y. In vivo reflectance-mode confocal microscopy in clinical dermatology and cosmetology. Int J Cosmet Sci. 2008;30:1–17.
50. Shahriari N, Rabinovitz H, Oliviero M, Grant-Kels JM. Reflectance confocal microscopy: melanocytic and nonmelanocytic. Clin Dermatol. 2021. https://doi.org/10.1016/j.clindermatol.2021.03.010.
51. Ardigo M, Cameli N, Berardesca E, Gonzalez S. Characterization and evaluation of pigment distribution and response to therapy in melasma using in vivo reflectance confocal microscopy: a preliminary study. J Eur Acad Dermatol Venereol. 2010;24:1296–303.
52. Kang HY, Bahadoran P, Suzuki I, Zugaj D, Khemis A, Passeron T, Andres P, Ortonne J-P. In vivo reflectance confocal microscopy detects pigmentary changes in melasma at a cellular level resolution. Exp Dermatol. 2009;19:e228–33.
53. Shoemark A, Frost E, Dixon M, et al. Accuracy of immunofluorescence in the diagnosis of primary ciliary dyskinesia. Am J Respir Crit Care Med. 2017;196:94–101.
54. de Broglie L. XXXV. A tentative theory of light quanta. Lond Edinb Dublin Philos Mag J Sci. 1924;47:446–58.
55. Busch H. Berechnung der Bahn von Kathodenstrahlen im axialsymmetrischen elektromagnetischen Felde. Ann Phys. 1926;386:974–93.
56. The Nobel Prize in Physics 1986 - Perspectives: Life through a Lens - NobelPrize.org. https://www.nobelprize.org/prizes/physics/1986/perspectives/. Accessed 28 Apr 2021.
57. Kruger DH, Schneck P, Gelderblom HR. Helmut Ruska and the visualisation of viruses. Lancet. 2000;355:1713–7.
58. Kruger DH, Mertens T. Classic paper: are the chickenpox virus and the zoster virus identical?: HELMUT RUSKA. Rev Med Virol. 2018;28:e1975.
59. Olszewski S. De broglie's velocity of transition between quantum levels and the quantum of the magnetic spin moment obtained from the uncertainty principle for energy and time. J Mod Phys. 2014;05:2022–9.
60. Danev R, Buijsse B, Khoshouei M, Plitzko JM, Baumeister W. Volta potential phase plate for in-focus phase contrast transmission electron microscopy. Proc Natl Acad Sci U S A. 2014;111:15635–40.
61. Franken LE, Grünewald K, Boekema EJ, Stuart MCA. A technical introduction to transmission electron microscopy for soft-matter: imaging, possibilities, choices, and technical developments. Small. 2020;16:e1906198.
62. Walcott B. Practical methods in electron microscopy. Volume 3. Part I: fixation, dehydration, and embedding of biological specimens. Audrey M. GlauertPractical methods in electron microscopy. Volume 3. Part II: Ultramicrotomy. Audrey M. Glauert. Q Rev Biol. 1976;51:207–8.
63. Biel SS, Gelderblom HR. Diagnostic electron microscopy is still a timely and rewarding method. J Clin Virol. 1999;13:105–19.
64. Schroeder JA. Ultrastructural pathology today – paradigm change and the impact of microwave technology and telemicroscopy. Diagnostic electron microscopy - a practical guide to interpretation and technique. 2013; 383–408.
65. Kapikian AZ. Overview of viral gastroenteritis. Arch Virol Suppl. 1996;12:7–19.
66. Zheng D-P, Ando T, Fankhauser RL, Beard RS, Glass RI, Monroe SS. Norovirus classification and proposed strain nomenclature. Virology. 2006;346:312–23.
67. Hedberg CW, Osterholm MT. Outbreaks of food-borne and waterborne viral gastroenteritis. Clin Microbiol Rev. 1993;6:199–210.
68. Dingle JH, Badger GF, Feller AE, Hodges RG, Jordan WS Jr, Rammelkamp CH Jr. A study of illness in a group of Cleveland families. I. Plan of study and certain general observations. Am J Hyg. 1953;58:16–30.
69. Kapikian AZ, Wyatt RG, Dolin R, Thornhill TS, Kalica AR, Chanock RM. Visualization by immune electron microscopy of a 27-nm particle associated with acute infectious nonbacterial gastroenteritis. J Virol. 1972;10:1075–81.
70. Cunney RJ, Costigan P, McNamara EB, Hayes B, Creamer E, LaFoy M, Ansari NA, Smyth NE. Investigation of an outbreak of gastroenteritis caused by Norwalk-like virus, using solid phase immune electron microscopy. J Hosp Infect. 2000;44:113–8.
71. Lewis DC. Three serotypes of Norwalk-like virus demonstrated by solid-phase immune electron microscopy. J Med Virol. 1990;30:77–81.
72. Lewis D, Ando T, Humphrey CD, Monroe SS, Glass RI. Use of solid-phase immune electron microscopy for classification of Norwalk-like viruses into six antigenic groups from 10 outbreaks of gastroenteritis in the United States. J Clin Microbiol. 1995;33:501–4.
73. Dolin R, Reichman RC, Roessner KD, Tralka TS, Schooley RT, Gary W, Morens D. Detection by immune electron microscopy of the Snow Mountain agent of acute viral gastroenteritis. J Infect Dis. 1982;146:184–9.
74. Kaplan JE, Feldman R, Campbell DS, Lookabaugh C, Gary GW. The frequency of a Norwalk-like pattern of illness in outbreaks of acute gastroenteritis. Am J Public Health. 1982;72:1329–32.
75. Light JS, Hodes HL. Studies on epidemic diarrhea of the new-born: isolation of a filtrable agent causing diarrhea in calves. Am J Public Health Nations Health. 1943;33:1451–4.
76. Flewett TH, Bryden AS, Davies H, Woode GN, Bridger JC, Derrick JM. Relation between viruses from acute gastroenteritis of children and newborn calves. Lancet. 1974;2:61–3.
77. Flewett TH, Woode GN. The rotaviruses. Arch Virol. 1978;57:1–23.
78. Mayor S. Flu experts warn of need for pandemic plans. BMJ. 2000;321:852.
79. Smith W, Andrewes CH, Laidlaw PP. A virus obtained from influenza patients. Lancet. 1933;222:66–8.
80. Taubenberger JK, Reid AH, Krafft AE, Bijwaard KE, Fanning TG. Initial genetic characterization of the 1918 "Spanish" influenza virus. Science. 1997;275:1793–6.
81. Tumpey TM, García-Sastre A, Taubenberger JK, et al. Pathogenicity of influenza viruses with genes from the 1918 pandemic virus: func-

tional roles of alveolar macrophages and neutrophils in limiting virus replication and mortality in mice. J Virol. 2005;79:14933–44.
82. Tumpey TM, Basler CF, Aguilar PV, et al. Characterization of the reconstructed 1918 Spanish influenza pandemic virus. Science. 2005;310:77–80.
83. Cohen J. Swine flu strain with human pandemic potential increasingly found in pigs in China. Science. 2020. https://doi.org/10.1126/science.abd5761
84. Pandemic flu and avian flu. PsycEXTRA Dataset. 2006. https://doi.org/10.1037/e433532008-006.
85. Wrigley NG. Electron microscopy of influenza virus. Br Med Bull. 1979;35:35–8.
86. Boktor SW, Hafner JW, Doerr C. Influenza (Nursing). StatPearls. 2020.
87. Kansas. Board of Emergency Medical Services (2020) KBEMS Guidance - 2019 Novel Coronavirus (COVID-19): Ambulance Staffing (Dated Material).
88. WHO Coronavirus Disease (COVID-19) Dashboard. Bangladesh Physiother J. 2020. https://doi.org/10.46945/bpj.10.1.03.01.
89. Melms JC, Biermann J, Huang H, et al. A molecular single-cell lung atlas of lethal COVID-19. Nature. 2021. https://doi.org/10.1038/s41586-021-03569-1
90. Laue M, Kauter A, Hoffmann T, Möller L, Michel J, Nitsche A. Morphometry of SARS-CoV and SARS-CoV-2 particles in ultrathin plastic sections of infected Vero cell cultures. Sci Rep. 2021;11:3515.
91. Pesaresi M, Pirani F, Tagliabracci A, Valsecchi M, Procopio AD, Busardò FP, Graciotti L. SARS-CoV-2 identification in lungs, heart and kidney specimens by transmission and scanning electron microscopy. Eur Rev Med Pharmacol Sci. 2020;24:5186–8.
92. Prasad S, Potdar V, Cherian S, Abraham P, Basu A, ICMR-NIV NIC Team. Transmission electron microscopy imaging of SARS-CoV-2. Indian J Med Res. 2020;151:241–3.
93. Hopfer H, Herzig MC, Gosert R, Menter T, Hench J, Tzankov A, Hirsch HH, Miller SE. Hunting coronavirus by transmission electron microscopy – a guide to SARS-CoV-2-associated ultrastructural pathology in COVID-19 tissues. Histopathology. 2021;78:358–70.
94. Möller L, Holland G, Laue M. Diagnostic electron microscopy of viruses with low-voltage electron microscopes. J Histochem Cytochem. 2020;68:389–402.

Physiomics and Phenomics

9

José Alberto Avila-Funes
and Virgilio Alejandro Hernández-Ruiz

Abbreviations

CHARGE	Cohorts for Heart and Aging Research in Genomic Epidemiology
DNA	Deoxyribunocleic Acid
IUPS	International Union of Physiological Sciences
MRI	Magnetic Resonance Imaging
NIH	National Institutes of Health
RNA	Ribonucleic Acid

Physiomics

The term physiome, as well as physiomics, first appeared in the 1990s, shortly after the boom of the other "omics." Physiomics can be defined as the systematic study of the physiome in biology. Likewise, the term physiome (coming from the roots "*physi*" meaning "life" and "*ome*" meaning "as a whole") intends to provide a quantitative description of the physiological dynamics and functional behavior of an intact organism and is built upon information and structure (genome, proteome, and the morpheme) [1, 2]. Stated differently, an organism reacts to its environment and its stimuli by numerous dynamic and intricately orchestrated responses (i.e., the genome and proteome, create the metabolome). So, the physiome could be seen as the complete system integrating those responses from a cellular to organism level, portending an inclusive framework of an organism's physiological processes [3].

In the last decades, the diverse branches of biology have provided an extremely detailed repository of the diverse components of a living human being, yet a limited understanding of how all those parts continuously interact and integrate. Thus, physiomics seek an understanding of the interaction between the physiological phenotypes of genes, expressed proteins, and their underlying networks [4]. For this task, physiomics need to rely on exhaustive databases and bioinformatics for the construction and analysis of the networks between diverse genes and its proteins, so, it may also be seen as a mean to integrate biology with informatics and its complex systems approach.

Thus, the ultimate objective of research in the field of physiomics can be the integration of biological and physiological information into complex mathematical models as a way to untangle an individual's physiology with the goal of surmounting disease [5].

This ambitious goal can only be envisioned by the progress that the diverse disciplines surrounding physiomics have had in the previous years, starting by computational science.

Innovations in computational sciences have profoundly impacted the fields of mathematics, engineering, and physics. Subsequently, those improvements (particularly in complex computational algorithms as they need to consider continuous multisystem interactions) are being collectively applied into biophysical models of human physiology that try to integrate the immense amount of data hierarchically derived from the rest of the "omics" (i.e., genomics, transcriptomics, proteomics, glycomics, etc.) and systems biology into the living structure of the human body [6]. As this colossal feat needs to access and integrate databases containing interactive models that span genes, proteins, functional cells and their structures, functional tissues and their structures, functional organs and their structures, and finally, the whole body, diverse publicly available models and open-source software repositories are being created [7].

J. A. Avila-Funes (✉)
Department of Geriatrics, Instituto Nacional de Ciencias Médicas y Nutrición Salvador Zubirán, Mexico City, Mexico
e-mail: alberto.avilaf@incmnsz.mx

V. A. Hernández-Ruiz
Univ. Bordeaux, Inserm, Bordeaux Population Health Research Center, UMR 1219, Bordeaux, France

J. C. Gomez-Verjan, N. A. Rivero-Segura (eds.), *Principles of Genetics and Molecular Epidemiology*,
https://doi.org/10.1007/978-3-030-89601-0_9

Practical Applications of Physiomics

While a whole-body physiological model of the human being may not be available yet, currently there are diverse practical applications for physiomic-based models.

Drug Research

Pharmaceutical research was one of the first disciplines that found applications for physiomic-based approaches. Particularly, potential oncological treatments have been studied using cell cycle models and the effects that the treatments will have on them [8]. This process differs from conventional drug research as the used cellular models may incorporate signature abnormalities in the cell cycle, and the expression of abnormal molecules and proteins, to estimate the effect of the potential treatment [9].

Organ Models

Physiomical-styled models (accounting for physical conservation laws) of diverse organ systems are being developed or are already available. As for other physiomic-based projects, the goals are to construct simulations that incorporate not only anatomically accurate structures but also their respective cellular components and expressed proteins [7]. Some of the more advanced projects include the heart, lungs, musculoskeletal, and digestive system models. To exemplify the intricacies of the physiomic approach in an organ, we can take one the collaboratively developed models of the heart. For the model to work properly, it needs to informatically integrate the geometrical and structural properties of the myocardium and connective tissue, membrane currents, ion channels, voltage changes, propagation of the electrical excitation, blood flow and oxygen delivery into the coronary vessels, etc. [10, 11]. These models hold an important promise for the better understanding and treatment of organic specific conditions (i.e., arrythmias, ischemia, or drug toxicity at heart level).

"The Human Physiome Project"

The "Physiome Project" was presented along with the working definition of *physiome,* as well as its intention to provide a quantitative description of physiological dynamics and functional behavior of the intact organism, as a report from the Commission on Bioengineering in Physiology to the International Union of Physiological Sciences (IUPS) Council at its 32nd World Congress in Glasgow, UK, in 1993. The rationale of the project can be understood by contrasting it with the Human Genome Project, in which the main objective was to dissect and describe each of the base pairs that conforms to human DNA. On the other hand, the Physiome Project intends to describe how every element on the human body works integrally in a complex, yet orchestrated fashion (the Human Physiome Project) [7].

Since its presentation, the Physiome Project has been striving to promote investigation that deepens the knowledge on how each component of the human body works as a component of a whole. In the same vein, diverse institutions are continuously contributing to develop computational and mathematical modelling frameworks that will integrate all those levels of human biology. With those tools one of the aims of the Physiome Project is to eventually produce an integral virtual physiological human (accounting for individual variability), with all the advantages that would bring to diverse fields of research including precision medicine.

This previous concept is important since diseases that affect the human body (e.g., cancer, hereditary diseases, neurological conditions) equally behave in a complex manner, affect diverse domains, and have numerous interactions with an individual's environment and lifestyle. However, clinical use of physiomic models remains limited as individual physiological systems have not been largely studied.

Phenomics

Our next *omic* of interest is phenomics, which is the acquisition of high-dimensional phenotypic data on an organism-wide scale, or the systematic study and analysis of qualitative and quantitative characteristics of the phenome [12]. The phenome is composed of the measurable traits that result from the often-complex interactions between genes, epigenetics, environmental, and stochastic factors [13]. Those measurable characteristics of physical, chemical, and biological phenotypes of an organism may span from mechanisms underlying genomic architecture and its regulatory pathways, the proteome, metabolome, and cellular features to the developed organs at the organism level.

Given its complex interests and objectives, phenomics (as well as physiomics) is considered a trans-discipline that relies on other fields as biology, physiology, epidemiology, computational and data sciences, as well as engineering. Those complementary disciplines allow us to consider and interpret the influence of the many potential sources of variation on the phenome.

Among the many potential advantages of phenomics is that it may assist, for example, in the tracing of complex causal links between a given genotype, a set of environmental factors, and a phenotype, which has been referred to as a *genotype-phenotype* map [13, 14].

Establishing potential causal explanations for a given phenotype is an alluring matter for different fields due to its relation to diverse important outcomes of interest like comorbidities in human beings, efficiency, adaptability, and resistance of seeds and plant life (or animals under domestication). Hence, in-species or populational variations of a phenotype that confer a certain risk or adaptation for a condition is one of the principal promises of phenomics. However, as previously mentioned this demands a transdisciplinary approach since potential causal interactions between a trait and an outcome can be complex in nature, for example, the effect of being obese on cancer risk.

For the purpose of a broad categorization of phenotypes, phenomics rely on some tools and technologies that already exist, yet that need to be improved for large-scale measurements, as they need to capture the whole spectrum of an organism (e.g., hierarchically spanning from DNA/RNA and molecular structures to physiological, morphological traits). These tools include transcriptomics, epigenomics (which are the most comprehensive sources for phenomic data), proteomics, metabolomics, and imagery techniques as spectroscopic imaging or MRI (even traditional microscopy). Among others, one of the currently needed improvements on phenomics-related measurements is the domain of gene expression. It still represents a great challenge to phenomics as it may change in function of cellular type and the temporal point on which it measured; hence, the construction of extensive datasets concerning this domain remains a test yet to overcome.

A phenomics project in humans is already under way, with phenotypes being longitudinally examined and measured in diverse studies. One of the pioneer fields interested in phenomic applications is cardiovascular medicine. Cohorts like the Framingham Heart Study and the Cohorts for Heart and Aging Research in Genomic Epidemiology (CHARGE) have already recorded a vast amount of phenomic data concerning thousands of their participants, as well as their medical history and environment-related variables. This, in conjunction with the use of data sciences and novel informatics techniques to integrate data from diverse *omics*, will provide new insights into cardiovascular disease [17].

Another example of an interdisciplinary and multicentered phenomics project is the Consortium for Neuropsychiatric Phenomics (CNP), funded by the NIH, in which genomic data and structural and functional information related to the central nervous system is being conducted in a case-control fashion for three psychiatric syndromes [18].

These projects, as well as the phenomic effort in general, may change current ways or visions on how to use experimental models to uncover the influence of genetic and environmental stimuli on a given phenotype. Likewise, they may represent an improvement on how to test causal hypothesis for different diseases and environmental exposures. In the long run, robust phenomics may even improve our understanding of complex longitudinal events, as aging, and their influence on health and disease.

Practical Applications of Phenomics

The agricultural sector and plant sciences share an interest in phenomics given the ever-growing food supply needs amidst a continuously changing global environment. As an increase in crop demand is expected over the coming years, crop performance and productivity are key factors to consider when assuring its supply [15]. However, as previously mentioned, a crop's phenotype is heavily influenced by its genotype and its surrounding environment. Hence, crop phenomic research combines agronomy, data sciences, mathematics, and engineering sciences to explore phenotypic information of crops and their complex environmental interactions in order to develop new methods of mining genes associated with relevant agronomic traits for precision breeding [16]. Diverse associations as the International Plant Phenotyping Network (IPPN) have been established to promote those purposes.

There are other phenome projects focused on insects (*Drosophila* Genome Reference Panel), small rodents and mammals as mice (Mouse Phenome Database), and dogs (Canine Phenome Project).

Closing Remarks

The physiome hierarchically entails the genome, the expressed proteome, metabolome, and their continuous interactions. Physiomics as one of the most recent *omics* benefits from diverse disciplines, including bioinformatics, to build physiological networks between the rest of those *omics* instead of focusing on a reduced scope or only one domain.

Phenomics work with the expressed phenotypes in organisms. The expressed phenotypes in organisms range from subcellular structures to a given physiological state and may change over time, and in relation to diverse stimuli (environment). However, causal relations between a given phenotype and the conditions surrounding it may be constructed with the appropriate input of diverse disciplines. Longitudinally, this may help to improve our understanding of the relation between diverse exposures and their potentially associated outcomes.

On the same vein, molecular epidemiology could be one of those disciplines whose input may help bridging the gap to causal relations. As one of its main interest is the analysis between given genetic and environmental risk factors at molecular level, the integration of powerful tool such as physiomics and phenomics to etiologic models of diseases may not only enhance the understanding of the subcellular origins of medical conditions but also the eventual development of preventive strategies on individual and populational level.

References

1. Hunter PJ, Borg TK. Integration from proteins to organs: the Physiome Project. Nat Rev Mol Cell Biol. 2003;4(3):237–43.
2. Bassingthwaighte JB. Strategies for the physiome project. Ann Biomed Eng. 2000;28(8):1043–58.
3. Welch GR. Physiology, physiomics, and biophysics: a matter of words. Prog Biophys Mol Biol. 2009;100(1–3):4–17.
4. Gomase VS, Tagore S. Physiomics. Curr Drug Metab. 2008;9(3):259–62.
5. Shim EB, Heldt T, Leem CH. Eds. Clinical applications of physiome models. Lausanne: Frontiers Media SA. 2020: https://doi.org/10.3389/978-2-88963-919-9
6. Karahalil B. Overview of systems biology and omics technologies. Curr Med Chem. 2016;23(37):4221–30.
7. Project P. Physiome Project. 2021. Available from: http://physiomeproject.org/data-repositories.
8. Chassagnole C, Jackson RC, Hussain N, Bashir L, Derow C, Savin J, et al. Using a mammalian cell cycle simulation to interpret differential kinase inhibition in anti-tumour pharmaceutical development. Biosystems. 2006;83(2–3):91–7.
9. Park IW, Reddy MV, Reddy EP, Groopman JE. Evaluation of novel cell cycle inhibitors in mantle cell lymphoma. Oncogene. 2007;26(38):5635–42.
10. Noble D. Modeling the heart--from genes to cells to the whole organ. Science. 2002;295(5560):1678–82.
11. Legrice I, Hunter P, Young A, Smaill B. The architecture of the heart: a data-based model. Phil Trans R Soc A [Internet]. 2001;359:1217–32.
12. Houle D, Govindaraju DR, Omholt S. Phenomics: the next challenge. Nat Rev Genet. 2010;11(12):855–66.
13. Bilder RM, Sabb FW, Cannon TD, London ED, Jentsch JD, Parker DS, et al. Phenomics: the systematic study of phenotypes on a genome-wide scale. Neuroscience. 2009;164(1):30–42.
14. Waddington CH. Towards a theoretical biology. Nature. 1968;218(5141):525–7.
15. Fahlgren N, Gehan MA, Baxter I. Lights, camera, action: high-throughput plant phenotyping is ready for a close-up. Curr Opin Plant Biol. 2015;24:93–9.
16. Zhao C, Zhang Y, Du J, Guo X, Wen W, Gu S, et al. Crop phenomics: current status and perspectives. Front Plant Sci. 2019;10:714.
17. Joshi A, Rienks M, Theofilatos K, Mayr M. Systems biology in cardiovascular disease: a multiomics approach. Nat Rev Cardiol. 2021;18(5):313–30.
18. Institute US. Consortium for neuropsychiatric phenomics. 2007. Available from: http://www.phenomics.ucla.edu.

Digital Health and Physiomics

10

Oscar Salvador Barrera-Vázquez,
Nadia Alejandra Rivero-Segura,
and Juan Carlos Gomez-Verjan

Abbreviations

COVID-19	Coronavirus Disease 19
DNA	Deoxyribonucleic Acid
GIS	Geographic Information Systems and Methods
ISF	Interstitial Fluid
SARS-Cov	Severe Acute Respiratory Syndrome Coronavirus
SARS-Cov2	Severe Acute Respiratory Syndrome Coronavirus 2
SM	Social Media
WBS	Wearable Biosensors
WHO	World Health Organization
WS	Wearable Sensors

Introduction

The classical health model is based primarily on the provision of medical services through the systems of hospitals and outpatient clinics. Several factors of the health services influence the quality of this model, such as the qualification of medical personnel, hospital facilities, and the availability of updated equipment [1]. It should be noted that although this model may vary from country to country, the basic principles are the same. The first principle is a "patient-oriented" approach and supporting infrastructure that provides optimal access to healthcare. In our days, platforms have experienced new challenges due to the rapid growth of technologies and the demand of the population for a high-quality medical service. These new digital technologies have offered the possibility of expanding the potential of various tools and of various diagnostic and therapeutic systems [2].

The implementation of digital medical technologies is intended to provide better access and flexibility to healthcare for the general public. This includes better availability of information about health, treatment, complications, and biomedical research on the Internet [1]. This chapter discusses recent trends and achievements in the field of digital medical technologies, social media, wearables and biosensors, their various applications, and their relationship to the physiome to solve real problems in healthcare.

Digital Health

Digital health is an emerging field of study at the intersection of healthcare and digital technologies, with high impact in the last decade. In 2019, the American Medical Association has reported that companies have made a large capital investment in new digital health endeavors [3]. Digital health is considered by the US Food and Drug Administration as a wide range of technologies, which include mobile health, wearable devices, telehealth and telemedicine, health information technologies, and personalized medicine [3]. Furthermore, the WHO emphasizes that digital health can be beneficial in achieving the Sustainable Development Goals by making health and wellness services accessible to high standards for all people around the world [3]; however, concept of digital health continues to evolve [4].

In the past decade, the potential for digital health intervention has grown rapidly through the use of devices such as

O. S. Barrera-Vázquez (✉)
Departamento de Farmacología, Facultad de Medicina, Universidad Nacional Autónoma de México (UNAM), Mexico City, Mexico

N. A. Rivero-Segura
Dirección de Investigación, Instituto Nacional de Geriatría (INGER), Instituto Nacional de Geriatría, Ciudad de México, Mexico
e-mail: nrivero@inger.gob.mx

J. C. Gomez-Verjan
Dirección de Investigación, Instituto Nacional de Geriatría (INGER), Ciudad de México, Mexico
e-mail: jverjan@inger.gob.mx

J. C. Gomez-Verjan, N. A. Rivero-Segura (eds.), *Principles of Genetics and Molecular Epidemiology*,
https://doi.org/10.1007/978-3-030-89601-0_10

laptops, tablets, smartphones, and wearable devices, and our understanding of how digital interventions can be theorized has improved [5]. Digital health technologies are now being applied more widely within medicine to improve diagnosis, treatment, clinical decision support, care management, and care delivery. Currently with regard to mobile health applications, there are around 3,000,000 health applications with more than 200 health applications added daily [5]. This makes it more apparent that there is an interest within medical care (patients, providers, payers, industry, and regulators) with respect to digital health, so there is a recurring challenge in finding solutions that provide real value [4].

Wearables and Biosensors

Due to the growing increase in the population, the increase in life expectancy, and as a consequence, aging and chronic diseases have had an economic impact with respect to medical care, and due to this reason, the medical care system has led to carry out a transformation from the traditional hospital-centered system to an individual-centered system [6]. Since the last century, portable sensors have been included within biomedical and healthcare monitoring systems, allowing continuous measurement of critical biomarkers to monitor the condition and health of the disease, medical diagnosis, and evaluation in biological fluids [6] like saliva, blood, and sweat. Currently, the developments of these devices have focused on electrochemical and optical biosensors, along with advances in the noninvasive monitoring of biomarkers, bacteria, and hormones, among others.

Technological advances in wearable sensors (WS) and wearable biosensors (WBS) have received great relevance due to their ability to collect useful information in real time about the health of an individual and their high specificity, portability, data acquisition speed, low costs, and low energy consumption that have improved over time [6].

WS are used to monitor various processes, such as body movements, or signals from the environment outside the body, such as exposure to vapors or environmental toxins. The WBS, on the other hand, have a much higher specificity and are characterized by having biological recognition sensors, allowing the specific detection of some type of ion, molecule, enzyme, cell receptor, antibody, or organelle in biological fluids [7], such as sweat, interstitial fluid (ISF), tears, or saliva through enzymatic, electrochemical, or colorimetric (optics) reactions [6].

Thanks to innovation and the latest advances in materials science and the development of mechanical engineering and wireless communication technologies, we have been able to develop portable devices (watches, straps, etc.) to simultaneously process and analyze biomarkers to improve health management [7, 8] . In addition, it has been seen that as the population has aged, the evidence of food safety and disease outbreaks has increased, so that the sale in the market of wearable technology is expected to increase to 70 billion US dollars in 2025 due to its ease of use [9].

A biosensor consists of two fundamental elements: a specific biological receptor in charge of the selective recognition of the analyte (enzyme, antibody, DNA, nucleic acid, peptide) and a transducer (such as optical, electrochemical, piezoelectric, and thermal) whose function is to convert the detected signal into a useful signal [10, 11]. Originally, the first biosensor devices were designed and developed for single-use or in vitro measurements, such as the glucometer and glucose test strips. In addition, advancement in biosensor technologies has paved the way to initiate improvements in modern portable biosensors for noninvasive monitoring in biomedical and healthcare applications [11].

In wearable devices, the main component is wearable sensors which have built-in functions for measuring identified markers in order to solve problems in the field of health, medicine, and sports [6]. According to their different measurement parameters, the WBS are classified into biophysical, biochemical, and state of motion sensors. Motion status sensors are used to measure human physical parameters such as gait, sleep, and tremor for real-time monitoring and collection of long-term information [12, 13]. With integrated laboratory-on-chip technology, portable biochemical sensors are used to measure the trace and run of different samples in parallel [6]. Portable biochemical sensors are used to accurately measure biomarkers in biological fluids in order to monitor health conditions and metabolism. The characteristics of portable biophysical sensors are determined by skin contact to provide real-time measurement of biophysical parameters such as blood pressure, heart rate, and temperature, which have significant values in healthcare applications [6]. It should be noted that of the different types of sensors, only the biophysical and state of motion are circulating in the market and are widely used by consumers, in contrast to biochemical biosensors, and due to their nature of having significant potential since biological fluids are complex and challenging matrices to detect the analyte of interest, they are not yet commercialized to the general public [14]. In addition, in recent years there has been an increase in the use of portable devices. In 2015, around 500 different healthcare-related wearable devices were found circulating in the market and more than 34.3 million of these devices were sold. This amount in devices sold is equivalent to three times the number sold in 2013 [15].

Social Media and Health

Since ancient times, humanity has thrived in social communities where each member shared so much knowledge, opinions, and experiences. As humanity advances in terms of technology, social media (SM), which are defined as “a group of Internet-based applications (apps) that allow the creation and exchange of user-generated content," have had an impact on the lives of millions of people [16]. However, the definition of SM is always evolving; the *Merriam-Webster Dictionary* defines it as "any form of electronic communication through which users create web-based communities to share information, personal messages, ideas, and other content such as photos and videos" [17].

The SM has become a tool for healthcare by allowing its users to acquire and share information, connect with others in the field, and communicate with colleagues, patients, or the public about health issues. Additionally, SM has supported the patient to expand their knowledge and place them in a position where they can take control of their own healthcare needs [18], including the recent use of SM during the COVID-19 pandemic. On the other hand, SM have gained relevance within the current scenario because the use of SM and social networking sites (SNS) is increasing around the world, especially in the healthcare industry [16].

Social Media in Healthcare

With the increasing orientation of the world toward digital, the healthcare industry sees SM as an important channel for promoting healthcare and employment, attracting new clients or patients, marketing for healthcare professionals, and building a captivating brand. On the other hand, health professionals have observed that SM goes much further than a platform for posting vacation photos and interacting with followers. Possibly the four most common areas in which SM has a strong influence in the healthcare industry are health promotion, research, marketing and branding for the public and practices, and recruitment. Also the SM has impacted on doctor-patient relationships as patients better understand health information and play a more active role in maintaining it [18]. SM have played an important role in public health surveillance in different aspects such as epidemiological surveillance and monitoring, awareness of the situation during emergency response, and communications surveillance.

Monitoring and Retrieval of Official Information

The use by public health officials has been seen to monitor official information disclosed by foreign authorities and to monitor national official accounts [19].

Disease Detection

Social networks can function as additional data sources for public health surveillance because they serve to detect disease outbreaks and estimate their incidence. This syndromic surveillance is carried out by detecting, through human readers or computer algorithms or through participatory epidemiology, the symptoms revealed by people on social networks for purposes not related to public health, where the applications allow participants to self-report their symptoms to patients, as has happened with the current COVID-19 pandemic [20]. On the other hand, the circulation of unofficial information or rumors about a new disease has served to detect it through surveillance based on events, as has happened with a patient with influenza A H7N9 uploaded to Weibo in 2013 [21], which was broadcast through an SM. On the other hand the official sites in SM of print media, radio, and television can be detected by surveillance systems based on events in order to find news about diseases [22]. In addition, digital data sources also provide epidemiologists with additional means to detect, investigate, and verify outbreaks.

Timely Estimates and Forecasts of Disease Incidence

Epidemiologists are currently conducting searches for the use of social media and other digital data to generate timely estimates and forecasts of disease incidence. This has been the case for influenza-related Twitter data which could generate timely incidence estimates, as it was found to correlate with seasonal influenza data in the United States of America (USA). The Wikipedia access log data also showed potential for the prognosis of certain infectious diseases in some countries. Advanced forecasting methods are being developed, and some use digital data as experimental inputs.

Situational Awareness During Emergency Response

Social media is also useful in situational awareness of humanitarian crises after natural or man-made disasters. In the event of dangerous situations, people have resorted to using social media to seek help and connect with family, friends, and first responders, and the authorities can use SM to identify people in danger and be able to help them. On the other hand, nongovernmental organizations have used SM for the purpose of tracking and mapping the needs of displaced people, as seen with the 2011 earthquake and tsunami in Japan [23] and the earthquake in Haiti in 2010 [24].

Communication Surveillance

Global Awareness

Social media data can also be used as a complement to more traditional methods for global awareness of disease outbreaks, as trends in SM can help quantify changes in awareness of the disease among people, users, and feelings toward treatments and preventive interventions [25].

Reaction to Public Health Campaigns and Messages

Analyses of SM data related to specific health promotion events have provided insights into useful information for public health professionals as they evaluate their campaigns [26].

Applications

According to the WHO in the upcoming years, the use of wearables monitoring our health will increase significantly and will revolutionize the way that people achieve higher standards of health. In fact, according to the WHO the so-called digital health provides the opportunity to improve the well-being since individuals will attend promptly to their health. Hence, in this section we will delve into the outstanding programs, initiatives, and clinical applications that are currently ongoing.

Recently, the WHO has developed a global strategy, named Global Strategy on Digital Health 2020–2024, which focuses on the promotion of healthy lives and well-being through the integration of technological resources [27]. In this context, several programs and initiatives have been powered by the WHO in collaboration with other institutions or corporations. For instance, the *Digital Health Technical Advisory Group and Roaster of Experts* developed by the WHO in collaboration with the public institutions, businesses, social enterprise, and investors aims to lead a strategy focused on advancing the universal digital health coverage (https://www.who.int/health-topics/digital-health/dh-tag--membership); *Be He@lthy/Be mobile* was develop by the WHO and the International Telecommunication Union (https://www.who.int/activities/Addressing-mobile-health), as well as *hearWHO* and *Google Fit*. For a more detailed review on the Global Strategy on Digital Health 2020–2024, we recommend referring to https://www.who.int/health--topics/digital-health. As well, in the clinic digital health has been used as an intervention in children with asthma [28] or in the oncology field [29] (https://www.physiomics.co.uk/).

Digital health interventions have been shown to have some clinical benefits in the treatment of musculoskeletal conditions [30], such as pain and functional disability; these conditions are the second largest contributor to disability worldwide and have important individual implications, social and economic [31]. Due to the increasing burden of musculoskeletal disability, there has been an urgent need for an integrated and strategic response, which is why digital health has been seen as an option because it provides high-scope, low-cost, easily accessible interventions and scalable for large patient populations that address time and resource constraints [30].

In a recent systematic review, the following databases were searched: Medical Literature Analysis and Retrieval System Online (MEDLINE), Excerpta Medica database (EMBASE), Cumulative Index to Nursing and Allied Health Literature (CINAHL), and Scopus since January 1, 2000. As of November 15, 2019, it showed that at a total of 19 studies evaluating musculoskeletal pain, nine reported statistically significant reductions after digital intervention. In total, 16 studies investigated functional disability, while ten studies showed statistically significant improvement; however, with heterogeneous results, it was not possible to perform a meta-analysis. Despite this, digital health interventions have the potential to positively contribute to reducing the multifaceted burden of musculoskeletal conditions for the individual, the economy, and society [30].

When the original SARS-CoV epidemic of 2002–2003 occurred and with seasonal influenza, geographic information systems (GIS) and methods allowed us to map online real or near real-time disease cases and reactions of social networks to the spread of diseases; in addition the population travel data provides us with a predictive mapping of risks and tracing of contact trajectories of super-spreaders in space and time [32]. Due to these characteristics that this technology offers, they make it a new source of information about diseases, dynamics, and epidemiology, allowing us to generate an effective response to them [33]. Modern GIS technologies focus on web-based tools, and they have been extremely popular for sharing and understanding the spread of SARS-CoV-2 [34]. On the other hand, communication through map-based dashboards has offered accessible information to many users, providing information about the places with the highest concentration of COVID-19 cases, helping to protect themselves and their communities. In addition, this type of tool improves the transparency of the data and helps the authorities to disseminate information [33].

Digital Health and Physiome

Physiomics is a branch of omics that uses large-scale databases and experimental databases together with computer algorithms to study and understand the physiological phenotypes of genes, proteins, their networks, and their relationships. It also uses bioinformatics to build networks of physiological characteristics associated with the network of genes and proteins. On the other hand, the total integration of these systems, from cells to organisms, can be called "Physiome" [35].

Physiome is a combinatorial word from "physio" and "ome" for "life" and "as a whole," respectively. It is the quantitative and integrated description of the functional behavior of the physiological state of an individual or species. It describes the physiological dynamics of the intact normal organism and is based on information and structure (genome, proteome, and morphome). In its broadest sense, physiology should define the relationships of the genome to the organism and from functional behavior to gene regulation. In the context of the Physiome Project, it includes integrated models of components of organisms, such as particular organs or cell systems or biochemical or endocrine systems (www.physiome.org) [36]. The Physiome Project was created with the purpose of understanding both human physiology and other eukaryotic organisms through a computational framework; this is carried out through the development of integrative models of all levels that are the biological organization of an organism. This project has established physiological databases accessible on the web that address different aspects such as data related to models and bibliographic information, at the level of cells, tissues, organs, and organ systems. However, a major problem for modelers today is the lack of standards for exchanging biological models, since the models published mathematically in journals were frequently incomplete and contained errors that made it difficult for anyone else to code the data [35].

The latest generations of portable biosensors have allowed us to obtain frequent measurement of health-related physiology. Studies have shown that these devices have served to determine the physiological changes of various users during various activities, resulting in better health management and diagnosis, in addition to disease analysis. Interestingly, different environments such as airline flights have been found that there is a decrease in peripheral capillary oxygen saturation [SpO2] and increased exposure to radiation; these processes were found to be associated with physiological macrophenotypes such as fatigue, thus being able to associate the reduction of pressure/oxygen and fatigue in high altitude flights [37].

These studies showed that wearable devices were helpful in identifying early signs of diseases such as Lyme disease and inflammatory responses; they were also able to distinguish physiological differences between insulin-sensitive and insulin-resistant individuals; therefore, these findings suggest that individuals have personal physiome and activity patterns that can be tracked using wearable sensors, so these devices could play an important role in health management, as well as allowing affordable access to healthcare for groups traditionally limited by socioeconomic class or remote geography [37].

Conclusion

The digital revolution has influenced healthcare systems around the world; it has changed from the principles to the fundamental approaches of medical service and education. Different investigations where this digital revolution was applied have shown to present an improvement in terms of accessibility, quality, and flexibility of medical care for the public not only in Western countries but also in developing countries.

However, currently the implementation of digital health platforms faces many limitations, including the clinical efficacy of the proposed technologies and their validation and also the question about the reliability and safety of these innovations. Therefore, extensive testing and clinical studies established in accordance with ethical principles are necessary. In addition, another problem that digital health faces is the lack of regulations and official recommendations, from interested parties, such as private and governmental organizations, as well as the appropriate validation and approval of new digital health technologies.

This chapter has shown that digital health interventions have some clinical benefits in the treatment of some diseases, in addition to showing the potential that digital health interventions have a positive effect on the personal, social, and economic impact of diseases. However, it should be noted that more research is needed to identify certain characteristics of the diseases, such as the identification of subgroups of patients who respond more positively to digital health interventions, and also to determine the pertinent characteristics of the interventions that are likely to achieve more successful patient outcomes. It is expected that as demand and technological improvements arise, the expansion of these devices will have a significant impact on daily life, being able to solve all current limitations and allowing a new interpretation and management mode of the health of its users to be obtained.

Acknowledgments This chapter is part of a registered project at the Instituto Nacional de Geriatria with number DI-PI-003/2018. Dr. Oscar Salvador Barrera-Vázquez receives a postdoctoral fellowship from the DGAPA-UNAM.

References

1. Senbekov M, Saliev T, Bukeyeva Z, Almabayeva A, Zhanaliyeva M, Aitenova N, Toishibekov Y, Fakhradiyev I. The recent progress and applications of digital technologies in healthcare: a review. Int J Telemed Appl. 2020; https://doi.org/10.1155/2020/8830200.
2. Mitchell M, Kan L. Digital technology and the future of health systems. Health Syst Reform. 2019;5:113–20.
3. Fatehi F, Samadbeik M, Kazemi A. What is digital health? review of definitions. Stud Health Technol Inform. 2020;275:67–71.
4. Mathews SC, McShea MJ, Hanley CL, Ravitz A, Labrique AB, Cohen AB. Digital health: a path to validation. NPJ Dig Med. 2019;2:38.
5. Gulliford M, Alageel S. Digital health intervention at older ages. Lancet Dig Health. 2019;1:e382–3.
6. Sharma A, Badea M, Tiwari S, Marty JL. Wearable biosensors: an alternative and practical approach in healthcare and disease monitoring. Molecules. 2021; https://doi.org/10.3390/molecules26030748.
7. Li G, Wen D. Wearable biochemical sensors for human health monitoring: sensing materials and manufacturing technologies. J Mater Chem B Mater Biol Med. 2020;8:3423–36.

8. Neethirajan S. Recent advances in wearable sensors for animal health management. Sens Bio Sens Res. 2017;12:15–29.
9. Ajami S, Teimouri F. Features and application of wearable biosensors in medical care. J Res Med Sci. 2015;20:1208–15.
10. Gray M, Meehan J, Ward C, Langdon SP, Kunkler IH, Murray A, Argyle D. Implantable biosensors and their contribution to the future of precision medicine. Vet J. 2018;239:21–9.
11. Kim J, Campbell AS, de Ávila BE-F, Wang J. Wearable biosensors for healthcare monitoring. Nat Biotechnol. 2019;37:389–406.
12. Song W, Gan B, Jiang T, Zhang Y, Yu A, Yuan H, Chen N, Sun C, Wang ZL. Nanopillar arrayed triboelectric nanogenerator as a self-powered sensitive sensor for a sleep monitoring system. ACS Nano. 2016;10:8097–103.
13. Xia S, Song S, Jia F, Gao G. A flexible, adhesive and self-healable hydrogel-based wearable strain sensor for human motion and physiological signal monitoring. J Mater Chem B Mater Biol Med. 2019;7:4638–48.
14. Imani S, Bandodkar AJ, Vinu Mohan AM, Kumar R, Yu S, Wang J, Mercier PP. A wearable chemical–electrophysiological hybrid biosensing system for real-time health and fitness monitoring. Nat Commun. 2016; https://doi.org/10.1038/ncomms11650.
15. Samydurai K. Technology: a key to patient satisfaction. Managed Health Care Executive. 2016.
16. Farsi D. Social media and health care, part I: literature review of social media use by health care providers. J Med Internet Res. 2021;23:e23205.
17. Definition of SOCIAL MEDIA. https://www.merriam-webster.com/dictionary/social%20media. Accessed 21 Apr 2021.
18. Antheunis ML, Tates K, Nieboer TE. Patients' and health professionals' use of social media in health care: motives, barriers and expectations. Patient Educ Couns. 2013;92:426–31.
19. Young SD. Social media as a new vital sign: commentary. J Med Internet Res. 2018;20:e161.
20. Gottlieb M, Dyer S. Information and disinformation: social media in the COVID-19 crisis. Acad Emerg Med. 2020;27:640–1.
21. Salathé M, Freifeld CC, Mekaru SR, Tomasulo AF, Brownstein JS. Influenza a (H7N9) and the importance of digital epidemiology. N Engl J Med. 2013;369:401–4.
22. Fung IC-H, Tse ZTH, Fu K-W. The use of social media in public health surveillance. Western Pac Surveill Response J. 2015;6:3–6.
23. Peary BDM, Shaw R, Takeuchi Y. Utilization of social media in the East Japan Earthquake and Tsunami and its effectiveness. J Nat Dis Sci. 2012;34:3–18.
24. Dugdale J, Van de Walle B, Koeppinghoff C. Social media and SMS in the Haiti earthquake. In: Proceedings of the 21st international conference on world wide web. New York: Association for Computing Machinery; 2012. p. 713–4.
25. Denecke K, Atique S. Social media and health crisis communication during epidemics. Participatory Health Through Social Media. 2016. pp. 42–66.
26. Edney S, Bogomolova S, Ryan J, Olds T, Sanders I, Maher C. Creating engaging health promotion campaigns on social media: observations and lessons from fitbit and garmin. J Med Internet Res. 2018;20:e10911.
27. Dhingra D, Dabas A. Global strategy on digital health. Indian Pediatr. 2020;57:356–8.
28. Ferrante G, Licari A, Marseglia GL, La Grutta S. Digital health interventions in children with asthma. Clin Exp Allergy. 2021;51:212–20.
29. Aapro M, Bossi P, Dasari A, Fallowfield L, Gascón P, Geller M, Jordan K, Kim J, Martin K, Porzig S. Digital health for optimal supportive care in oncology: benefits, limits, and future perspectives. Support Care Cancer. 2020;28:4589–612.
30. Hewitt S, Sephton R, Yeowell G. The effectiveness of digital health interventions in the management of musculoskeletal conditions: systematic literature review. J Med Internet Res. 2020;22:e15617.
31. Briggs AM, Woolf AD, Dreinhöfer K, Homb N, Hoy DG, Kopansky-Giles D, Åkesson K, March L. Reducing the global burden of musculoskeletal conditions. Bull World Health Organ. 2018;96:366–8.
32. Kamel Boulos MN, Geraghty EM. Geographical tracking and mapping of coronavirus disease COVID-19/severe acute respiratory syndrome coronavirus 2 (SARS-CoV-2) epidemic and associated events around the world: how 21st century GIS technologies are supporting the global fight against outbreaks and epidemics. Int J Health Geogr. 2020;19:8.
33. Liu N, Chee ML, Niu C, et al. Coronavirus disease 2019 (COVID-19): an evidence map of medical literature. BMC Med Res Methodol. 2020;20:177.
34. Franch-Pardo I, Napoletano BM, Rosete-Verges F, Billa L. Spatial analysis and GIS in the study of COVID-19. A review. Sci Total Environ. 2020;739:140033.
35. Gomase VS, Tagore S. Physiomics. Curr Drug Metab. 2008;9:259–62.
36. Leem CH. Perspectives of physiome research. Integr Med Res. 2016;5:37–40.
37. Li X, Dunn J, Salins D, et al. Digital health: tracking physiomes and activity using wearable biosensors reveals useful health-related information. PLoS Biol. 2017;15:e2001402.

Principles of Imaging for Epidemiologists

11

Omar Yaxmehen Bello-Chavolla, Arsenio Vargas-Vázquez, Mónica Itzel Martínez-Gutiérrez, Enrique C. Guerra, Carlos Alberto Fermín-Martínez, and Alejandro Márquez-Salinas

Abbreviations

AD Alzheimer's Disease
BOLD Blood Oxygen Level Dependent
CT Computed Tomography
DWI Diffusion Weighted Imaging
MRI Magnetic Resonance Imaging fMRI
Functional MRI
PET Positron Emission Tomography
SPECT Single-Photon Emission Computed Tomography

Introduction to Imaging Methods in Epidemiology

Imaging methods have been consistently used in biomedical research given that they have become invaluable resources for diagnosis, monitorization, and prognosis of many diseases. Imaging has been adopted by several clinical and basic research models and plays a fundamental role in many translational research studies [1, 2]. Particularly, imaging methods have had a strong presence in clinical trials and observational studies, where they can be used as a diagnostic method to confirm a clinical condition, whether it be as an inclusion criterion or as a clinical endpoint for the study [3, 4]. However, imaging methods are vastly helpful beyond a clinical setting, as they have been utilized on basic research on animal models and even in tissue preparations [4, 5]. The noninvasive nature of imaging methods has been exploited to a great extent in preclinical and clinical research focusing on countless areas of biomedical research.

In the last few years, medical image analysis has grown exponentially largely due to the development of high-throughput computing. The increased number of pattern recognition tools, increase in dataset sizes, and the development of processes for high-throughput extraction of innumerable quantitative features result in the conversion of images into mineable data and with the subsequent analysis permit support the epidemiological and clinical decisions. Advanced technologies that include complex imaging methods, such as computed tomography (CT) or magnetic resonance imaging (MRI), can provide information on subclinical and clinical diseases in a very short time. In addition, application of these methods to epidemiological studies is increasingly frequent. The large-scale acquisition of medical images in controlled population-based cohort is known as population imaging. These approaches can be used to identify persons at risk of developing specific diseases or may aid in disease-specific outcome prediction [3].

Representativity is the main assumption required in imaging studies to establish reference values that are generalizable to the entire source population. Reference values' importance lies on a method's capacity of distinguishing "normal" and "abnormal" or between "healthy" and "unhealthy." Therefore, large-scale population-based studies allow the definition of reference ranges dependent on sex, age, or body weight. The resulting dataset provides an opportunity to interrelate between risk factors, imaging phenotypes, and clinical outcomes resulting in the identification of subgroups of patients based on their individual risk factors and subclinical phenotypes. Typically, large-scale population-based studies are designed as cohort studies which include multiple examinations and follow-ups for morbidity and mortality; these studies allow the identification of individual risk factors, development of predictive models for incident disease, and/or risk scores that are useful tool for clinical practice patient's individual risk stratification.

O. Y. Bello-Chavolla (✉)
Research Division, Instituto Nacional de Geriatría, Mexico City, Mexico

A. Vargas-Vázquez · M. I. Martínez-Gutiérrez · E. C. Guerra
C. A. Fermín-Martínez · A. Márquez-Salinas
(PECEM) Program, Faculty of Medicine, National Autonomous University of Mexico, Mexico City, Mexico

J. C. Gomez-Verjan, N. A. Rivero-Segura (eds.), *Principles of Genetics and Molecular Epidemiology*,
https://doi.org/10.1007/978-3-030-89601-0_11

Interestingly, comprehensive phenotyping is usually supplemented by in-depth characterization of the genome, the metabolome, and other omic approaches. These approaches allow the identification of pathophysiological mechanisms and provide useful information for clinical practice. *Radiomics* is defined as the conversion of images to higher-dimensional data and the subsequent mining of this data to improve decision support. Radiomics is a relatively new field with substantial challenges for its implementation in a clinical setting; however, radiomics is motivated by the concept that biomedical images contain information which reflects underlying pathophysiological information, and these relationships can be revealed via quantitative image analysis. Radiomics offers a supply of imaging biomarkers that could aid in detection, diagnosis, assessment of prognosis prediction to treatment response, and monitoring of disease status. Furthermore, the correlation of radiomic data with genomic patterns is known as *radiogenomics*; this field has elicited especially great interest within the research community [6].

Table 11.1 Differences between advantages and disadvantages of clinical and population imaging and its potential applications for epidemiological studies

	Advantages	Weakness/disadvantage
Clinical imaging	Sequences applied are specific to define a disease Immediate clinical relevance Tailored to particular patient profile	Not easily reproducible Relevance is context-specific Limited knowledge extraction for other uses
Population imaging	It can establish reference values generalizable Coupled with other types of data (e.g., risk factors, clinical outcomes, and *omics*), it can serve for identification of subgroups of patients based on their individual risk factors or subclinical phenotypes and increase understanding of molecular pathogenesis of diseases It can be used to identify disease markers or risk factors associated with adverse outcomes	Sequences applied are too nonspecific to define diseases Usually less importance given to external validity which leads to reduced comparability among studies (e.g., imaging devices may have company-specific software not readily available everywhere) Generalizability requires huge sample size that includes healthy participants Underpowered to identify small high-risk groups Less suitable for investigating the individual effects of a given drug on imaging-based diagnosis Relatively high investment and costs that may limit the number of studies using high-end imaging technologies

Source: Adapted from Gillam et al. [3]

Finally, it is important to understand that population imaging differs from medical imaging as applied in clinical practice; these differences are highlighted in Table 11.1. In clinical settings a single imaging examination can provide insights for a certain diagnosis, whereas images acquired in large-scale population-based studies are regarded in much broader context. Population imaging combines imaging data with a variety of information collected from the participants that include questionnaires, physical examination, and/or laboratory measurement, and these factors permit transferring results from epidemiological studies to clinical practice.

In this chapter, we sought to evaluate how imaging is currently used in epidemiological research in several disciplines in medical research to understand their applications and limitations as currently used. Furthermore, we discuss how imaging is currently used in large-scale population-based epidemiological studies and imaging biobanks and finalize discussing the challenges and innovations required to meaningfully integrate imaging findings onto epidemiological and clinical research.

Imaging Applications in Neurological and Psychiatric Research

The field of neurology comprises a multitude of degenerative diseases with a steadily changing disease progression, where imaging follow-up might prove to be a useful tool to monitor disease progression and evaluate prognosis [7]. Neuroepidemiology studies how risk and protective factors may exert their effect directly on the underlying neuropathology of diseases or how this may influence the clinical expression of signs and symptoms in the presence of progressive neurological damage. The implementation of imaging to neuroepidemiology has improved the identification of subclinical neurological disorders and study of the natural course of the neurological diseases [8]. Similarly, neuroimaging has given to us the ability to study brain structure and function in vivo generating great excitement over an opportunity to address many of the scientific questions about brain development, damage, human cognition, and emotion in living humans [9].

There are numerous imaging techniques to study neurological disorders, such as MRI to assess brain atrophy and positron emission tomography (PET) to detect amyloid-beta aggregation. PET can also be used to visualize the presence of dopamine transporters, which can be beneficial in the evaluation of Parkinson's disease [10]. Tissues at greater risk of stroke can be detected with [^{15}O]-PET, which can be used to calculate the oxygen extraction fraction (OEF). Furthermore, diffusion-weighted MR imaging (DWI) offers vascular data that can reveal a mismatch between perfusion and oxygen consumption [1]. Regarding epilepsy, glucose consumption rates are elevated during seizures, and a decrease in its metabolism between episodes can be used to obtain an anatomical guide for treatment of the epileptic foci [1].

These advantages often make it difficult to recognize that brain imaging is a technological tool like many others in the research field, with their own strengths and weaknesses and with a fundamental limitation clearly described by Horga et al. as "No technology alone can generate valid scientific findings. Rather it is only technology coupled with a strong experimental design that can generate valid and reproducible findings [...]" [12]. Most of the neuropsychopathologies have one characteristic in common which makes them difficult to study: uncertainty in clinical diagnosis. Psychiatric disorders including depression, schizophrenia, or autism are still diagnosed primarily using behavioral signs and symptoms, diagnostic criteria which do not seem to have clear relations to the biological processes involving their pathogenesis. Similarly, these disorders have several neural systems throughout the brain, and patients' lack of focal damage makes it harder to classify, study, and diagnose the illness. Imaging has a key role as a "window" on the brain without physical invasive techniques, but despite this incredible approach, it still has one important problem to solve: relationship between structure, function, and behavior [9, 10, 13].

Illustrative Example: Population Imaging in Alzheimer's Disease

Neuroimaging in Alzheimer's disease (AD) has moved from an assistive tool in research to a notable place in diagnosis and temporal and spatial evolution of the illness. Multiple imaging studies have shown characteristic traits in brain function and structure in patients with AD. The biological, functional, and topographical information imaging can quantify has been notable in recent years and has shown its potential for improving correlation between clinical AD and its biological aspects, helping in the recognition of the prevalence of mixed pathology in mental diseases. Here, we continue describing the most relevant neuroimaging modalities used in AD, with observations about their advantages and limitations that conform their complementary roles.

Structural MRI AD has a characteristic topographic pattern, and the earliest changes are found in the medial temporal lobe, entorhinal and perirhinal cortex, and the hippocampus. These characteristics, previously observed with computed tomography, allowed for the exploration of more anatomical patterns linked with preclinical features and progression of AD [14]. In atrophy evaluation, one of the most notable advantages of MRI is the visualization of changes described by histopathology in the living brain. Neuronal counts at autopsy are closely related to volumes that can be measured with MRI, and since neuronal damage in certain regions (e.g., hippocampal region) is interpreted as decreased volume, volumetric MRI scans have been accepted as an accurate method for assessing AD progression. MRI findings include a pattern of loss described as insidious and progressive atrophy that has its first occurrence in the medial temporal lobe. Typically, the entorhinal cortex is the earliest zone of atrophy followed by the hippocampus, amygdala, and parahippocampus. Nevertheless, the accuracy of entorhinal cortex measurements has high variability due to anatomic ambiguity in its cortex boundaries, so over time measurement of the hippocampus has been established as one of the best biomarkers for AD [15]. Longitudinal MRI studies of individuals who were initially asymptomatic and subsequently develop AD or that were measured during their progression from amnesic mild cognitive impairment to clinically diagnosed AD support the idea that before the clinical diagnosis is made, atrophy is established [16–18]. In fact, one study of Sluimer et al. found that higher rate of brain atrophy was associated with an elevated risk of developing dementia [18].

Despite these promising results, MRI has some disadvantages. Most of the studies have small sample sizes often below a hundred participants, despite the wider availability of MRI compared to other imaging tools. This situation may be related with the high cost of repeated MRI measurements and technological complexity that often requires a multidisciplinary team, which is the case for longitudinal studies that gather repeat measurements over time to address the gap in knowledge of AD progression [19]. Similarly, atrophy is not entirely AD specific; patterns overlap with other diseases and atypical presentations of AD. Volume changes on MRI could be produced by more factors related or not with neuronal loss, like changes that could be attributable to the multidimensional aspect of AD as we commented before. Decreased hippocampal volume is also not AD-specific, as other diseases and conditions have been characterized with this anatomical feature such as Parkinson's disease, epilepsy, Huntington's disease, cardiac arrest, and chronic alcohol abuse [15]. Finally, as we previously pointed out, structure does not necessarily assess function. It remains difficult to distinguish whether anatomical findings, even if they are strongly related with mental functions observed in patients during the progression of the disease, are a consequence or a cause of the medical condition. MRI cannot explain completely why the presence of atrophy in early stages does not correlate with the clinical diagnosis of the disease as the signs and symptoms appear years later. It appears that the clinical state could be related not with the presence but with the extent of neurodegeneration.

Functional MRI (fMRI) fMRI is a noninvasive imaging technique which provides an indirect measure of neural activity using blood-oxygen-level-dependent (BOLD) signals and its changes. This imaging tool has the advantage of measuring changes in blood oxyhemoglobin/deoxyhemoglobin ratio, which reflects the integrated synaptic activity of neurons, without contrast agents. Therefore, fMRI is an attempt to assess the function-structure relationship that structural MRI cannot. fMRI can be assessed using two broad perspectives: (1) resting state, which measures changes in BOLD signals during "inactivity" or without a specific task, so specific brain networks can be observed, and (2) task-related fMRI in which patients perform different cognitive tasks that produce certain BOLD signals that could be compared with a control condition (e.g., encoding new information or activity compared with viewing familiar information) [20].

Having said that, we can highlight some important points:

1. *Assumption in the BOLD signal:* The concentration of deoxyhemoglobin depends on the metabolic activity within the studied area. The assumption of fMRI is that neuronal activity increases oxygen consumption, showing an increase in deoxyhemoglobin concentration which is measured with the fMRI scan, and the overall measured hemodynamic response function reflects the effects of changes in metabolic rate of oxygen, cerebral blood flow, and cerebral blood function. According to Schleim and Roiser, there are influential studies which showed a strong correlation of the BOLD signal with focal potentials reflecting synaptic activity; nevertheless caution should be taken when concluding from this activity and its correlation with the BOLD signal because it is only an indirect indicator of neuronal activity [20].
2. *Correlation in time:* Images representing BOLD signals must be processed in basically three forms: (1) First, they are *realigned* to make them lie in the same space because image slices were collected at different times. (2) Then, images often are *co-registered* with the findings in anatomical scans and *spatially normalized*. (3) Finally, images are *smoothed* slightly. The BOLD responses needed for this procedure occur slowly, so each event is modified to match the form of an average hemodynamic response function (*convolution*) [20]. Convolution process reflects the fact that fMRI cannot have the same temporal resolution of other functional studies as electroencephalography, for example. This is one of the most noticeable limitations.
3. *Group-level analysis:* The analysis of the fMRI in groups of participants or patients conducted in most studies shows the results in an image of statistical values, and voxels (or 3D pixels, spatial unit of measurement of fMRI) are colored according to t-values. Therefore, colors on the final images do not represent activity per se but represent statistical values based on BOLD signals.

As fMRI compares the performance of cognitive task studies that include patients with an advanced condition of AD or other neurological disorders, we are losing one of the most important advantages of this technique if participants are unable to perform the cognitive activity adequately, particularly in resting-state fMRI. Another challenge in interpreting fMRI data is the variability of the results. BOLD response is variable across subjects and even across the circumstances as BOLD measurements can be influenced by age and disease. Similarly, BOLD signal is also questionable since there are studies that show lower neural activation associated with stronger BOLD signals or patterns that suggest adaptation in the case of patients clinically diagnosed with AD who exhibit decreased hippocampal activity during encoding of new information but increased prefrontal cortical activity [20, 21].

AD is itself a challenge because there are a relatively small number of fMRI studies that have been realized with AD patients and mild cognitive impairment and genetically at-risk individuals; studies often count with a small sample size as structural MRI studies and the effect of medication in BOLD response of psychiatric patients has not been clarified, representing similar limitations that we assessed before, in structural MRI. So, it will not be plausible to affirm fMRI can completely differentiate function and specific processing or can assess the complexity of the interaction between networks and neuromodulation. References to neural aspects as "activation, processes, cognitive function," and so on required discretion.

Imaging Applications in Cardiovascular Research

Over the last few decades imaging methods in cardiovascular research have provided very detailed information about the structure and function of the heart and its vasculature. Imaging cardiology has provided unique ways to assess ischemic heart disease and a myriad of other heart conditions through the use of pharmacologic stress testing, SPECT and PET viability, vascular plaque imaging, and other technological breakthroughs [22]. The current frontiers of imaging cardiology research are directed toward the use of targeted molecular imaging methods to further assess cardiovascular conditions and the evaluation of inflammation in various disease states [23].

Several population-based cohort studies have used imaging methods for cardiovascular risk stratification and prediction of all-cause and cardiovascular mortality. In this context, coronary artery calcium (CAC) scanning is a rapid non-contrast CT of the heart used to identify calcification within epicardial coronary arteries. This imaging method has been repeatedly demonstrated to be the most effective predictor of coronary events in asymptomatic subjects. Several large-scale studies have shown that CAC is superior to risk factor scores in the reclassification of patients, and its prognostic value has been replicated in many studies, including MESA, Dallas Heart, and Rotterdam, among others.

Other imaging method useful in cardiovascular screening of asymptomatic populations is ultrasonography, used for the measurement of carotid artery intima-media thickness and carotid plaque. This approach for detection of subclinical atherosclerosis is attractive as it requires no ionizing radiation, is highly reproducible, and can be done in an office setting with appropriately trained personnel. Also, large-scale cohort studies have assessed its prognostic value to predict relevant cardiovascular outcomes, with the most important including MESA, Rotterdam, Three-City study, Tromso study, MDCS, ARIC, CAPS, and CHS [24].

Nuclear medicine is a relatively new area of knowledge concerned with the use of artificial radionuclides for clinical, therapeutical, and biomedical research. Radionuclides are typically bound to other chemical substances constituting radiopharmaceuticals, which after administration are metabolized by the body physiological pathways. The energy emitted by the decay of radiopharmaceuticals can be registered by special devices (e.g., γ-camera, SPECT or PET systems) and generate an image which can be coupled with other imaging methods (e.g., CT or MRI). Because of the intrinsic relationship of radiopharmaceuticals with physiologic pathways, one of the main advantages of nuclear imaging methods over traditional radiological images is that they can directly assess function rather than just anatomical structures [22]. In this way, the Hamburg City Health Study (HCHS) is a single-center, prospective, and population-based cohort which aims at identifying novel risk factors for major cardiovascular diseases, such as coronary artery disease, atrial fibrillation, heart failure, dementia, and stroke. The cohort includes about 45,000 volunteers between 45 and 74 years old with a baseline cardiovascular and neurological examination. HCHS focuses on evaluating the prognostic value of cardiovascular MRI that is performed in a sub-cohort with increased risk of cardiovascular disease; the imaging protocol includes stress perfusion MRI and radionuclide perfusion imaging. The cardiac MRI objective is to evaluate measurement of cardiac volumes, mass, and function. Therefore, the excellent reproducibility of quantitative measurements compared with other techniques, such as ultrasonography, positioned cardiac MRI as an attractive imaging method for large-scale population-based studies. Other cohorts that include the cardiac MRI examination are UK Biobank, German National Cohort, and the Canadian Alliance for Healthy Hearts and Minds [25].

Characteristics of this technique make it a very useful tool not only for clinical medicine but also for research endeavors. Depending on the radiopharmaceutical employed, various organ systems such as the central nervous system, the endocrine system, the respiratory system, and many others can be studied with novel unique perspectives. An interesting and novel research approach that does not require the need of implementing new technologies or radiopharmaceuticals is the use of radiomic methods to further improve the precision of diagnosis and risk stratification [23]. Radiomics in imaging cardiology may supersede the current semiquantitative approach in interpreting study results and could enhance the reliability and reproducibility of studies. One example in which radiomics has been useful is in the assessment of coronary artery calcification using myocardial perfusion studies [26, 27]. Furthermore, the correlation between cardiac imaging data and genomic data could provide a better understanding of subclinical disorders and improve the performance of cardiovascular scores for risk stratification in asymptomatic population. Further advances in the field of radiomics may prove useful to translate the insights of clinical utility of cardiovascular imaging into significant research advances and implications for epidemiological research in cardiovascular health.

Imaging Applications in Cardio-Metabolic Research

The current obesity and diabetes epidemic is a major challenge in developed and developing countries. Insulin resistance and impaired insulin secretion are key features in the physiopathology of these diseases that lead to well-known adverse cardiovascular outcomes [28, 29]. These phenomena interact with adipose tissue function to increase risk of adverse outcomes depending on the localization of adipose tissue depots and its specific function [30, 31]. To assess this latter function, advance imaging modalities that include whole-body composition assessed by bioelectrical impedance analysis (BIA), dual-energy x-ray densitometry (DXA), and MRI as the gold standard have provided the opportunity to analyze parameters of subclinical disease and body composition independent of BMI.

Body composition is important in order to analyze trends in obesity, adipose tissue function, sarcopenia, and other weight-related health conditions. The National Health and Nutrition Examination Survey (NHANES) using a national representative sample of the US population has included body composition analysis to estimate total body water (TBW), fat-free mass (FFM), total body fat (TBF), visceral adipose tissue (VAT), and percentage of body fat (%BF).

These assessments had been evaluated using BIA and DXA. This approach has permitted for the establishment of the percentiles of body composition parameters across the US population by age, sex, and ethnicity. Also, large-scale population-based studies as NHANES have evaluated the sex-specific body composition and determined that males had higher mean of TBW and FFM than did females, whereas females had higher mean of TBF and %BF than males, regardless of age and ethnic status [32, 33].

VAT is the most deleterious fat deposition in the body that it is strongly related to insulin resistance and increases the risk of developing cardio-metabolic disease and certain types of cancer [34, 35]. Large cohort studies that have evaluated VAT using DXA, CT, or MRI have demonstrated that the association between VAT and all-cause mortality is an independent risk factor of several variables and that the use of reduction of VAT as a primary target for obesity-reduction strategies even in the absence of weight loss is associated with less risk of developing cardio-metabolic diseases [36, 37]. In the KORA-MRI study, metabolic findings associated with diabetes and prediabetes included ectopic fat accumulation, particularly of hepatic lipids, and differential adipose tissue distribution compared to healthy subjects. Similarly, in the Framingham Heart Study using CT, VAT was strongly associated with an adverse metabolic risk profile even according to standard anthropometric indexes [38]. The use of imaging techniques in the evaluation of ectopic fat accumulation in subjects with subclinical or clinical risk has allowed for the identification of fatty liver disease as an important risk factor for morbidity and mortality [39].

Whole-body imaging examination in large-scale studies has shown the presence of white matter lesions, presence of carotid plaque, impaired function left ventricle, and significantly increased myocardial mass in patients with diabetes, even in subjects with prediabetes [40]. These findings have been reported previously in different studies using ultrasound, echocardiography, and CT, and all these large studies confirm diabetes as a major cardiovascular risk factor [41]. However, using a population imaging approach, epidemiological research has made possible the identification of subclinical phenotype at high cardiovascular risk in diabetes [42].

In summary, the implementation of the population imaging to the study of cardio-metabolic diseases offers clear evidence of subclinical changes associated with high cardiovascular risk and shows substantial variability of subclinical and clinical phenotypes between healthy and unhealthy subjects [43]. This could explain the very heterogeneous relative risk observed in longitudinal studies that depend on the observed subclinical and clinical manifestations. Additionally, implementation in population-based cohort studies of the radiomic approach which includes volume, shape, intensity, and texture of metabolic tissues assessed with imaging features will allow more accurate risk stratification across cardio-metabolic phenotypes [6].

Imaging Applications in Cancer Research

Implementation of imaging in the study of cancer is primarily focused on evaluating strategies for early cancer detection, stratification, and management and follow-up for prognosis. Correlation between bio-specimen-derived biomarkers and imaging biomarkers is common in the oncology field. In this way, imaging biomarkers offer available, cost-effective, and noninvasive tools for screening, staging, prognosis, therapy planning, and serial monitoring of patients to evaluate therapy response, recurrence, and palliation [44]. Imaging methods vary in physical characteristics, such as sensitivity and temporal and spatial resolution. In this case, X-ray is the less sensitive clinical imaging technique, in contrast to PET and nuclear medicine which are the most sensitive imaging techniques. Worldwide, the most important population-based imaging studies for cancer screening involve breast, lung, colorectal, and prostate cancer. For example, the population-based Cancer Screening Program in Urban China is an ongoing national program initiated in 2012, and about 72,000 participants have a valid ultrasound screening. With this approach adjuvant screening value of the ultrasonography in women with dense breast has been evaluated and determined that this imaging method is a helpful tool for designing large-scale effective cancer screening strategies [45].

However, large-scale studies for breast cancer screening have used mammography and MRI methods for the detection and follow-up of patients at risk. In Cote d'Or, France, a population-based study of breast cancer screening investigated factors affecting the adequacy of breast cancer screening rounds using mammography and the clinical implications of this approach. Comparatively, in the Breast Cancer Surveillance Consortium, using background parenchymal enhancement on breast MRI showed that this method is a strong predictor of breast cancer risk, independent of breast density and another established risk factor [46, 47]. Similarly, CT is useful for the evaluation of tumor size and density, whereas dynamic CT with contrast perfusion can further aid in the evaluation of blood flow, blood volume, and capillary permeability in patients with cancer. MRI allows for a more detailed examination of soft tissues compared to CT, and contrast-enhanced MRI using gadolinium has been employed to detect tumor angiogenesis and to monitor antiangiogenic drugs such as bevacizumab [1]. The combination of CT and MRI with nuclear medicine has additionally increased diagnostic sensitivity, reproducibility, and reliability, and these technologies have been used for cancer staging and monitoring of post-therapy tumor response [1, 48, 49]. Diffusion-weighted MR imaging (DWI) can also be used to differentiate tumors with a high cellularity from those with necrosis, swelling, and cell lysis. As discussed above, imaging mass spectrometry (IMS) can be used in the detection of a myriad of molecular biomarkers on a specific tissue; moreover, these biomarkers can be combined to determine specific metabolic dysregulations. In the oncology area this can be helpful to assess tumor-specific molecules to inform prognosis, diagnosis, and treatment response [50]. One important drawback from this methodology is the great influence of sample preparation on the quality and reproducibility of results, but in recent years significant progress has been made in this regard [50, 51].

Optical imaging can also be utilized in cancer research as there is a wide range of chromophores available to assess neoplasia, both endogenous including deoxyhemoglobin, water, amino acids, nicotinamide, flavins, porphyrins, collagen, and elastin and exogenous including drugs, small peptides, and antibodies targeting specific surface proteins overexpressed in tumors [52]. For example, these probes have been used to discriminate malignant breast tumors by assessing deoxyhemoglobin concentrations and water content within the tissue and to detect up-regulation of HER-2/neu to assess tumor growth and metastasis using fluorescent probes conjugated with trastuzumab, a drug that has high affinity for the receptor [53, 54]. Depending on the radiopharmaceutical employed, various organ systems such as the central nervous, endocrine, respiratory, and many other systems can be studied with novel imaging perspectives. Since 1970, nuclear medicine has been deeply intertwined with oncology, and it currently is vital for screening, diagnosis, characterization, stratification, and performing special procedures in various types of cancers [2]. The IP1-PROSTAGRAM study is a population-based prostate cancer screening that compared the prostate-specific antigen, ultrasonography, and non-contrast MRI to screen prostate cancer. The study concluded that non-contrast MRI screening may be a useful tool for community-based screening compared with ultrasonography [55]. Similarly, screening with MRI is a useful method for lung cancer detection in asymptomatic individuals [56].

Although radiomics is applied to many fields, in oncology it is most developed because of support from many institutions that include the National Cancer Institute, Quantitative Imaging Network, and other initiatives from the Cancer Imaging Program. This approach offers quantitative image features based on texture, shape, size, and volume information from the tumor phenotype and microenvironment. The correlation with clinical and biochemical information could help in clinical decisions and potentially cancer detection, diagnosis, prognosis assessment, prediction of response to treatment, and disease status monitoring. Recently, PET radiomics has been the subject of matter in oncology research in order to determine the extent by which these methods can provide a comprehensive quantification of neoplasm phenotype and to fine-tune outcome predictions of patients. While valuable results have been reported, there is still much more to be clarified and it is a worthwhile area of opportunity. Particularly in the field of oncology, radiomics has been investigated to great extent. For instance, it has been applied for the differentiation of pheochromocytoma and other adenomas in the workup of adrenal incidentalomas [57], for discrimination between lung-invasive adenocarcinoma and other noninvasive lesions in the management of lung nodules [58], for differentiation between benign and malignant renal tumors [59], and many other cancers with very good results. The main advantage of radiomics over traditional interpretation is that the deep learning approaches used to analyze the data can provide insights that may be omitted by the naked eye or by traditional statistical methods [60, 61].

Use of Imaging in Population-Based Epidemiological Studies

Population-based epidemiological studies are conducted from a sample randomly selected from population registry databases. These studies allow the generalizability of their results to the rest of the population in contrast to studies carried out in specialized centers. Application of large-scale population-based epidemiological studies using imaging techniques can be both logistically and technically challenging as the heterogeneity of methods and its applications exposed before show that there is no single effective method or approach to design, conduct, develop, evaluate, and analyze studies based on imaging data. Below, we will describe some examples of epidemiological studies which have primarily focused on imaging using population-based approaches to obtain more grounded statistical inferences.

Study of Health in Pomerania (SHIP)

The SHIP is a population-based epidemiological cohort study which includes two independent projects: SHIP and SHIP-TREND, conducted in the northeast of Germany. Between 1997 and 2001, the first SHIP cohort enrolled 6265 subjects followed in 5 years from 2002 to 2006, then again during 2008–2012, and finally in 2014. The second cohort (SHIP-TREND) was conducted between 2008 and 2012 and enrolled 8016 subjects, and the first follow-up was scheduled in 2015. The main aims include the evaluation of common risk factors, subclinical disorders, and manifest diseases using highly innovative noninvasive methods, involving the collection and assessment of data relevant to prevalence and incidence of common diseases and their risk factors. The

SHIP was the first population-based cohort study worldwide which includes whole-body MRI. The inclusion of MRI evaluation allowed for the establishment of population-based MRI reference parameters for certain systems. Furthermore, this study provided prevalence, incidence, and progression estimates for different MRI findings and their correlation with clinical examinations, metabolomics, and genome-wide analysis to help elucidate the complex interactions between risk factors and diseases [62].

Cooperative Health Research in the Region Augsburg (KORA)

KORA is a research platform for population-based surveys and subsequent follow-up studies in the south of Germany. This project was started in 1996 to continue and expand the MONICA project. In total, KORA cohort comprises about 18,000 subjects drawn from the population registry as a random sample of all 25–74-year-old residents of south of Germany. Examinations have been conducted at 5-year intervals since 1984. However, in 2004 an extensive biobank was set up. Between 2013 and 2014, a subgroup of subjects was included to undergo a whole-body MRI, KORA-FF4 MRI sub-study; in total 400 subjects were enrolled to design a case-control study to evaluate metabolic and cardiovascular disorders [63].

Generation R

Generation R is a population-based prospective cohort study initiated in Rotterdam. The main aims include the identification of early environmental and genetic factors and causal pathways leading to normal and abnormal growth, development, and health during fetal life until adulthood. In total, 9778 mothers with delivery date between 2002 and 2006 were enrolled. The cohort included data from mothers, fathers, and children and includes questionnaires, physical and ultrasound examinations, behavioral observations, and biological samples. The inclusion of MRI measurements in about 4000 children aims to assess cardiac, pulmonary, body composition, and liver parameters [64, 65].

Imaging Biobanks for Epidemiological Research

Imaging biobanks are organized databases of medical images. These imaging biomarkers are linked to other biorepositories with the aim to give researchers access to large collection of imaging datasets from healthy subjects or patients with specific diseases integrated with clinical, demographic, and biospecimen data. Due to development of modern radiology and nuclear medicine, the access to huge imaging biomarker datasets from all sources of digital imaging, such as CT, MRI, PET, SPECT, and US, has increased exponentially, opening up the possibilities of developing large-scale imaging biobanks. The European Society of Radiology established a working group aimed at monitoring and implementing imaging biobanks in Europe. Imaging biobanks allow storage of image data and metadata and storage of associated non-imaging data. In Europe, many countries count with imaging biobanks; however, most of them are for research and clinical reference and are disease-oriented, primarily in oncology and cardiovascular research. Currently, access to most imaging biobanks is restricted to local department/hospital personnel [66].

The UK Biobank and the German National Cohort (GNC) are two of the largest repositories of imaging data. UK Biobank is one of the largest ongoing population studies that includes about 100,000 volunteers examined using medical imaging starting in 2014. Imaging methods include ultrasonography, DXA, and whole-body MRI covering neuroradiological, cardiovascular and musculoskeletal assessment. These imaging data can be correlated with demographic, biometric, and functional data as well as biological samples and genetic data. In the last few years, genetic associations performed in large populations such as UK Biobank have been receiving particular attention. Radiomic approaches as an emergent field might make it possible to combine imaging datasets and genetic data for stratification of metabolic risk phenotypes or evaluation of subclinical and clinical conditions. Access to the UK Biobank is granted to scientific community members who meet the criteria of scientific quality and public interest [67].

The GNC is a large-scale epidemiological study that enrolled 200,000 individuals between 20 and 69 years old from different regions of Germany. The main aim is to identify genotypic and phenotypic features associated with health and disease. However, in a sub-cohort of 30,000 volunteers, whole-body MRI examinations were performed; the imaging protocol includes neurological, cardiovascular, and musculoskeletal evaluation. In addition to the imaging examination, the protocol includes biometrical and functional test as well as biological samples that are stored in a biobank in Southern Germany. The GNC-MR study in the beginning was planned as a cross-sectional study, but follow-up studies are planned and underway [68].

Analysis of Imaging Data from Population-Based Studies

Big Data Challenges

Because a large amount of biological, clinical, genetic, metabolomic, and imaging data have been generated and collected at an exceptional speed and scale, high-throughput computing has made possible the efficient analysis of large-scale data, as well as enabled the access of information to the medical and scientific community. In epidemiology, imaging data represent a big data challenge in multiple ways, with significant technical challenges ranging from data acquisition, data management, and data analysis [69]. The combination of traditional data and new forms of data at the individual and population level has permitted the incorporation of imaging data into epidemiological research. Datasets from a multitude of sources have improved and supported faster and more reliable research and discovery. However, the statistical analysis can be a big challenge which limits the fast growth of the research. In this field, the characteristics of the big data are defined by the four major "Vs":

1. **V**olume which implies the enormous volumes of data, and depending on the machine, it is possible to capture a huge volume of data.
2. **V**ariety refers to the sources and different types of structured and unstructured data. Currently, biomedical data can be gathered from emails, photos, videos, and imaging, among other modalities, and this variety of unstructured data can be a challenge for storing, mining, and analyzing data.
3. **V**elocity implies the speed of production and processing of the data.
4. **V**eracity, which refers to the biases, noise, and abnormalities in medical data, which is vital for the taking decisions in medicine [69–71].

Although big data offers exciting new opportunities for research, in the imaging field images must be processed in order to get an enhanced image or to extract useful information. In summary, imaging processing includes importing via image acquisition tools, analysis and manipulation of the image, and output in which results can be altered images or reports based on image analysis. Five major challenges which involve the image processing are speckle noise, computation time, feature dimensionality, retrieval accuracy, and semantic gap [69, 71]:

- *Speckle reduction* – A major problem for handling medical images is the presence of various granular structures such as speckle noise. Some real-time algorithms can remove speckles and allow smooth regions where no features or edges exist while maintaining and enhancing edges and borders. These algorithms do not eliminate information, and this improves image quality and can increase consistency in diagnosis, reducing patient and operator dependence. In this way, speckle reduction imaging is a real-time algorithm available exclusively on ultrasonography.
- *Image registration* – This process consists of combining two or more images obtained from different modalities for providing more information. For example, combination of information from an MRI and CT modalities provides more information than each individual modality separately [72].
- *Image segmentation* – Tissues and body organs are analyzed as delimited images; therefore, the procedure to segment images for extracting the region of interest (ROI) through an automatic or semiautomatic process is called image segmentation, for example, border detection in angiograms of coronary arteries, tumor detection and segmentation, brain studies that include functional mapping, and heart segmentation, among others. This procedure can be used for separating different tissues from each other, through extracting and classifying features. In this context, image classification in pixels may be useful in extracting bones, muscles, and blood vessels from specific anatomical regions. Currently, there are many medical image segmentation methods, algorithms, and applications that can be useful tools for imaging segmentation [69].
- *Image classification and retrieval* – The main challenge of image processing is the retrieval accuracy, which means achieving meaningful mappings between the high-level semantic concepts and the low-level features including color, shape, and texture, known as the semantic gap. For example, in the body composition analysis in images from DXA, CT, or MRI scans, other tissues such as cysts can be content in the fat deposits and the results might be biased. However, retrieval accuracy of images might improve correct classification and quantification.

If image acquisition and processing appear to be a great challenge, statistical analysis for classification or simply for reporting the characteristics for populations may be the greatest challenge for researchers. However, the development of multivariate techniques and other fields of statistical analysis like artificial intelligence and Bayesian statistics has permitted the study of asymptomatic population in more detail to differentiate them into different risk groups.

Multivariate Analysis in Imaging

Multivariate analysis (MVA) includes statistical and pattern recognition techniques which involve the processing of data that contains multiple measurements per sample. MVA is ideal for imaging tasks and can be used to analyze the correlation within the entire image dataset in order to provide valuable guides for the unveiling and understanding of related processes. MVA techniques are quite varied in their potential applications [73, 74]:

- Principal component analysis (PCA) is the most popular unsupervised MVA which computes orthogonal components that maximize the variance captured from the underlying measurements. PCA relies implicitly on two assumptions: Large variances have important dynamics and data has a high signal-to-noise ratio, while the independent component analysis is a useful technique based on discovering non-Gaussian distributions in datasets resulting in the exhibition of mixed signals.
- Clustering methods like K-means is an expectation-maximization algorithm which iteratively alternates. For example, K-means iteratively alternates between assigning pixels to the closest available cluster center and recalculating cluster centers as the mean spectrum of all the pixels in the cluster; this permits data to be grouped in different clusters. This unsupervised method can be refined using a supervised methods as linear discriminant analysis (LDA) that is a classification method for the observation of a dataset into groups using regression equations which maximize between group variances and minimize within group variance. This method has been applied to imaging mass spectrometry to improve the data classification more than clustering methods.

Overall, dimensionality reduction methods have permitted modeling the relationship between statistical analysis techniques and computational technologies that are used automatically and semiautomatically [73, 74].

Artificial Intelligence: Machine Learning and Deep Learning

Artificial intelligence (AI) deals with all aspects of mimicking cognitive functions for real-world problem-solving and building systems that learn. This field has been able to build causal models which support explanations and understanding as well as solving recognition patterns which often occur in imaging analyses [75, 76]. AI techniques play an important role in imaging analysis in the big data era because of the applications that include imaging processing, computer-aided diagnosis, imaging interpretation, fusion, registration, segmentation, retrieval, and analysis [77]. Machine learning and deep learning algorithms have been developed to improve imaging analysis, such as triaging screening mammograms, reducing or eliminating gadolinium-based contrast media for MRI, reducing the radiation dose of CT imaging to improve image noise reduction, etc. However, the use of supervised learning requires large and heterogeneous training, validation, and testing datasets. In this context, if the images are from a specific population or are measured by specific machines, the performance will be affected, and the algorithm capacity will be biased. Furthermore, the validation phase is very important as it allows the algorithm to be tuned until the final performance of the model is evaluated with a test dataset. Currently, multiple internal validation methods are available; however, independent validation on an external dataset is preferred to internal validation to assess model generalizability [75].

AI techniques are composed of conventional algorithms without learning such as support vector machine (SVM) and deep learning algorithms which include convolutional neural networks (CNN), recurrent neural networks (RNN), long short-term memory (LSTM), and extreme learning model (ELM), among others. These algorithms are fed by raw data and try to learn multiple levels of abstraction, representation, and information automatically from large imaging datasets. However, these techniques are limited by the processing of images in their raw form; are computationally intensive and time-consuming, based on expert knowledge; and require considerable time for feature tuning. Although AI algorithms in imaging analysis are a fascinating and exponentially growing research area, there are several barriers which slow down its progress. One of the biggest issues is the so-called black-box problem, which refers to the fact that the mathematical concepts used to construct the models are straightforward, but the output is exceedingly complicated and the understanding how the model works could be a huge issue [75].

Bayesian Statistics in Imaging

The Bayesian approach permits the incorporation of prior knowledge into data analysis and revolves around estimating posterior probabilities, which summarizes the degree of one's certainty concerning a given situation. Therefore, Bayesian approaches have multiple applications in image analysis and interpretation because it permits the use of prior knowledge concerning the situation under study. In the image analysis, the number of variables can range from thousands to millions. Although the same Bayesian principles apply, the computational burden is obviously magnified [78]. One of the major applications of these methods is imaging reconstruction. For example, some authors have used the Bayesian approach to interpret medical image adapting the prior of Geman and Geman which smooths images inside identified regions but avoids smoothing across region boundaries. The application to bone-scan images results in improvements in images and analysis quality. In the analysis of positron emission tomography and MRI, Bayesian analysis permits reconstruction using these line processes, as well as freely adaptable ones [79].

Bayesian inference is also a useful tool for the analysis of imaging datasets. Because many advanced methods for a variety of imaging techniques have been developed, such as

BOLD fMRI, diffusion, and ASL perfusion, Bayesian inference provides a mathematical framework which makes possible to take an approach to the complex problems in imaging analysis.[81] This approach offers a consistent way to handle the uncertainty and then to quantify the resulting uncertainty in the estimates. However, when we use a Bayesian approach, it is important to define the prior information as well as the analysis goal since the final interpretation can be affected by an excessive use of prior information. Therefore, the incorporation of prior information to image processing and analysis aims to provide a unified framework that incorporates external information and that can be adjusted through computational techniques such as simulation [80].

Limitations and Biases in Imaging Studies

To understand some of the limitations of imaging studies in epidemiological settings, we will retrieve the case study of Alzheimer's disease from Sect. 2 of this chapter. Alzheimer's disease is challenging to study because of its complex pathophysiology but also because many of the imaging findings which have been identified in epidemiological studies have offered limited insight into pathophysiology, management, and monitoring of the disease [81]. The following assessments are applicable to many disease models but are particularly relevant for studies related to Alzheimer's disease and cognitive impairment.

Anatomical and Incidental Findings

One of the major challenges of all imaging studies is distinguishing which findings represent pathophysiological processes, which represent adaptive changes derived for the presence of illness and which are simply anatomical variations due to age, ethnicity, sex, and so on [82]. For example, in neurological research the best understanding of the human mind requires the relationship between brain structure and function, coupled with mental processes. The brain of patients with psychiatric disorders could look similar to those of controls, and anatomical findings often overlap between diseases, so at the end they will not be as specific as we would like. This problem is particularly relevant of studies that include already-affected individuals. Structural, emotional, physiological adaptations and cognitive changes associated with the development of disease, especially in those which have chronic evolution, are thus likely to induce new alterations that coexist with the primary disturbances that caused the illness in the first place. If neuroimaging analysis is designed to identify similarities between these already-affected individuals, it will be more difficult to discern causes and signs of an illness from its consequences. Having said that, it would be reasonable to be cautious when an anatomical finding appears even as a common pattern in the group of study because this does not necessarily translate into saying that every patient will show the same sign or saying that this finding can separate entirely this pathology from others [19, 83].

Nonrepresentative Samples

Having a nonrepresentative sample implies that there are limitations to what can be concluded from the differences detected between groups. In case-control imaging studies there are at least three important considerations:

- *Samples of convenience:* Recruited patients come from local clinics which admit certain subtypes of an illness or patients that fit on certain diagnostic criteria, advocacy organizations, and particular socioeconomic groups. Differences detected across the cases and controls likely represent their sociodemographic variability and not the relative risk.
- *Select participants based on their diagnosis:* This has led to a bias toward studying diseases in patients who have different states of the illness. For example, in Alzheimer's research individuals who are mildly cognitively impaired compared with patients who have progressed to the disease state. Having in mind that only a fraction of patients with mild cognitive impairment progress to clinical Alzheimer's disease over 5–10 years or even later, the problem of discerning between findings linked to the pathogenic process, aging, and casual abnormalities becomes bigger.
- *Ethnically representative sample:* For example, in the Alzheimer's Disease Neuroimaging Initiative (ADNI) study, over 90% of participants were Caucasian, meaning that Afro-Americans, Asians, and Hispanics were underrepresented in this sample because people with white Caucasian ancestry in that time represent 63% of the total population [19].

Establishing Causal Relationships

Observed associations between neural and physiological traits do not necessarily imply a causal relationship; in fact, these associations could result from an unmeasured third variable that independently influences the other measures as a confounding factor. One of the best methods to infer causal effects is a systematic manipulation of an independent variable and the subsequent observation of its effects on the dependent variable. This allows for the interpretation of any observed relationships as causal, and imaging studies have tried to achieve this goal with different designs [15, 19].

Most imaging studies experimentally manipulate psychological processes by presenting a stimulus or task to the subject so investigators can infer that brain activity, or a part of it, was caused by performing these psychological functions, but they can infer with confidence that this brain features are causally responsible for the psychological process under study. Another interesting strategy is the adoption of randomized controlled trial designs to study the casual effects of therapeutic interventions, where the outcome measure is a brain imaging measure instead of changes in symptoms. The random assignment allows the inference that the treatment caused the observed change in imaging measure without saying, of course, that these changes can absolutely have a causal relationship with the illness being treated. For example, it seems quite reasonable to infer that not because a medication with an affinity for a particular neurotransmitter receptor can change the severity, frequency, or manifestation of an abnormal activity in an illness, the illness itself is a consequence of abnormal activity of that neurotransmitter or receptor [13, 15].

Brain structure linked with function is also another important point that has been assessed with imaging studies; its importance rests on the fact that structure and function influence each other, and the description of that interaction could sustain the biological component on the diagnosis, treatment, and understanding in this kind of pathologies. Nevertheless, this point remarks the importance of integrating different imaging modalities in imaging studies as each one has their own weaknesses that could be compensated by another technique. For example, structural abnormalities or differences between patients and controls could be related with findings in the same region using other modalities, improving their interpretations and neurobiological validity [13].

Costs in Population Imaging

Costs associated with imaging can represent a limitation when developing these large-scale studies and especially in those with serial evaluations in prospective cohort studies. These costs include those related to data acquisition, use of a hospital-based scanner, development and continuation of a well-functioning research infrastructure that includes personnel, and storage facilities of large-scale datasets and imaging analysis. However, the costs can vary considering the imaging method and the number of the participants included in the study. Despite this, some large-scale studies are in follow-up, such as UK Biobank and German National Cohort. Although there are some population-based studies, development of large-scale imaging studies in developing countries could result in a great limitation, and the sub-representation of these populations in other large-scale studies might limit the generalizability of the findings to developing countries with populations with specific characteristics.

Concluding Remarks

Besides the ethical issues, the standardization of processes, analysis, and costs, most large-scale epidemiological studies strive to ensure high internal validity, that is, that the equipment and software used to obtain the images remains unchanged for a certain period and during this period the personnel receives intensive training to achieve the standardization of obtaining the images. However, external validity is complex since the use of a specific methodology which includes the imaging method used and the protocol for obtaining the image may prevent comparability between studies. These problems are clearly exemplified when comparing different imaging methods and software, even inside of each imaging method. For example, there are many devices and software for magnetic resonance imaging and computed tomography, which limits the comparison between populations or even studies. However, cohort studies should keep the imaging methods used stable throughout follow-up. For example, the Rotterdam study in 1990 started using a 1.5 tesla MRI method and 22 years after continued to use the same method [24], even considering that now three Tesla scanners have improved the imaging process and are gaining ground in clinical implementation. Thus, implementing the latest in technology when large-scale studies are started is essential.

References

1. Wehrl HF, Sauter AW, Judenhofer MS, Pichler BJ. Combined PET/MR imaging — technology and applications. Technol Cancer Res Treat. 2010;9(1):5–20.
2. Rudin M. Noninvasive structural, functional, and molecular imaging in drug development. Curr Opin Chem Biol. 2009;13(3):360–71.
3. Gillam LD, Leipsic J, Weissman NJ. Use of imaging endpoints in clinical trials. JACC Cardiovasc Imaging. 2017;10(3):296–303.
4. Murphy P, Koh D-M. Imaging in clinical trials. Cancer Imaging. 2010;10(1A):S74–82.
5. Wehrl HF, Judenhofer MS, Wiehr S, Pichler BJ. Pre-clinical PET/MR: technological advances and new perspectives in biomedical research. Eur J Nucl Med Mol Imaging. 2009;36(1):56–68.
6. Yip SSF, Aerts HJWL. Applications and limitations of radiomics. Phys Med Biol. 2016;61(13):R150–66.
7. Albanese E. Chapter 3 – Advanced epidemiologic and analytical methods. En: Aminoff MJ, Boller F, Swaab DF, editores. Handbook of clinical neurology [internet]. Elsevier; 2016 [citado 2 de abril de 2021]. p. 39–52. (Neuroepidemiology; vol. 138). Disponible en: https://www.sciencedirect.com/science/article/pii/B9780128029732000033.
8. Cohen AB, Klein JP, Mukundan S. A guide to imaging for common neurological problems. BMJ. 2010;341:c4113.
9. Kerr JND, Denk W. Imaging in vivo : watching the brain in action. Nat Rev Neurosci. 2008;9(3):195–205.

10. Schwarz CG. Uses of human MR and PET imaging in research of neurodegenerative brain diseases. Neurotherapeutics [Internet]. 15 de marzo de 2e021 [citado 2 de abril de 2021]; Disponible en: https://doi.org/10.1007/s13311-021-01030-9.
11. Tsushima Y, Taketomi-Takahashi A, Endo K. Prevalence of abnormal findings on brain magnetic resonance (MR) examinations in adult participants of brain docking. BMC Neurol. 2005;5(1):18.
12. Horga G, Kaur T, Peterson BS. Annual research review: current limitations and future directions in MRI studies of child- and adult-onset developmental psychopathologies. J Child Psychol Psychiatry. 2014;55(6):659–80.
13. Poldrack RA, Farah MJ. Progress and challenges in probing the human brain. Nature. 2015;526(7573):371–9.
14. Chandra A, Dervenoulas G, Politis M. Alzheimer's Disease Neuroimaging Initiative. Magnetic resonance imaging in Alzheimer's disease and mild cognitive impairment. J Neurol. 2019;266(6):1293–302.
15. van Oostveen WM, de Lange ECM. Imaging techniques in Alzheimer's disease: a review of applications in early diagnosis and longitudinal monitoring. Int J Mol Sci. 2021;22(4):2110.
16. Chan D, Janssen JC, Whitwell JL, Watt HC, Jenkins R, Frost C, et al. Change in rates of cerebral atrophy over time in early-onset Alzheimer's disease: longitudinal MRI study. Lancet. 4 de octubre de 2003;362(9390):1121–2.
17. Jack CR, Weigand SD, Shiung MM, Przybelski SA, O'Brien PC, Gunter JL, et al. Atrophy rates accelerate in amnestic mild cognitive impairment. Neurology. 6 de mayo de 2008;70(19 Pt 2):1740–52.
18. Sluimer JD, van der Flier WM, Karas GB, Fox NC, Scheltens P, Barkhof F, et al. Whole-brain atrophy rate and cognitive decline: longitudinal MR study of memory clinic patients. Radiology. 2008;248(2):590–8.
19. Lawrence E, Vegvari C, Ower A, Hadjichrysanthou C, De Wolf F, Anderson RM. A systematic review of longitudinal studies which measure Alzheimer's disease biomarkers. J Alzheimers Dis. 2017;59(4):1359–79.
20. Schleim S, Roiser JP. FMRI in translation: the challenges facing real-world applications. Front Hum Neurosci. 2009;3:63.
21. Johnson KA, Fox NC, Sperling RA, Klunk WE. Brain imaging in Alzheimer disease. Cold Spring Harb Perspect Med [Internet]. abril de 2012 [citado 2 de abril de 2021];2(4). Disponible en: https://www.ncbi.nlm.nih.gov/pmc/articles/PMC3312396/.
22. Zaret BL. Nuclear cardiology: history and milestones [Internet]. Nuclear Cardiac Imaging. Oxford University Press; [citado 2 de abril de 2021]. Disponible en: https://oxfordmedicine.com/view/10.1093/med/9780199392094.001.0001/med-9780199392094-chapter-1.
23. Di Carli MF. Challenges and opportunities for nuclear cardiology. J Nucl Cardiol. 2019;26(4):1043–6.
24. Ikram MA, van der Lugt A, Niessen WJ, Krestin GP, Koudstaal PJ, Hofman A, et al. The Rotterdam Scan Study: design and update up to 2012. Eur J Epidemiol. 2011;26(10):811–24.
25. Bohnen S, Avanesov M, Jagodzinski A, Schnabel RB, Zeller T, Karakas M, et al. Cardiovascular magnetic resonance imaging in the prospective, population-based, Hamburg City Health cohort study: objectives and design. J Cardiovasc Magn Res. 2018;20(1):68.
26. Bogowicz M, Vuong D, Huellner MW, Pavic M, Andratschke N, Gabrys HS, et al. CT radiomics and PET radiomics: ready for clinical implementation? Q J Nucl Med Mol Imaging. 2019;63(4):355–70.
27. Ashrafinia S, Dalaie P, Yan R, Ghazi P, Marcus C, Taghipour M, et al. Radiomics analysis of clinical myocardial perfusion spect to predict coronary artery calcification. J Nucl Med. 2018;59(Suppl 1):512.
28. Yaribeygi H, Farrokhi FR, Butler AE, Sahebkar A. Insulin resistance: review of the underlying molecular mechanisms. J Cell Physiol. 2019;234(6):8152–61.
29. Yazıcı D, Sezer H. Insulin resistance, obesity and lipotoxicity. Adv Exp Med Biol. 2017;960:277–304.
30. Antonio-Villa NE, Bello-Chavolla OY, Vargas-Vázquez A, Mehta R, Fermín-Martínez CA, Martagón-Rosado AJ, et al. Increased visceral fat accumulation modifies the effect of insulin resistance on arterial stiffness and hypertension risk. Nutr Metab Cardiovasc Dis. 2021;31(2):506–17.
31. Fernández-Chirino L, Antonio-Villa NE, Vargas-Vázquez A, Almeda-Valdés P, Gómez-Velasco D, Viveros-Ruiz TL, et al. Elevated serum uric acid is a facilitating mechanism for insulin resistance mediated accumulation of visceral adipose tissue. medRxiv. 2020;2020:09.20.20198499.
32. Chumlea WC, Guo SS, Kuczmarski RJ, Flegal KM, Johnson CL, Heymsfield SB, et al. Body composition estimates from NHANES III bioelectrical impedance data. Int J Obes. 2002;26(12):1596–609.
33. Hinton BJ, Fan B, Ng BK, Shepherd JA. Dual energy X-ray absorptiometry body composition reference values of limbs and trunk from NHANES 1999–2004 with additional visualization methods. Plos One. 2017;12(3):e0174180.
34. Després J-P. Body fat distribution and risk of cardiovascular disease: an update. Circulation. 2012;126(10):1301–13.
35. Jean-Pierre D. Body fat distribution and risk of cardiovascular disease. Circulation. 2012;126(10):1301–13.
36. Katzmarzyk PT, Mire E, Bouchard C. Abdominal obesity and mortality: the Pennington Center Longitudinal Study. Nutr Diabetes. 2012;2:e42.
37. Kay SJ, Fiatarone Singh MA. The influence of physical activity on abdominal fat: a systematic review of the literature. Obes Rev. 2006;7(2):183–200.
38. Fox CS, Massaro JM, Hoffmann U, Pou KM, Maurovich-Horvat P, Liu C-Y, et al. Abdominal visceral and subcutaneous adipose tissue compartments: association with metabolic risk factors in the Framingham Heart Study. Circulation. 2007;116(1):39–48.
39. Rinella ME. Nonalcoholic fatty liver disease: a systematic review. JAMA. 2015;313(22):2263–73.
40. Bamberg F, Hetterich H, Rospleszcz S, Lorbeer R, Auweter SD, Schlett CL, et al. Subclinical disease burden as assessed by whole-body MRI in subjects with prediabetes, subjects with diabetes, and normal control subjects from the general population: the KORA-MRI study. Diabetes. 2017;66(1):158–69.
41. Schram MT, Henry Ronald MA, van Dijk Rob AJM, Kostense Piet J, Dekker Jacqueline M, Giel N, et al. Increased central artery stiffness in impaired glucose metabolism and type 2 diabetes. Hypertension. 2004;43(2):176–81.
42. Fox ER, Sarpong DF, Cook JC, Samdarshi TE, Nagarajarao HS, Liebson PR, et al. The relation of diabetes, impaired fasting blood glucose, and insulin resistance to left ventricular structure and function in African Americans: the Jackson heart study. Diabetes Care. 2011;34(2):507–9.
43. Grundy SM. Pre-diabetes, metabolic syndrome, and cardiovascular risk. J Am Coll Cardiol. 2012;59(7):635–43.
44. O'Connor JPB, Aboagye EO, Adams JE, Aerts HJWL, Barrington SF, Beer AJ, et al. Imaging biomarker roadmap for cancer studies. Nat Rev Clin Oncol. 2017;14(3):169–86.
45. Wang Y, Chen H, Li N, Ren J, Zhang K, Dai M, et al. Ultrasound for breast cancer screening in high-risk women: results from a population-based cancer screening program in China. Front Oncol. 2019;9:286.
46. Arasu VA, Miglioretti DL, Sprague BL, Alsheik NH, Buist DSM, Henderson LM, et al. Population-based assessment of the association between magnetic resonance imaging background parenchymal enhancement and future primary breast cancer risk. J Clin Oncol. 2019;37(12):954–63.
47. Ouedraogo S, Dabakuyo TS, Gentil J, Poillot M-L, Dancourt V, Arveux P. Population-based study of breast cancer screening in Côte d'Or (France): clinical implications and factors affecting screening round adequacy. Eur J Cancer Prev. 2011;20(6):462–74.

48. Daamen LA, Groot VP, Goense L, Wessels FJ, Rinkes IHB, Intven MPW, et al. The diagnostic performance of CT versus FDG PET-CT for the detection of recurrent pancreatic cancer: a systematic review and meta-analysis. Eur J Radiol. 2018;106:128–36.
49. Han S, Choi JY. Impact of 18F-FDG PET, PET/CT, and PET/MRI on staging and management as an initial staging modality in breast cancer: a systematic review and meta-analysis. Clin Nucl Med. 2021;46(4):271–82.
50. McDonnell LA, Angel PM, Lou S, Drake RR. Chapter eleven – mass spectrometry imaging in cancer research: future perspectives. En: Drake RR, McDonnell LA, editores. Advances in cancer research [internet]. Academic press; 2017 [citado 3 de abril de 2021]. p. 283-90. (applications of mass spectrometry imaging to cancer; vol. 134). Disponible en: https://www.sciencedirect.com/science/article/pii/S0065230X16300835.
51. Karlsson O, Hanrieder J. Imaging mass spectrometry in drug development and toxicology. Arch Toxicol. 2017;91(6):2283–94.
52. Solomon M, Liu Y, Berezin MY, Achilefu S. Optical imaging in cancer research: basic principles, tumor detection, and therapeutic monitoring. MPP. 2011;20(5):397–415.
53. Soliman H, Gunasekara A, Rycroft M, Zubovits J, Dent R, Spayne J, et al. Functional imaging using diffuse optical spectroscopy of neoadjuvant chemotherapy response in women with locally advanced breast cancer. Clin Cancer Res. 2010;16(9):2605–14.
54. He K, Zeng S, Qian L. Recent progress in the molecular imaging of therapeutic monoclonal antibodies. J Pharm Anal. 2020;10(5):397–413.
55. Eldred-Evans D, Burak P, Connor MJ, Day E, Evans M, Fiorentino F, et al. Population-based prostate cancer screening with magnetic resonance imaging or ultrasonography: the IP1-PROSTAGRAM study. JAMA Oncol. 2021;7(3):395.
56. Wu N-Y, Cheng H-C, Ko JS, Cheng Y-C, Lin P-W, Lin W-C, et al. Magnetic resonance imaging for lung cancer detection: Experience in a population of more than 10,000 healthy individuals. BMC Cancer. 2011;11(1):242.
57. Yi X, Guan X, Zhang Y, Liu L, Long X, Yin H, et al. Radiomics improves efficiency for differentiating subclinical pheochromocytoma from lipid-poor adenoma: a predictive, preventive and personalized medical approach in adrenal incidentalomas. EPMA J. 2018;9(4):421–9.
58. Zhao W, Xu Y, Yang Z, Sun Y, Li C, Jin L, et al. Development and validation of a radiomics nomogram for identifying invasiveness of pulmonary adenocarcinomas appearing as subcentimeter ground-glass opacity nodules. Eur J Radiol. 2019;112:161–8.
59. Zhou L, Zhang Z, Chen Y-C, Zhao Z-Y, Yin X-D, Jiang H-B. A deep learning-based radiomics model for differentiating benign and malignant renal tumors. Transl Oncol. 2019;12(2):292–300.
60. Biswas M, Kuppili V, Saba L, Edla DR, Suri HS, Cuadrado-Godia E, et al. State-of-the-art review on deep learning in medical imaging. Front Biosci (Landmark Ed). 2019;24:392–426.
61. Avanzo M, Stancanello J, Pirrone G, Sartor G. Radiomics and deep learning in lung cancer. Strahlenther Onkol. 2020;196(10):879–87.
62. Völzke H, Alte D, Schmidt CO, Radke D, Lorbeer R, Friedrich N, et al. Cohort profile: the study of health in pomerania. Int J Epidemiol. 2011;40(2):294–307.
63. Hetterich H, Bayerl C, Peters A, Heier M, Linkohr B, Meisinger C, et al. Feasibility of a three-step magnetic resonance imaging approach for the assessment of hepatic steatosis in an asymptomatic study population. Eur Radiol. 2016;26(6):1895–904.
64. Jaddoe VWV, van Duijn CM, Franco OH, van der Heijden AJ, van Ilzendoorn MH, de Jongste JC, et al. The Generation R Study: design and cohort update 2012. Eur J Epidemiol. 2012, 27;(9):739–56.
65. Jaddoe VWV, Bakker R, van Duijn CM, van der Heijden AJ, Lindemans J, Mackenbach JP, et al. The generation R study biobank: a resource for epidemiological studies in children and their parents. Eur J Epidemiol. 2007;22(12):917–23.
66. European Society of Radiology (ESR). ESR position paper on imaging biobanks. Insights Imaging. 2015;6(4):403–10.
67. Sudlow C, Gallacher J, Allen N, Beral V, Burton P, Danesh J, et al. UK biobank: an open access resource for identifying the causes of a wide range of complex diseases of middle and old age. PLOS Med. 2015;12(3):e1001779.
68. The German National Cohort: aims, study design and organization. Eur J Epidemiol. 2014;29(5):371–82.
69. Big data in meedical image processing [internet]. CRC Press; 2018 [citado 4 de abril de 2021]. Disponible en: https://www.taylorfrancis.com/https://www.taylorfrancis.com/books/mono/10.1201/b22456/big-data-medical-image-processing-suganya-rajaram-sheik-abdullah
70. Ezhilraman SV, Srinivasan S. State of the art in image processing & big data analytics: issues and challenges. Int J Eng Technol. 2018;7(2.33):195–9.
71. Kharat AT, Singhal S. A peek into the future of radiology using big data applications. Indian J Radiol Imaging. 2017;27(2):241–8.
72. Brown LG. A survey of image registration techniques. ACM Comput Surv. 1992;24(4):325–76.
73. Levman J, Takahashi E. Multivariate analyses applied to healthy neurodevelopment in fetal, neonatal, and pediatric MRI. Front Neuroanat. 2015;9:163.
74. Muir ER, Ndiour IJ, Goasduff NAL, Moffitt RA, Liu Y, Sullards MC, et al. Multivariate analysis of imaging mass spectrometry data. En: 2007 IEEE 7th international symposium on bioinformatics and bioengineering. 2007. p. 472–9.
75. Willemink MJ, Koszek WA, Hardell C, Wu J, Fleischmann D, Harvey H, et al. Preparing medical imaging data for machine learning. Radiology. 2020;295(1):4–15.
76. Holzinger A, Langs G, Denk H, Zatloukal K, Müller H. Causability and explainability of artificial intelligence in medicine. WIREs Data Min Knowl Discov. 2019;9(4):e1312.
77. Tahmassebi A, Ehtemami A, Mohebali B, Gandomi AH, Pinker K, Meyer-Baese A. Big data analytics in medical imaging using deep learning. En: Big data: learning, analytics, and applications [internet]. International Society for Optics and Photonics; 2019 [citado 4 de abril de 2021]. p. 109890E. Disponible en: https://www.spiedigitallibrary.org/conference-proceedings-of-spie/10989/109890E/Big-data-analytics-in-medical-imaging-using-deep-learning/10.1117/12.2516014.short.
78. Hanson KM. Introduction to Bayesian image analysis. En: Medical imaging 1993: image processing [Internet]. International Society for Optics and Photonics; 1993 [citado 4 de abril de 2021]. p. 716-31. Disponible en: https://www.spiedigitallibrary.org/conference-proceedings-of-spie/1898/0000/Introduction-to-Bayesian-image-analysis/10.1117/12.154577.short.
79. Ma J, Feng Q, Feng Y, Huang J, Chen W. Generalized Gibbs priors based positron emission tomography reconstruction. Comput Biol Med. 2010;40(6):565–71.
80. Zhang L, Guindani M, Vannucci M. Bayesian models for functional magnetic resonance imaging data analysis. WIREs Comput Stat. 2015;7(1):21–41.
81. Illes J, Kirschen MP, Edwards E, Stanford LR, Bandettini P, Cho MK, et al. Incidental findings in brain imaging research. Science. 2006;311(5762):783–4.
82. Frisoni GB. Structural imaging in the clinical diagnosis of Alzheimer's disease: problems and tools. J Neurol Neurosurg Psychiatry. 2001;70(6):711–8.
83. Völzke H, Schmidt CO, Hegenscheid K, Kühn J-P, Bamberg F, Lieb W, et al. Population imaging as valuable tool for personalized medicine. Clin Pharmacol Ther. 2012;92(4):422–4.

Bioinformatics and Genomics for Epidemiologists

12

Omar Yaxmehen Bello-Chavolla, Luisa Fernández-Chirino, Neftali Eduardo Antonio-Villa, Marco Antonio Delaye-Martinez, and Alejandro Sicilia-Andrade

Abbreviations

EWAS	Epigenome-wide association studies
GWAS	Genome-wide association studies
HGP	Human Genome Project
HWE	Hardy-Weinberg equilibrium
LD	Linkage disequilibrium
PRS	Polygenic risk scores
SNP	Single nucleotide polymorphisms

Introduction: Bioinformatics Applied to Modern Epidemiology

The shape and focus of clinical and epidemiological research have had significant advances over the last few years. Traditional research approaches based on exposures and relation to outcomes for clinical characterization of human diseases are significantly limited given the vast amount of data available at hand. Similarly, the scope and speed at which data are being gathered and analyzed is growing fast, and the use of traditional statistical approaches may be undermined by the curse of multidimensionality and multiple hypothesis testing, which may complicate the identification of true positive findings, which is pivotal to translate genomic findings into significant clinical and epidemiological findings. Furthermore, characterization of disease processes has led to the identification of multidimensional factors involved in highly incident diseases such as diabetes, cancer, allergies, rheumatic, or neurodegenerative diseases [1].

Epidemiological designs for human research during the last century have been primarily focused on the characterization of human diseases using a traditional epidemiology approach, which can be defined as "the study of the distribution and determinants of health-related states or events in specified populations, and the application of this study to the control of health problems" [2]. While this approach makes it possible to study the general effects of a disease process on the population, in a broad sense, the current medical trend demands a more integrative, systems biology-driven perspective [3]. In this context, the Human Genome Project (HGP) has been considered one of the most important and ambitious steps in medical history. From its proposal in May 1985 until its approval and initiation in October 1990, the HGP has attracted the participation of the Wellcome Trust Sanger Institute, the Broad Institute of MIT and Harvard, the Genome Institute of Washington University, the Joint Genome Institute, and the Whole Genome Laboratory at Baylor College of Medicine; overall, the HGP had brought this systems biology perspective as an attractive and applicable approach to epidemiology. The HGP has also brought forth technical advances in sequencing technologies, mathematical, computational, and statistical tools for the management and storage of genetic information, which overall represented the most significant achievement for medical sciences to comprehend human diseases in the last century [4].

The need for dedicated investigators who could perform and maintain the complex needs of large-scale data analysis has been of interest to academia and industry, but most importantly for human research. With a growing need for experts in this field, most laboratories seek to have a department in bioinformatics among their toolbox and main areas of work. Beyond data acquisition and analysis, the individual researcher would certainly need external advice for any complex analysis derived from modern epidemiological studies [5]. The challenge is amped up when investigators aim to

O. Y. Bello-Chavolla (✉)
Research Division, Instituto Nacional de Geriatría, Mexico City, Mexico

L. Fernández-Chirino · N. E. Antonio-Villa
Faculty of Chemistry, National Autonomous University of Mexico, Mexico City, Mexico

M. A. Delaye-Martinez · A. Sicilia-Andrade
(PECEM) Program, Faculty of Medicine, National Autonomous University of Mexico, Mexico City, Mexico

J. C. Gomez-Verjan, N. A. Rivero-Segura (eds.), *Principles of Genetics and Molecular Epidemiology*,
https://doi.org/10.1007/978-3-030-89601-0_12

integrate patient information, clinical data, laboratory evaluations, imaging, and genetic data to solve a clinically relevant problem. The integration of multidimensional datasets enables the potential to identify unique biological signatures, providing a unique platform for advances in clinical and translational sciences.

In recent years, the term "omics" has spread across the fields of genomics, proteomics, metabolomics, and others to provide an insight view for understanding the systems biology mechanisms within human diseases. This unified vision allows the traditional epidemiology approach to have a novel interaction between other disciplines such as computational biology, genetics, and most importantly, bioinformatics [1, 6]. Specialized areas, such as clinical bioinformatics, have been created to analyze the massive amount of data that modern epidemiology generates within current populational studies. The field of clinical bioinformatics includes the analysis of not only genetic microarrays and other *omics* data but also an interplayed combination of medical, demographic, and psychometric information extracted from clinical research and human databases. Overall, the current systems biology perspective allowed for the integration of clinical bioinformatics in modern epidemiology to better comprehend human diseases.

In this chapter, we focus on evaluating three main applications in the use of genomics and bioinformatics in modern epidemiological approaches. We evaluate the necessary bioinformatics tool to engage in genomics discovery through genome- or exome-wide association studies, the development and use of genetic risk scores to predict outcomes and disease processes, and the integration of information from multiple omics technologies to identify pathways which may have biological relevance to model disease processes using a systems biology approach. We aim to provide a comprehensive understanding on the potential applications of genomics into epidemiology and refer for further reading to gain a more in-depth understanding of the application of genomics to inform epidemiological studies.

Applications of Bioinformatics in Modern Epidemiology

The Need for Clinical Bioinformaticians in the Era of Big Data in Epidemiology

The implementation of new techniques that allow for a more robust sequencing of the human genome has been introduced in public health laboratories worldwide. Methods such as pulse-field gel electrophoresis have been displaced by techniques related to whole-genome sequencing that could provide a high load of information and definition at the nucleotide level for a specific human disease. Furthermore, this type of technique could establish causality mechanisms through the analysis of nearby genetic structures [6, 7]. The tremendous amount of data that can be extracted from a single patient can now be translated into new information and that new information into new knowledge which, in turn, will lead to action on how to treat human disease and ideally how to prevent it, opening a new era where personalized medicine shifts from wishful thinking to a tantalizing and feasible application of bioinformatics.

Bioinformatics' importance in modern epidemiology relies on integrating multimodal patients' data obtained from sources at different biological levels, including population, demographic, psychometrics, clinical, tissue-specific, and cellular data to identify physiological routes for the timely diagnosis and treatment of diseases even in preclinical stages. Hence, bioinformatics has a unique domain within this current epidemiology perspective [8]. Moreover, novel specialties such as translational bioinformatics have been created to focus on "the development of storage, analytic, and interpretive methods to optimize the transformation of increasingly voluminous biomedical data, and genomic data, into proactive, predictive, preventive, and participatory health." It has been emphasized that the integration of clinical bioinformatics into research could improve medical care and provide an integration of multidisciplinary research with the unique objective to provide relevant information for personalized medical care [9, 10]. If there is data in medicine, there will be an area for bioinformatics.

Technical and Conceptual Limitations in Bioinformatics

Bioinformatics continues to be a growing area of application in epidemiology, and its recent use has led to the identification of various challenges and difficulties in its systematization in various research and clinical centers. The great benefit that is expected to be obtained from the techniques and knowledge of bioinformatics within personalized medicine underlies its various limitations. Despite the advances presented for the applied genomic analysis, some limitations have been identified on technology necessary to provide sufficient technical and computational capacity to analyze the information provided by novel sequencing techniques. This has also provided exciting opportunities for the development and advancement of the computer science field and has accelerated its integration into health-related research.

First, a great deal of computational power is required to manipulate all the components extracted from genomic sequencing. It has been recognized that desktop and laptop computers may not have the computational capacity to process and analyze large datasets in short periods of time with the level of technical complexity required to achieve causal inference using systems biology approaches. With so-called workstations and graphics processing units (GPU), along with appropriately trained bioinformatics staff, hundreds of genomes can be generated and analyzed in approximately 2–3 days, providing near real-time results for surveillance and monitoring of emerging disease outbreaks and deep genotyping for the characterization of complex chronic diseases.

Second, appropriate knowledge in bioinformatics is required to guide the hypothesis-focused analysis plan. An appropriate analysis of the data generated from sequencing starts by choosing the best algorithm tool to compare our query sequences with database sequences. It has been proposed a wide variety of analyzing tools, like the Basic Local Alignment Search Tool (BLAST), that can be chosen according to the purposes of the study. This requires an expert in bioinformatics to know the possibilities for the best approach to the analysis. Moreover, programming languages are widely used in data processing (e.g., Python, Perl, R, Julia, or Shell). A good domain of this programming language allows a more efficient analysis.

Finally, conclusions must be interpreted by a multidisciplinary team dedicated to the area of interest in the problem to be limited. Given the large amount of information that is managed and the possible confounders that may be implicated, investigators should have experience in the area of interest for a correct interpretation of the result. This needs to be assessed from the point of view of statistical and biological significance. Whatever a genome sequence variation is characterized in a group of individuals, its physiological and pathological implications must be established to determine the real impact on health's influence.

Genome-Wide Association Studies

Fundamentally, epidemiology aims to establish a relationship between outcomes and exposures, as well as pursuing understanding of disease processes to improve public health and healthcare in general [11]. Genomic technology and data integration into public health research has highlighted its usefulness and common goals [12]. Genetic epidemiology continues to be a growing field; it is predicted that this approach will soon replace the role of family studies when investigating heritability mechanisms and the identification of how risk factors may influence disease risk at an individual level. Focus on single nucleotide polymorphisms (SNP), which are substitution or variations of single nucleotides, inside the dense amount of genome data that has become more common, along with microarray technology progress, has allowed to better map genetic relationships which diverge from Mendelian mechanisms [13]. Throughout the following segment, we will discuss how two of the most common genomic studies, GWAS (genome-wide association studies) and EWAS (epigenome-wide association studies), could be applied for epidemiological research and how these can be useful to inform research designs.

A Primer on GWAS and its Applications to Epidemiology

Genome Wide Association Studies (GWAS) is a genomic analysis technique at a population level centered in the study of SNP, and sometimes copy number variants (CNVs), the linkage disequilibrium (LD) associated to them, and their relationship with a phenotype, be it a disease or a specific disease trait. For instance, in the locus where a group of a population could have a cytosine (C), the other group could have a thiamine (T). Even though there are four different theoretic possibilities for nucleotides in a locus, in practice, the number of alleles commonly found is primarily reduced to two. When two or more substitutions of this type have a population frequency of 1% or greater, they are called SNPs. This type of genetic change accounts for approximately 75% of all genetic variation [14].

When SNPs are studied, a concrete combination of them is expected to be found depending on their prevalence among the population. As an illustration, if the SNP "A" has a prevalence of 80% and the SNP "B," which resides at another locus, has a prevalence of 30%, the expected proportion of subjects with both SNPs present at the same time would be 24% (0.8*0.3 = 0.24). Linkage disequilibrium is the difference between the expected prevalence of two SNPs and the one observed or obtained through an experimental measurement, being positive in the case it is greater than the one expected or negative, if it is lower. Both concepts, SNPs and LD, are fundamental in order to be able to understand GWAS. Using LD, dependence patterns between previously characterized allele groups can be inferred, and correlation analysis can be done, grouping them in LD blocks. These LD blocks allow for genealogical and familial tracking, as well as database building, from which polymorphisms that will not be measured in a posterior study can be predicted or imputed. Data imputation in GWAS has been proven useful and gives the opportunity to perform an extensive analysis of the obtained genetic code [15].

The study of SNPs is important because in classic Mendelian genetics, mutations could be followed for generations due to the great effect that a single gene had in a trait or a disease; however, not every trait has a clear monogenetic inheritance, nor every gene has a strong effect on the phenotype. In consequence, the study of multiple genes with moderate or mild effects was necessary if these more complex inheritance and diseases were to be explained. This is where GWAS comes into play, as it allows the simultaneous observation of the whole genome when searching for these variants, and their cohesive analysis, making the association between the genetic code and the observed phenotype possible. The rarer a variant, the larger the sample size will need to be to reduce the possibility of false-positive or false-negative findings. It is for this reason that cooperation among researchers and institutions is frequent and that more common variants are the ones studied.

How Is the GWAS Approach Applied to Epidemiology?

Methodologically, a cases-control study design is often used for GWAS. A few hundred samples are genotyped, which could belong to a cohort, such as the Rotterdam or Framingham cohorts, or to a biobank. A ratio of 1:1 control to cases is regarded as the gold standard, but a ratio of 1:4 is often preferred [16]. Once the genetic material is extracted, the genotype is obtained through hybridization to quartz chips with oligonucleotides of approximately 50 base pairs that are flanking the desired SNP, then, DNA amplification is performed, along with washing steps, and finally, an ultrahigh-definition laser that detects the fluorochromes is used in the process. This allows for many SNPs to be rapidly genotyped, as current SNP arrays range in size from 200,000 to 2,000,000 [15]. A large amount of computational power is required for the process and storage of obtained information, as a 10,000 sample database could be as large as 15 terabytes [16].

GWAS can also be done in two stages. First, relevant or clinically significant variants are identified and followed by a technical and statistical quality control. Second, another independent analysis is performed to verify the findings. For a SNP to be genome-wide significant, a p value of 10^{-8} is often used [17]. After getting the genotype, the stored information is analyzed. The objective is to associate known phenotypes to the identified variants. The minor allele frequency (MAF) is the proportion of the second most common allele in a population. When this frequency is $<1\%$, the assumptions could no longer hold and should be studied with a larger sample size [16].

For those SNPs that are statistically significant, a posterior analysis is made to find its relationship with adjacent or near genes and LD blocks. A great benefit from GWAS is that associations and correlations for variants that had no previous information are found, and this fact opens a new opportunity for future research and therapeutic targets. On the other hand, GWAS suffers from the limitation of not offering an explanation of the mechanisms which underlie the variant's effect on the phenotype, which makes the selection of a new variant to study difficult, as many loci could have newly identified associations and characterizations.

Linkage analysis was the main method used to study the causal implication of certain variants. It was specifically useful for Mendelian traits, but it was not as effective for complex traits and common diseases. For this reason, a method for analyzing mutations that had a small effect on the phenotype and that could not be identified through linkage analysis was needed. The International HapMap Project, or Haplotype Map, was an international collaboration project focused on mapping the most relevant SNPs in such a way that out of the ten million in total, 500,000 could be enough to be representative of the entire genome, and haplotypes or groups of these SNPs could be formed. This allowed for an efficient approach to genetic variation and made possible the calculation of linkage disequilibrium.

Advances in metabolic, psychiatric, and autoimmune diseases have happened as a direct result from GWAS, ranging from the discovery of a new therapeutic target to medication repurposing. There is no doubt that GWAS has been a powerful resource since its first implementation and that it will continue to be in the foreseeable future. It is highly likely that GWAS will continue to allow for discoveries of new SNPs and their association with known diseases, being supported with larger sample sizes and bigger databases to work with. Almost 10,000 genome-wide significant associations between genetic variants and complex traits have been reported [18]. From the data that GWAS has provided, it has been clear that nearly every complex trait has multiple loci contributing to the genetic variation and that the same genetic variants can be associated with multiple traits or diseases, a characteristic called pleiotropy.

Incorporating GWAS Findings into Epidemiology

Incorporation of genetic information into epidemiological findings is a relatively new approach. However, with the newest advancements in sequencing and microarray technologies, the challenge has shifted from the performance of individual GWAS to the management of the enormous amount of data that has originated from these analyses and in identifying meaningful biological and clinical ways to interpret and integrate the data into current frameworks [19]. Genetic epidemiology, per se, involves a mixture of both complex epidemiologic models and additional statistical designs which have evolved from physical mapping to Hardy-Weinberg equilibrium and, further on, for their understanding. GWAS data and epidemiology should aim to be studied in an intersectional manner, as a more accurate statistical modelling may help understand genetic multicausality in hereditable diseases. Nevertheless, it should be noted that one of the main limitations of GWAS analysis is its lack of reproducibility and further inconsistency, which makes strong case for a need to systematize and optimize quality of GWAS in epidemiology [20].

GWAS focuses on the identification of common SNP alleles for an evaluation of disease risk and is a popular approach to investigate associations between genetic information and diseases [21]. Originating from the premise of *penetrance,* the most basic probabilistic genetic model, which describes the probability of a phenotype for a given genotype [22], altogether with Mendelian (or chromosomic) inheritance, genetic models have evolved to adapt with the immense amount of information supplied by GWAS. As of February 2020, there were more than 175,000 SNP and disease outcome associations [21], and more than a million SNPs were identified from every individual sample, capturing human genome variability, based on the idea that common alleles in a given population would explain, at least partly, disease heritability [23]. Today, complementary to the first wave of studies that explored if and which specific DNA sequences or genes were associated with a single disease, GWAS is focusing more and more toward comparing the buildup of the transcriptome and RNA composition in affected and unaffected individuals, pretending to answer what could be going on in the cell from a dynamic and time-dependent perspective during the physiopathology of a disease [24].

Population studies from GWAS data have shifted from focus into familial or close-relation cases where the studied disease was common into massive numbers of unrelated individuals for case-control or cohort stratified studies. Associations have been found as commonly in coding regions as they have in genetic deserts. However, the detection of rare variants remains more likely in extended pedigree analysis; although they co-segregate with family traits, the effect size is still relevant despite the small sample size [25]. Family information remains relevant for study and statistical analysis design. The ultimate goal of a case-control study with GWAS data is to identify allelic frequency differences between controls and cases in order to identify particular SNPs that could be related to disease susceptibility [26]. For GWAS analysis, both findings in independent cohorts and a large sample functional analysis should be completed as gold standard.

Analyzing GWAS Data

There are several considerations to be made when performing statistical analyses on the observations from a SNP array, regarding experimental design, data cleaning, and further analysis. Although GWAS has become more and more common since its introduction, there are still several challenges regarding its analysis [27], which will be covered throughout this section.

The first thing to consider, as in any statistical analysis one aims to perform, is the quality of the data. Most of the variants that show relationship with disease outcomes that are detected during the analysis have small effects, accounting for a small fraction of the sample's genetic variance. Given this, special attention should be paid to data quality control in order to ensure the observed variability is not caused by random or systematic error [26]. A large data size is critical for this analysis, but quality control and assurance may be even more relevant. However, numerous quality control strategies have been developed, and there are different software packages [28] (and R packages) which can take care of this procedure before taking on any analysis. It is relevant to point out that it has been observed that SNP calling is closely related with the software used for the analysis [25]. Common problems in analysis quality arise from experimental factors which relate to allelic frequency, such as differences in population structures, related individuals under independence assumptions, or non-random missing data [26].

Missing data in GWAS is measured with the *missing call rate*. This metric refers to "the fraction of missing calls per SNP over samples or the fraction per sample over SNPs" [26], which refers overall to genotyping efficiency. When samples have a low genotyping efficacy, they should be removed from the analysis. The threshold for the missing call rate should be established considering the specific efficacy of the genotyping of the SNPs of interest for each individual [29]. When there is a high missing call rate for a certain SNP, for example, >5%, the SNP should not be considered for the analysis. Missing data can be dealt with using imputation techniques based on reference datasets, like the HapMap Project [30], or with known haplotype data [31]. SNP alleles with a very low frequency (<1%) also should be excluded from the analysis, given that they are likely representative of genotyping error and on themselves have low statistical power. Outliers should also be dealt with nearest-neighbor analysis, considering ~4 standard deviations. There are no established thresholds for quality data control, but they should always be specified and determined according to the current sample characteristics [32].

As preliminary data quality control, Hardy-Weinberg equilibrium (HWE) should be evaluated. If a deviation exists for SNPs of interest in case groups, it is normally regarded as a signal of true association [32]. Normally, after a Chi-squared evaluation, a statistical significance of $p < 5x10^{-8}$ is considered, given that there is multiple hypothesis testing going on. Other corrections, such as Bonferroni or false discovery rate, could also be performed to determine optimal statistical significance, as well as the use of Bayes factor [33]. When a "normal" p-value limit of 0.05 is considered, it means that there is five in a hundred chances that the null hypothesis is being wrongly rejected. However, when the analysis includes around a million different SNPs, there could be up to 50,000 false-positive carriers. That is why, to try to minimize error related to multiple testing, significance is established at such threshold [20]. It is recommended to ensure HWE in control groups; if there is a deviation in this group, there probably exists genotyping error or even genetic association [27].

Another relevant aspect to consider for epidemiologic GWAS application is population structure. Preexisting differences in ancestry contrary to disease could lead to spurious observations, so population stratification should be checked within groups as part of quality data control [29]. This condition is one of the most common reasons for the lack of reproducibility of GWAS analysis. Population stratification is defined as when the difference in allele frequencies between cases and controls "is due to systematic ancestral differences" [32]. Different approaches, such as genomic control based on Chi-squared tests [34], structured association involving clustering based on a Bayesian algorithm [35], principal component analysis (PCA) [36], and multidimensional scaling [28], have been proposed to both correct and detect population structure.

Once quality control has been guaranteed, the statistical analysis for association in GWAS data has as main objective the comparison of allele or phenotype frequency between cases and controls [32]. One of the most common tests in single genetic association studies is the Cochran-Armitage test, which follows the gene-dosage model and is distributed asymptotically as a Chi-squared distribution. However, the gene-dosage assumption tends to deviate from reality, so other optimal trend tests for basic genetic association, especially for heterogeneous diseases, have been developed [37]. These tests are improved by considering HWE inside the test. This provides a more sensitive testing alternative for each case. Other methods as logistic and linear models, t-tests, or survival analysis can also be used depending on the viability of the data. When considering complex diseases, penalized regression models are often considered after a selection screening process. After relevant SNP identification, further independent validation should be carried on in order to discard spurious relationships in replication samples. However, biological evaluation and analysis should be considered before discarding any association [32]. The final step for any basic association analysis then consists in data visualization in order to understand GWAS observations. The most common and used tool for this is the Manhattan plot, which graphs p-values in the log-scale against SNP physical position in order to distinguish possible SNPs related with the studied condition [32].

Epigenome-Wide Association Studies

The term genetic epidemiology has been defined as "the integration of epigenetic analyses into population-based epidemiological research with the goal of identifying both the causes and the phenotypic consequences" [38]. Genetic epidemiology also includes the consideration of data coming from epigenome-wide association studies or EWAS, whose main goal is to inform on environmental factors which influence gene expression in genetic studies [39]. With the same aim to that of a GWAS, EWAS seek to replace familial genetic studies and look for association for common conditions instead of rare variants. Tissue specificity is key in these studies; however, ideal tissues for analyses might not always be readily available. Therefore, common samples are drawn from peripheral blood, and white blood cells are selected depending on research interest [40] or drawn from easily accessible sources, such as saliva, nasal swabs, or urine samples. Tissue heterogeneity is another important issue to be considered for sample selection for EWAS, as also the ideal sample choice depends on the pathological nature of the disease under study. For more information regarding sample selection, readers are suggested to consult Rakyan et al. and Michels et al.

The epigenome has the peculiarity that it is dynamic; it can show hereditable changes in chromatin that modify genetic expression without modifying its sequence, and these same modifications could be caused by environmental factors. Studying the epigenome is of epidemiologic interest given that it is inheritable across cell generations, so it represents a reliable way to study long-term environmental exposure and its outcomes for disease processes [41]. Its study has been of special interest for cancer genomics and heritability, especially regarding hypomethylation in CpG islands, local hypermethylation, and silencing of DNA repair-related genes. The epigenome regulates dynamic gene expression that determines cellular phenotypes, with genomic imprinting being one of the most important mechanisms regulated by the epigenome.

The most common epigenetic marker is DNA methylation, whose regulation is essential for the correct functioning of the cell, and it occurs most commonly in CpG islands. Although EWAS is centered on cytosine methylation analysis, it is the most accessible epigenetic modification to be analyzed [42]. The epigenome involves other modifications, such as histone modification (methylation, ubiquitination, acetylation), or enzymatic modifications in either DNA or histones [43]. DNA methyltransferases (DNMTs) are responsible for the maintenance of methylation patterns in the cell; these methylation patters are plastic, enzymatically reversible, cell-type specific, and can be affected by SNPs [11]. This is why, unlike GWAS, EWAS associations can be both causal and consequential for a determined phenotype [42], making its analysis relevant not only for disease comprehension but for its prevention, as it includes a time-varying component [44]. EWAS might be a very relevant tool to find epidemiologically significant biomarkers.

Methods for Conducting EWAS

EWAS were first introduced in 2011 in a review, as a new proposal for a novel analysis in identification of loci associated with common diseases [42], but one of the first reported EWAS was performed in 2010 by Patel et al. for type 2 diabetes mellitus (T2D), in which methodology from GWAS was adapted to identify relevant environmental factors which could influence T2D etiology [39]. Before this, epigenetic studies were performed with limitations such as inadequate genome coverage or inadequate sample size, considering microarray technology (e.g., Illumina array) to identify methylation signatures in patients with ovarian cancer, bladder cancer, and breast cancer, among others [43]. These studies did not present independent validation, therefore lacking reproducibility. Today, 450 K array are used to perform EWAS, and although it covers less than 2% of the CpG sites in the genome, it tries to cover all known genes. Some of the most robust EWAS findings have been the relationships between methylation and smoking status and methylation and increasing age. EWAS-based sequencing is a promising approach which could influence future studies, perhaps using Bis-Seq data [43].

Methods for EWAS have passed through several iterations. The first one to be used included arrays, such as MethylScope and CHARM, and methylation-sensitive restriction enzymes, which were aimed to detect differentially methylated sequences. The limitations of this method are the large volume of starting DNA required (100 ng to 1000 ng), the susceptibility to false-positives, and the qualitative or semiquantitative nature of the results. Chip-based methods are largely in disuse. Affinity enrichment uses 5-methylcytosine (5mC)-specific antibodies or MBD cap proteins (methyl-CpG-binding domain) to detect methylated DNA segments and is another method for epigenomic analysis. Methyl-DNA immunoprecipitation (MeDIP-chip) is an example of this. MeDIP-chip targets preferentially zones of DNA with low CpG concentration and can work with approximately 5 ng of DNA. It is a good option for large-scale EWAS and has been used to study cell-free DNA for cancer detection. The results are semiquantitative, it has a low resolution, and it is susceptible to batch effects.

The current gold standard for epigenetic analysis are the bisulfite conversion-based methods. These are based on the differential conversion of unmethylated cytosines to uracil by sodium bisulfite, which are then converted to thiamine during the following polymerase chain reaction (PCR) and are finally compared to the reference sequence for analysis. Bisulfite conversion-based techniques allow for single nucleotide definition and quantitative results. Among this group, there are DNA methylation arrays and bisulfite sequencing. An example of the former is Illumina's DNA MethylationEPIC array, with a coverage of 850,000 CpG sites, a 3% of the total in the human genome. It has the advantage of being relatively inexpensive, having a simple setup, and allowing for analysis of large volumes of data. It is the most popular method for large quantities of samples. This method requires ~500 ng of DNA, making its application in liquid biopsies or needle samples difficult.

The latter is a wider group of techniques. Bisulfite sequencing could be classified by the coverage of the epigenome they have. Whole-genome bisulfite sequencing (WGBS) has the largest coverage, with 28 million CpG sites, and is the most expensive and time-consuming, which makes it unviable for a large volume of samples, and it is excellent at characterizing epigenomes of specific cell types and is the gold standard for fine-mapping of CpG sites. Reduced representation bisulfite sequencing (RRBS) has a smaller coverage ranging from 3 to five million CpG sites. It relies on methylation-sensitive restriction enzymes and has proven useful to analyze DNA from single cells and cell-free DNA. Bisulfate sequencing can also be done targeting specific desired sections by probe hybridization. It has an approximate coverage of five million CpG sites, and it enables tens of thousands of probes to be utilized simultaneously. This method has been used to approach hepatocellular carcinoma.

Application of EWAS in Epidemiological Studies

Epidemiology constantly faces the challenge of distinguishing causal inferences from statistical associations with confounded causality, and in the case of genomic data, this becomes more relevant. GWAS look for direct causality only, given that DNA remains constant throughout life; however, mechanisms cannot be ascertained as expression of certain variants is not guaranteed. EWAS, on the other hand, allow for more flexibility, being that different epigenomic markers may indicate both either cause or consequence of several disease processes [45]. Some considerations should be made before starting any EWAS analysis, including study design, study population, sample size, and consideration of possible confounders.

After tissue and epigenetic markers had been chosen, the best study design must be chosen to ensure the possibility of identifying meaningful causal associations. Missing data and population stratification should be dealt with in a similar manner as if data were analyzed using a GWAS approach. However, EWAS data presents more challenges compared to GWAS data. The epigenome is widely modified by environmental factors, which do not have any known matrices or theories as to how they behave, being additionally heterogeneous along the life course [11]. Similarly, lack of consensus and quality control in environmental confounders is still a major challenge for EWAS data analysis [44]. Methylation states are tissue-specific, and significant measurement error may impair precise identification of causal associations. Common epigenetic confounders include aging, specific allele methylation, and other environmental factors to which individuals are constantly exposed. There is a great way to go still on information regarding EWAS data; there is still work in progress mapping the human epigenome and cataloguing the inter-individual epigenetic variable regions within the genome [38]. The information provided by epigenomic analysis is invaluable in understanding the mechanics of disease, and it should be pursued alongside statistical and epidemiological rigor for the greatest benefit of public health studies.

Polygenic Risk Scores

Polygenic Risk Scores in Personalized Medicine

Over the past decade, the decreasing cost of whole-genome sequencing (WGS) has made increasingly accessible for its application in large genome-wide association studies (GWAS). More and more single nucleotide polymorphisms (SNPs) extracted from GWAS of high prevalent diseases such as diabetes, cancer, and cardiovascular disease have been derived [46–49]. However, the question remained as to how these SNPs could be combined to predict a relevant clinical outcome. This approach has led to the development of polygenic risk scores (PRS) as a way of integrating genetic information onto epidemiological or clinical applications in the new era of personalized medicine. Today the terms genetic risk scores, polygenic scores, and PRS are used indistinctively and consider the incorporation of genetic markers for the prediction of specific disease traits or its relevant outcomes [50].

Genetic risk scores (GRS) have two main goals: 1) to predict the likelihood of developing a disease or disease trait and 2) to estimate the predictive capacity that is captured by the associated genetic variants onto a specific outcome or trait [51]. The proposed applications of PRS range from assisting disease diagnosis, informing the selection of therapeutic interventions, improving risk prediction, and reporting disease detection even in a preclinical state [49]. Furthermore, the implementation of GRS has aided the evaluation of the cumulative effect of genetic factors taken together with clinical indicators over the outcome of specific disease traits [52], and given that genomic profiling alone is still debatable to be able to predict complex diseases for routine clinical use, PRS may offer an attractive alternative [52]. Development of PRS has increased in recent years, but it has also led to a reframing of its current approaches regarding their methodological and pragmatical application within personalized medicine.

First Approaches within the Estimation of GRS

GRS found their very first primitive origins in marker-assisted selection (MAS), which was first applied for animal breeding in 1998 [53]. This technique originated from the use of molecular markers to find individual quantitative trait loci that control important traits and therefore combining the effects the loci have on them to develop a ranking to aid development of lines or populations. These effects were determined through population genetics and linkage disequilibrium (LD), and although these methodologies were not applied into human populations until 2001 [54], they set a first approach toward what we know today as genetic risk scores. Furthermore, common GRS have evolved to PRS from the premise that heritable traits are not due to the transmission of one relevant loci, but to the combination of the smaller effect of other genes [50]. For most diseases, a single variant is not enough for the assessment of disease risk, so a PRS is formed from a set of independent risk variants associated with a disorder, based on evidence from GWAS data and weighted by the number of alleles the subject is carrying and the specific risk for the variant [49]. A PRS then is defined as a "single value estimate of an individual's genetic liability to a phenotype, calculated as a sum of their genome-wide genotypes, weighted by corresponding genotype effect size estimates derived from GWAS data" [55].

Determination of Genetic Risk Scores

The creation of PRS has been improving over the past few years. As originally conceptualized, PRS are created using the weighted sum of the risk alleles of single nucleotide polymorphisms (SNPs) extracted from GWAS of a specific disease. Numerous examples of PRS had been created using thousands of SNPs for several diseases including diabetes, breast cancer, and coronary artery disease [48]. The advances in statistical methods, computational capacity, and the availability of large population datasets have led to the rapid development of PRS applied to modern epidemiology over the past few years.

There are many different approaches for GRSs' determination. They range over a simple sum of odds ratios derived from logistic regression multiplied by the number of risk alleles [28] to a Bayesian mixture for additionally complex traits for PRS [55]. One of the main issues concerning GRSs and PRSs' determination is missing data from GWAS or EWAS, specially missing measurements on SNPs of interest [51]. Nevertheless, it should be noted that GRSs and PRSs' accuracy is normally assessed by the area under the receiver operating characteristic (ROC) curve using a cutoff of >0.75 for informative screening and > 0.99 for use as a diagnostic aid [56]. Other metrics are also used to estimate the proportion of trait variability established by the determined markers, such as the population attributable risk (PAR) which expresses the fraction of cases attributable to a single exposure [51]. All these metrics have important statistical considerations, along with the need for calibration curves to evaluate if there is overfitting. Even though GRS have become popular in use, until recently there had not been a solid delimitation of steps and methodology to perform and interpret them [55]. Readers are recommended to follow methods proposed within the TRIPOD statement for reporting on the development of predictive models using population data, which can be applicable to PRS [57].

PRS require two datasets in order to be performed: the base data (composed of GWAS/EWAS data with summary statistics corresponding to SNPs and genotype-phenotype associations) and the target data (which is composed of an independent sample from the base data used for PRS analysis). Similarly, as in GWAS/EWAS analysis [32], shrinkage algorithms should be performed to ensure the ideal number of SNPs are included considering effect size. PRS will tend to predict heritability and phenotype as GWAS data size increases [58]. Next, data should be submitted to a strict quality control process, starting from an adequately large sample size and with several considerations such as the effect allele and the genotyping method; these should be known and equal for both sets of data in order to be able to ensure trustworthy results [55], as well as avoidance of sample overlapping to avoid result inflation. It should be remembered that data mishandling at this stage could result in false positives and observations, as described for GWAS/EWAS. To perform any PRS analysis, there should be LD in the selected SNP(s) in order to ensure that the observations are due to polygenicity and not confounding [59]. GWAS should be performed on standardized protocols to ensure reproducibility, as previously described in this chapter. For more details on this, the work of Marees et al. proposes an additional guidance in the matter [60].

After ensuring that selected data is appropriate for analysis, the corresponding PRS should be calculated. Normal methodologies rely heavily on shrinking or clumping algorithms similar to GWAS/EWAS, which could include regularization such as penalized regression using Least Absolute Shrinkage and Selection Operator (LASSO) or ridge penalization [61] or other Bayesian approaches, considering p-values as explained in GWAS section. These algorithms are also useful to control for LD inside observations and must be optimized based on biological observations as well in order to avoid overfitting; furthermore, the selected method will go in accordance with what the researcher is aiming to observe. It should be remembered that PRS units will be concordant with those observed during GWAS. Finally, a regression using the calculated PRS as a predictor for the outcome must be performed in the target data, adjusting by covariates if necessary. Adequate graphic representation and the aforementioned goodness-of-fit metrics should then be estimated [55].

Limitations of Polygenic Risk Scores when Applied to a Real-World Scenario

The limitations that have arisen to the application of GRS in a real-world context have been discussed. First, there is great heterogeneity attributable to the type of cohorts that are used for the creation of PGS [51]. Most cohorts in which RMPs are developed may not be representative of the high ethnic diversity, suggesting that RMPs can only be applied in populations of origin and opening a public health gap in minority populations which are often underrepresented in genetic research [62]. Second, biobanks which have been used for the creation of PGR (e.g., UK Biobank) may not have enough individuals for a particular disease, limiting the size effect substantially. Third, the creation of a PGR has generated a debate regarding the ethical, legal, and social (ELSI) aspects of its application in clinical practice. It has been pointed out that for some health conditions, such as neurodegenerative and psychiatric diseases, the use of PGR has to be advised from an ethical-medical point of view [63].

For epidemiology, the utility on PRS lies on its interpretation and its applicability. There are four stated considerations for this:

- Previous known information for the individual.
- What is not known for the individual, such as the subject's environment or missing data in sequencing.
- Potential for incorrect information, such as possible bias at the time of sample selection.
- The ultimate goal for the PRS for the individual [49].

These four considerations encompass much of the unknowns that epidemiologists are facing when dealing when GWAS, EWAS, or genomic data in general. However, clinical utility of PRS is yet to be established. There are still some challenges, such as the basal risk for many of these scores has not been yet determined, and there are still disadvantages in individuals that were not subjected to familial evaluation. Sample size also presents a challenge in applicability and transferability for these observations, including ethnic considerations [64]. The great majority of genomic analyses have been performed in individuals of European ancestry, and this also presents another issue regarding study transferability, as not all polygenic risk scores may not be equally useful for all ethnicities [65].

Perspectives for Bioinformatics and Genomics Using Systems Biology

Nowadays, approaches to study of disease processes have evolved from trait-driven toward including a more comprehensive view of their components and the interactions they hold between each other. Although the study of the entire genome in a population has made it possible to establish associations between the presence of gene variants and the development of diseases or specific traits, many of these associations have not resulted in the determination of specific disease mechanisms or have made their way into personalized medicine approaches. Pathophysiological processes comprise a more complex set of elements in which the influence that transcription has on certain regions of the genome contributes to the regulation of the expression of other genes; similarly, expression of said transcripts on specific protein patterns or its influence on metabolic profiles also has specific influences on the course of diseases and its understandings. Thus, the presence of Gene Regulation Networks (GRN) represents one of the fundamental paradigms in the relationship between genomics and systems biology.

Systems biology is defined as the "study of interactions between parts of a system through the use of experimental and computational methods." This involves the integration of both structural elements, which included genomics and transcriptomics analyses, and functional elements, which consider expression levels of transcripts ranging from protein expression, metabolic profiles, and cross talks between tissues and its environment. Sciences such as physiology, genomics, metabolomics, proteomics, and population biology are the fundamental pillars of systems biology. In its essence, systems biology aims to provide a more comprehensive view of epidemiological findings at different levels in order to reconstruct disease processes to identify biomarkers, targets for the development of therapeutic interventions, and predictors of clinically meaningful outcomes. Therefore, systems biology is a multidisciplinary field and has two main approaches: 1) the top-down approach, which uses omics data in an integrative fashion, and 2) the bottom-up approach, which uses detailed specific information, like channel physiology, to construct a model describing the overall system and specific interactions among its components [18].

Emerging technologies, such as genome-wide association studies (GWAS) and epigenome-wide association studies (EWAS), allow for population-level analysis of the genome and epigenome, which are known to continuously influence each other and to have extensive networks of interactions that shape their impact on the phenotype. They also produce vast amounts of data that require considerable computational power and storage capacity to allow for casual inference of the impact of identified associations into disease processes and specific disease traits. These technologies have benefited from the analysis brought forth by modern bioinformatics and continue to make discoveries on the influence of SNPs and methylation patterns on specific diseases, such as diabetes mellitus and schizophrenia [66]. Approaches which integrate omics data to infer biological pathways and disease patterns require multivariate techniques which are often coupled with machine learning algorithms to increase feasibility for causal inference and reduce the possibility of false-positive findings. These have led to the development of systems medicine, which integrate systems biology findings onto clinical practice by means of innovative ways to model complex diseases processes, drug discovery, and biomarker identification for prevention, prediction, diagnosis, and treatment as a means of furthering the field of personalized medicine [3]. Among the main applications that this interrelation between genomics and systems biology keeps is the development of personalized medicine. For example, flow balance analysis has been used to model the metabolism of patients with hereditary hemorrhagic telangiectasia in order to identify alterations in the metabolic pathways. Likewise, other applications revolve around developing new drugs with broader safety profiles in less time and costs. The role of pharmacogenomics is particularly important in identifying interactions between drugs and the genotype of the individual to facilitate individualized treatment and monitoring and increased the likelihood of therapeutic benefit while minimizing adverse reactions.

Another very promising area is the study of the composition and interactions of bacterial communities with their environment, which is considered in the study of the microbiome [67]. A variety of sequencing projects have revealed that many undefined microbial species interact cooperatively with the environment in every imaginable ecological niche. This includes microbial communities associated with specific niches of the human body that cause a wide variety of diseases such as inflammatory bowel disease and obesity; furthermore, these microbial species modify the influence of genetic and environmental associations in disease processes and have a relevant role in predicting the development of complications, response to treatment, and prognosis in conjunction with mechanistic insights [68]. While we accumulate large sets of metagenomic data and catalog the bacterial genomes that make up the human microbiome, under both normal and pathological conditions, it is still very difficult to connect the presence or change in the frequency of specific bacterial species with the associated phenotypes.

These previous examples only demonstrate the increasing need for an approach which considers more and more variables at different biological levels into consideration. Systems biology considers the assessment of different aspects of a specific process which are identified through a wide array of methods and techniques and which individually provide unique insights on the particularities of a disease process and the interaction of all involved components [69]. Due to the ever-growing size of data collected from multiple fields of biological sciences, an integrated analytic technique or which that can measure, predict, or model the interactions among all the components of a system will need to be developed, and its development is currently an area of active bioinformatics research. Although systems biology brings a more holistic approach that is different from the historic reductionist view of epidemiology and has made important contributions, like chronotherapeutics, that take into account the endogenous biological rhythms to treat diseases, theories that integrate all these different components are yet to be confirmed and its translation into personalized medicine requires further validation studies [70]. This is an area of opportunity for complexity sciences which seek to explain the general principles that underlie complex systems, and their common properties, so they can be understood as a whole and beyond the explanation of a single element. Promising contributions of systems biology to epidemiological research require that further research efforts aim for multidisciplinary approaches to target these findings [71].

Conclusions

As discussed throughout this chapter, the applications of bioinformatics and genetics to epidemiological research are vast. Epidemiological associations identified using GWAS, EWAS, GRS, PRS, and systems biology offer insights onto how many of them are modified by complex interactions between genetics, the environment, and intrinsic disease processes. It should be considered that these different techniques are like a chain of knowledge, or multiple levels of data refinement, which offer insights onto different levels of biological complexity but ultimately require integrative multidisciplinary approaches to meaningfully unravel their contribution to the understanding of pathophysiology and its implications for clinical and epidemiological practice. Genomics brings the fundamental and basic physiology, the building blocks of a vast system that affects the phenotype, and bioinformatics builds networks from the multiple discovered elements, identifying interactions among them, and complexity sciences will attempt to find general rules which apply to genomics and the organism, taking into consideration the nonlinear nature of many of those systems. However, in order to make more accurate models, which meaningfully represent the processes they sought to represent, more information is required, and as more technologies that contribute large amounts of information are developed, the application of the proposed models will increase and become more useful to predict future outcomes and to explain the whole network of systems which integrate them.

References

1. Maojo V, Martin-Sanchez F. Bioinformatics: towards new directions for public health. Methods Inf Med. 2004;43(3):208–14.
2. Centers for Control Diseases and Prevention. An introduction to applied epidemiology and biostatistics [Internet]. 2019 [citado 21 de marzo de 2021]. Disponible en: https://www.cdc.gov/csels/dsepd/ss1978/Lesson1/Section1.html#_ref1.
3. Schleidgen S, Fernau S, Fleischer H, Schickhardt C, Oßa A-K, Winkler EC. Applying systems biology to biomedical research and health care: a précising definition of systems medicine. BMC Health Serv Res. 2017;17(1):761.
4. Hood L, Rowen L. The human genome project: big science transforms biology and medicine. Genome Med. 2013;5(9):79.
5. Bayat A. Science, medicine, and the future: bioinformatics. BMJ (Clinical Research ed). 2002;324(7344):1018–22.
6. Kulikowski CA, Kulikowski CW. Biomedical and health informatics in translational medicine. Methods Inf Med. 2009;48(1):4–10.
7. Maljkovic Berry I, Melendrez MC, Bishop-Lilly KA, Rutvisuttinunt W, Pollett S, Talundzic E, et al. Next generation sequencing and bioinformatics methodologies for infectious disease research and public health: approaches, applications, and considerations for development of laboratory capacity. J Infect Dis. 2020;221(Suppl_3):S292–307.
8. Trent R. Clinical bioinformatics. New York: Humana Press Inc.; 2007.
9. Chang PL. Clinical bioinformatics. Chang Gung Med J. 2005;28(4):201–11.
10. Butte AJ. Translational bioinformatics: coming of age. J Am Med Inform Assoc. 2008;15(6):709–14.
11. Tanić M. Chapter twenty-nine – epigenome-wide association study (EWAS): methods and applications. En: Tollefsbol T, editor Epigenetics Methods [Internet] Academic Press; 2020. p. 591–613. Disponible en: https://www.sciencedirect.com/science/article/pii/B9780128194140000029X
12. O'Leary P, Zimmern RL. Genomics and public health: translating research into public benefit. Public Health Genomics. 2010;13(4):193–6.
13. Traynor BJ. The era of genomic epidemiology. Neuroepidemiology. 2009;33(3):276–9.
14. Strachan T, Goodship J, Chinnery P. Genetics and genomics in medicine. Boca Raton: CRC Press; 2014.
15. Visscher PM, Wray NR, Zhang Q, Sklar P, McCarthy MI, Brown MA, et al. 10 years of GWAS discovery: biology, function, and translation. Am J Human Genet. Cell Press. 2017;101:5–22.
16. Uitterlinden AG. An introduction to genome-wide association studies: GWAS for dummies. Semin Rep Med. Thieme Medical Publishers, Inc. 2016;34:196–204.
17. Mullen SA, Crompton DE, Carney PW, Helbig I, Berkovic SF. A neurologist's guide to genome-wide association studies. Neurology. 2009;72(6):558–65.
18. Visscher PM, Brown MA, McCarthy MI, Yang J. Five years of GWAS discovery. Am J Hum Genet. 2012;90:7–24.
19. Fallin MD, Duggal P, Beaty TH. Genetic epidemiology and public health: the evolution from theory to technology. Am J Epidemiol. 2016;183(5):387–93.
20. Peter I, Seddon JM. Genetic epidemiology: successes and challenges of genome-wide association studies using the exam-

ple of age-related macular degeneration. Am J Ophthalmol. 2010;150(4):450–452.e2.
21. Cao X, Xing L, He H, Zhang X. Views on GWAS statistical analysis. Bioinformation. 2020;16(5):393–7.
22. Östensson, Malin. Statistical Methods for Genome Wide Association Studies [Internet]. [Göteborg, Sweden]: Chalmers University of Technology and University of Gothenburg; 2012. Disponible en: https://core.ac.uk/download/pdf/70596623.pdf.
23. Moore JH, Asselbergs FW, Williams SM. Bioinformatics challenges for genome-wide association studies. Bioinformatics. 2010;26(4):445–55.
24. Frelinger JA. Big data, big opportunities, and big challenges. J Invest Dermatol Symposium Proc. 2015;17(2):33–5.
25. Ziegler A, Sun YV. Study designs and methods post genome-wide association studies. Hum Genet. 2012;131(10):1525–31.
26. Laurie CC, Doheny KF, Mirel DB, Pugh EW, Bierut LJ, Bhangale T, et al. Quality control and quality assurance in genotypic data for genome-wide association studies. Genet Epidemiol. 2010;34(6):591–602.
27. McCarthy MI, Abecasis GR, Cardon LR, Goldstein DB, Little J, Ioannidis JPA, et al. Genome-wide association studies for complex traits: consensus, uncertainty and challenges. Nat Rev Genet. 2008;9(5):356–69.
28. Purcell S, Neale B, Todd-Brown K, Thomas L, Ferreira MAR, Bender D, et al. PLINK: a tool set for whole-genome association and population-based linkage analyses. Am J Hum Genet. 2007;81(3):559–75.
29. Turner S, Armstrong LL, Bradford Y, Carlson CS, Crawford DC, Crenshaw AT, et al. Quality control procedures for genome-wide association studies. Curr Protoc Hum Genet. 2011;Chapter 1:Unit1.19-Unit1.19.
30. Howie B, Fuchsberger C, Stephens M, Marchini J, Abecasis GR. Fast and accurate genotype imputation in genome-wide association studies through pre-phasing. Nat Genet. 2012;44(8):955–9.
31. Balding DJ. A tutorial on statistical methods for population association studies. Nat Rev Genet. 2006;7(10):781–91.
32. Zeng P, Zhao Y, Qian C, Zhang L, Zhang R, Gou J, et al. Statistical analysis for genome-wide association study. J Biomed Res. 2015;29(4):285–97.
33. Stephens M, Balding DJ. Bayesian statistical methods for genetic association studies. Nat Rev Genet. 2009;10(10):681–90.
34. Dadd T, Weale ME, Lewis CM. A critical evaluation of genomic control methods for genetic association studies. Genet Epidemiol. 2009;33(4):290–8.
35. Falush D, Stephens M, Pritchard JK. Inference of population structure using multilocus genotype data: dominant markers and null alleles. Mol Ecol Notes. 2007;7(4):574–8.
36. Price AL, Zaitlen NA, Reich D, Patterson N. New approaches to population stratification in genome-wide association studies. Nat Rev Genet. 2010;11(7):459–63.
37. Lee W-C. Optimal trend tests for genetic association studies of heterogeneous diseases. Sci Rep. 2016;6(1):27821.
38. Mill J, Heijmans BT. From promises to practical strategies in epigenetic epidemiology. Nat Rev Genet. 2013;14(8):585–94.
39. Patel CJ, Bhattacharya J, Butte AJ. An Environment-Wide Association Study (EWAS) on type 2 diabetes mellitus. Plos One. 2010;5(5):e10746.
40. Michels KB. Considerations in the design, conduct, and interpretation of studies in epigenetic epidemiology. En: Michels KB, editor. Epigenetic epidemiology [internet]. Dordrecht: Springer Netherlands; 2012 [citado 19 de marzo de 2021]. p. 21-35. Disponible en: http://link.springer.com/10.1007/978-94-007-2495-2_3.
41. Birney E, Smith GD, Greally JM. Epigenome-wide association studies and the interpretation of disease -omics. PLOS Genet. 2016;12(6):e1006105.
42. Rakyan VK, Down TA, Balding DJ, Beck S. Epigenome-wide association studies for common human diseases. Nat Rev Genet. 2011;12(8):529–41.
43. Flanagan JM. Epigenome-wide association studies (EWAS): past, present, and future. Methods Mol Biol. 2015;1238:51–63.
44. Zheng Y, Chen Z, Pearson T, Zhao J, Hu H, Prosperi M. Design and methodology challenges of environment-wide association studies: a systematic review. Environ Res. 2020;183:109275.
45. Latvala A, Ollikainen M. Mendelian randomization in (epi)genetic epidemiology: an effective tool to be handled with care. Genome Biol [Internet]. 14 de julio de 2016 [citado 2 de abril de 2021];17. Disponible en: https://www.ncbi.nlm.nih.gov/pmc/articles/PMC4944517/.
46. Weedon MN, Jackson L, Harrison JW, Ruth KS, Tyrrell J, Hattersley AT, et al. Use of SNP chips to detect rare pathogenic variants: retrospective, population based diagnostic evaluation. BMJ. 2021;372:n214.
47. Deng N, Zhou H, Fan H, Yuan Y. Single nucleotide polymorphisms and cancer susceptibility. Oncotarget. 2017;8(66):110635–49.
48. Fiatal S, Ádány R. Application of single-nucleotide polymorphism-related risk estimates in identification of increased genetic susceptibility to cardiovascular diseases: a literature review. Front Public Health. 2018;5:358.
49. Lewis CM, Vassos E. Polygenic risk scores: from research tools to clinical instruments. Genome Med. 2020;12(1):44.
50. Maher BS. Polygenic scores in epidemiology: risk prediction, etiology, and clinical utility. Curr Epidemiol Rep. 2015;2(4):239–44.
51. Igo RP Jr, Kinzy TG, Cooke Bailey JN. Genetic risk scores. Curr Protoc Hum Genet. 2019;104(1):e95.
52. Schork AJ, Schork MA, Schork NJ. Genetic risks and clinical rewards. Nat Genet. 2018;50(9):1210–1.
53. Xie C, XU S. Efficiency of multistage marker-assisted selection in the improvement of multiple quantitative traits. Heredity. 1998;80(4):489–98.
54. Meuwissen THE, Hayes BJ, Goddard ME. Prediction of Total genetic value using genome-wide dense marker maps. Genetics. 2001;157(4):1819.
55. Choi SW, Mak TS-H, O'Reilly PF. Tutorial: a guide to performing polygenic risk score analyses. Nat Protoc. 2020;15(9):2759–72.
56. Janssens ACJW, Moonesinghe R, Yang Q, Steyerberg EW, van Duijn CM, Khoury MJ. The impact of genotype frequencies on the clinical validity of genomic profiling for predicting common chronic diseases. Genet Med. 2007;9(8):528–35.
57. Collins GS, Reitsma JB, Altman DG, Moons KGM. Transparent reporting of a multivariable prediction model for individual prognosis or diagnosis (TRIPOD): the TRIPOD statement. BMJ. 2015;350:g7594.
58. Evans LM, Tahmasbi R, Vrieze SI, Abecasis GR, Das S, Gazal S, et al. Comparison of methods that use whole genome data to estimate the heritability and genetic architecture of complex traits. Nat Genet. 2018;50(5):737–45.
59. Bulik-Sullivan BK, Loh P-R, Finucane HK, Ripke S, Yang J, Patterson N, et al. LD Score regression distinguishes confounding from polygenicity in genome-wide association studies. Nat Genet. 2015;47(3):291–5.
60. Marees AT, de Kluiver H, Stringer S, Vorspan F, Curis E, Marie-Claire C, et al. A tutorial on conducting genome-wide association studies: quality control and statistical analysis. Int J Methods Psychiatr Res. 2018;27(2):e1608.
61. Mak TSH, Porsch RM, Choi SW, Zhou X, Sham PC. Polygenic scores via penalized regression on summary statistics. Genet Epidemiol. 2017;41(6):469–80.
62. Rosenberg NA, Edge MD, Pritchard JK, Feldman MW. Interpreting polygenic scores, polygenic adaptation, and human phenotypic differences. Evol Med Public Health. 2019;2019(1):26–34.

63. Lewis ACF, Green RC. Polygenic risk scores in the clinic: new perspectives needed on familiar ethical issues. Genome Med. 2021;13(1):14.
64. Torkamani A, Wineinger NE, Topol EJ. The personal and clinical utility of polygenic risk scores. Nat Rev Genet. 2018;19(9):581–90.
65. Duncan L, Shen H, Gelaye B, Meijsen J, Ressler K, Feldman M, et al. Analysis of polygenic risk score usage and performance in diverse human populations. Nat Commun. 2019;10(1):3328.
66. Systems Chronotherapeutics | Pharmacological Reviews [Internet]. [citado 2 de abril de 2021]. Disponeible en: https://pharmrev.aspetjournals.org/content/69/2/161.
67. Kumar M, Ji B, Zengler K, Nielsen J. Modelling approaches for studying the microbiome. Nat Microbiol. 2019;4(8):1253–67.
68. Awany D, Allali I, Dalvie S, Hemmings S, Mwaikono KS, Thomford NE, et al. Host and microbiome genome-wide association studies: current state and challenges. Front Genet [Internet]. 2019 [citado 2 de abril de 2021];9. Disponible en: https://www.frontiersin.org/articles/10.3389/fgene.2018.00637/full.
69. Oulas A, Minadakis G, Zachariou M, Sokratous K, Bourdakou MM, Spyrou GM. Systems Bioinformatics: increasing precision of computational diagnostics and therapeutics through network-based approaches. Briefings Bioinform. 2019;20(3):806–24.
70. Ma'ayan A. Complex systems biology. J Royal Soc Interface. 2017;14(134):20170391.
71. ten Bosch L, Hamme HV, Boves L. Discovery of Words: towards a Computational Model of Language Acquisition. Speech Recognition [Internet]. 1 de noviembre de 2008 [citado 2 de abril de 2021]; Disponible en: https://www.intechopen.com/books/speech_recognition/discovery_of_words__towards_a_computational_model_of_language_acquisition.

Spatial Statistics and Health Sciences: Methods and Applications

13

Ricardo Ramírez-Aldana

Introduction

Nowadays, the availability of more and more information and sources to obtain it requires the use of the most sophisticated analyses. This can be seen in the exponential increase in the number of jobs and research papers concerning data science and statistics, particularly concerning Big Data. Additionally, there is a lot of georeferenced data, a term commonly used in the Geographic Information Systems (GIS) framework, see for example [1], meaning the association of maps or images with spatial locations, that is, positions on the earth surface. GIS corresponds to technology joining information and informatics tools for the analyses of spatial data, thus organizing and visualizing these data, producing maps, and allowing spatial consultations and analyses or even the creation of models.

The availability of spatial data is usually of two types: raster (or image files) or vectorial. The latter corresponds to points, lines, and polygons, for instance representing locations of trees in an area, rivers, and states, respectively. Perhaps the most known vectorial format is a shapefile, but there are other formats depending on the software used. In terms of the shapefile, at first it can be confusing that the format does not consist of one, but at least three files, a .shp file containing the geometric characteristics of the objects, a .shx file including indexes of the spatial data, and a .dbf including the data set or attributes associated with the objects, and all of them should be contained in the same directory. The availability of software in which different aspects concerning spatial analyses are available is broad, but some examples are ArcGIS [2], QGIS [3], GeoDa [4], and R in specific packages [5].

Now, considering that we already have software available to analyse spatial information, the question is: What type of analyses are possible? It depends on the data type and the aim of our study. For instance, in a more geographical study, they may be interested in representing specific zones and transportation available between them, identifying through images the orographic characteristics of the zones. However, for spatial statistics analyses, the topic we are interested in this chapter, we usually have vectorial data, frequently just points or polygons (Fig. 13.1), and the questions are closer to those related with statistics and data science, for instance, mapping, clustering, predicting, and explaining a variable through others.

Descriptive Statistics

As in classical statistics, the first type of analysis we could perform is descriptive. This mainly consists of mapping or, more specifically, representing our data through figures. This could seem easy at first, but the required process could vary according to the data and software available. In terms of the data, for instance, we might get a shapefile consisting of the states of a country containing as information (in the .dbf file) only the names of each state. However, our research perhaps concerns the analysis of unemployment; thus we would have to get that information from other sources and join our shapefile with this information. Another possibility is that we do not want all the states in our country, only a zone; thus we would have to select first that zone and then work from this selection. Also, someone would want to represent the states including also the rivers in all the country, both files obtained separately; and thus, the person should have to join the two corresponding layers and, what is more, must be sure that the scale in both is the same, this being a challenge by itself. Since the earth is spherical-shaped, we require a spherical coordinate system, usually latitude and longitude; however, it is difficult to obtain measures in that system, and thus planar coordinates systems are used by projecting the data from the sphere to a plane, see for example [6]. There are several types of projections, according to their type, references used,

R. Ramírez-Aldana (✉)
Dirección de Investigación, Instituto Nacional de Geriatría (INGER), Mexico City, Mexico

J. C. Gomez-Verjan, N. A. Rivero-Segura (eds.), *Principles of Genetics and Molecular Epidemiology*,
https://doi.org/10.1007/978-3-030-89601-0_13

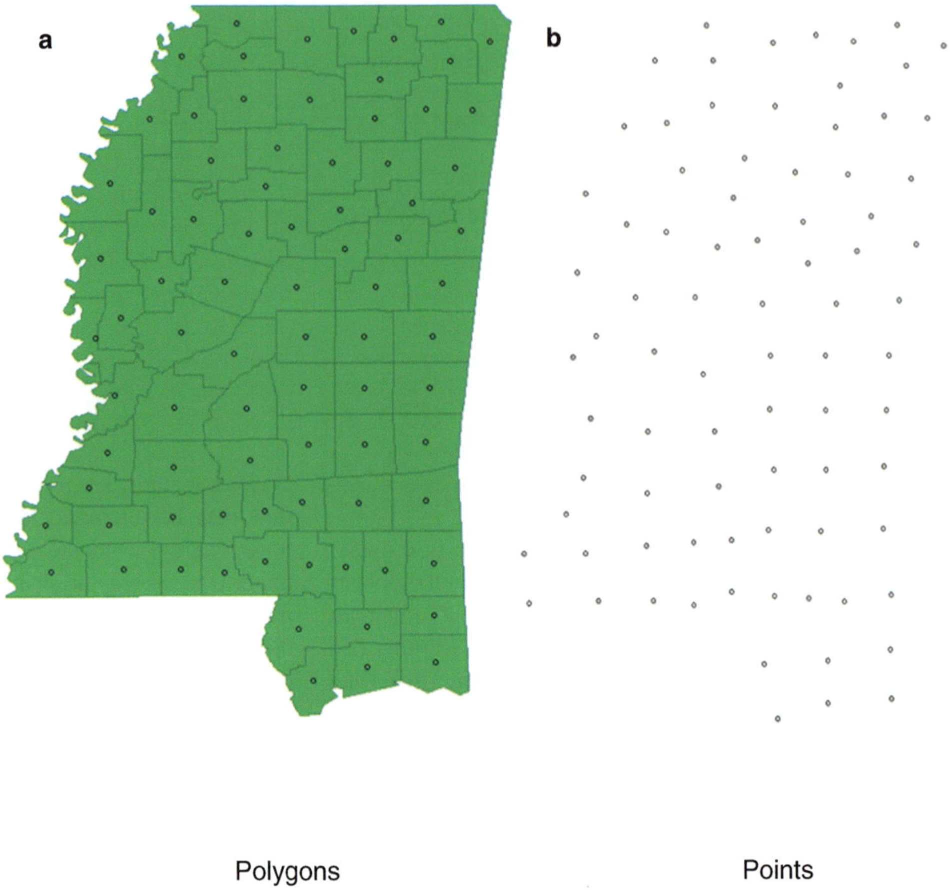

Fig. 13.1 Maps associated with two shapefiles: (**a**) A polygon shapefile and (**b**) A point shapefile corresponding to the centroids of the polygons in (**a**)

etc., and those decisions should be made. This is not the proper place to further introduce these topics; thus, in the following we will assume that all our data is in an adequate projection system, and that if we wanted to add layers concerning different geographical information, they are in the same scale.

Let us return to our study concerning descriptive analyses in spatial data and consider that we calculated all the variables we need, defined an appropriate zone of study, and obtained the adequate layers if we wanted to add geographical information. Assume that our interest is to geographically represent one variable and to understand how it is distributed along a territory. For instance, considering unemployment by states in a country, we might be interested to identify places in which it is greater. Of course, we could represent all possible values with a different colour or even colour them according to a colour gradient; however, we usually try to group the information, for instance, by calculating quantiles and associated groups representing each one with a different colour, for instance, from darkest to lightest. We could also identify the states that are possibly outliers when considering interquartile ranks, or any other method, or map the standard deviation of our variable, among other possibilities.

When mapping risks or rates, something common in epidemiological studies, an additional discussion is needed. In that case, we can calculate these measures and represent them as any other variable; however, we could also represent other measures as standardized mortality risks (SMR) or risks adjusted for sex or other variables using certain standardization processes. For instance, the SMR can be calculated by obtaining the expected value that our variable can take in a state considering that the rates are the same as those associated with all the country and comparing the true values by state with these expected values. There are also some smoothing procedures, for instance, using Bayesian procedures, that improve these measures, considering that in geographical units with lower risks there is a greater variability. And, of course, we could use any other conventional descriptive analysis such as histograms, box plots, dispersion diagrams, etc. to have a better understanding of our data. In Fig. 13.2, we show a map concerning SMR associated with COVID-19 in Mexico and a map of quartiles associated with a variable for the same population. There is software, for instance, GeoDa, that even allows an interaction between a map and one of these classical descriptive representations; for instance, we could identify a point corresponding to an

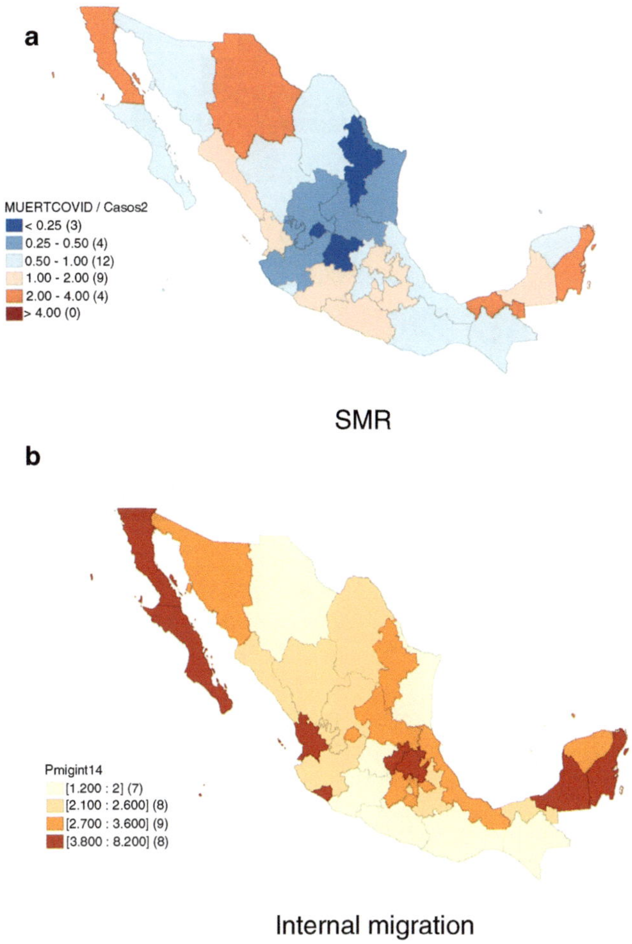

Fig. 13.2 Two descriptive representations: (**a**) Standardized mortality risks or SMR (standardization compared with national values) from COVID-19 among tested individuals in Mexico at the beginning of the pandemic and (**b**) Quartiles associated with the proportion of internal migration (rate of people moving between states in Mexico) for the same population

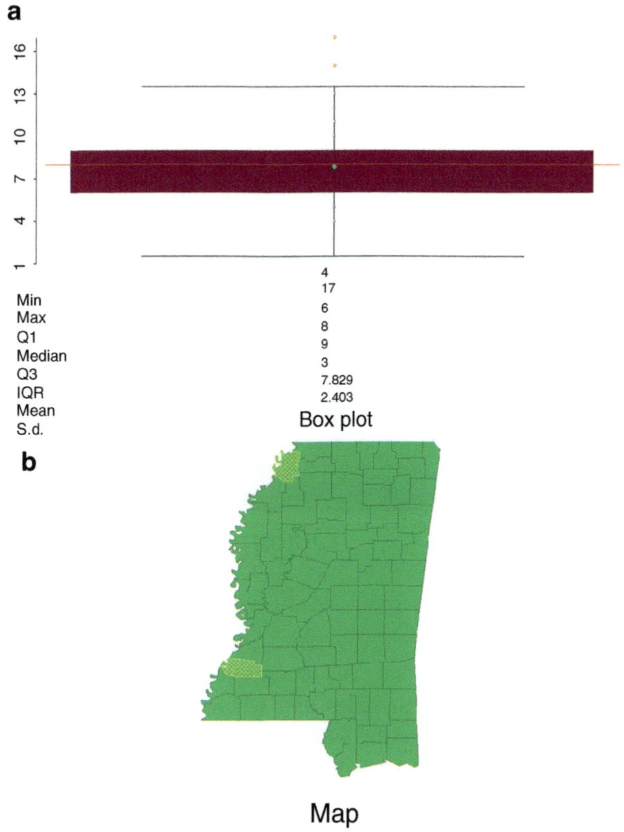

Fig. 13.3 Linking, in yellow, between possible outliers according to (**a**) A box plot and (**b**) A map

outlier in a box plot and identify in real time to which spatial unit it corresponds (Fig. 13.3).

As in classical statistics, a descriptive analysis corresponds to a first step in the analyses we perform to understand a phenomenon. Unfortunately, perhaps due to the effort that is required to obtain the information and maps, or the lack of knowledge concerning the availability of other analyses, some studies end in this step.

Global and Local Spatial Autocorrelation

Two aspects we are interested in when studying spatial data is whether one variable is spatially associated; that is, whether we expect that nearby places have similar values and if there is spatial clustering. To be able to analyse these aspects, we are required to define when spatial units are neighbours and from there define a spatial weight matrix, see for example [7]. This matrix is used to calculate measures of spatial autocorrelation, as well as in some spatial linear models.

Neighbours

When polygons are used, the neighbours can be defined according to two criteria:

1. *Queen*: Neighbours correspond to polygons having a common vertex as in Fig. 13.4a in which neighbours are shown in yellow.
2. *Rook*: Neighbours correspond to polygons having a border (line) in common as in Fig. 13.4b.

A problem with these criteria is when there are islands, polygons not connected with the others, which according to the criteria would not have neighbours, or when we have a

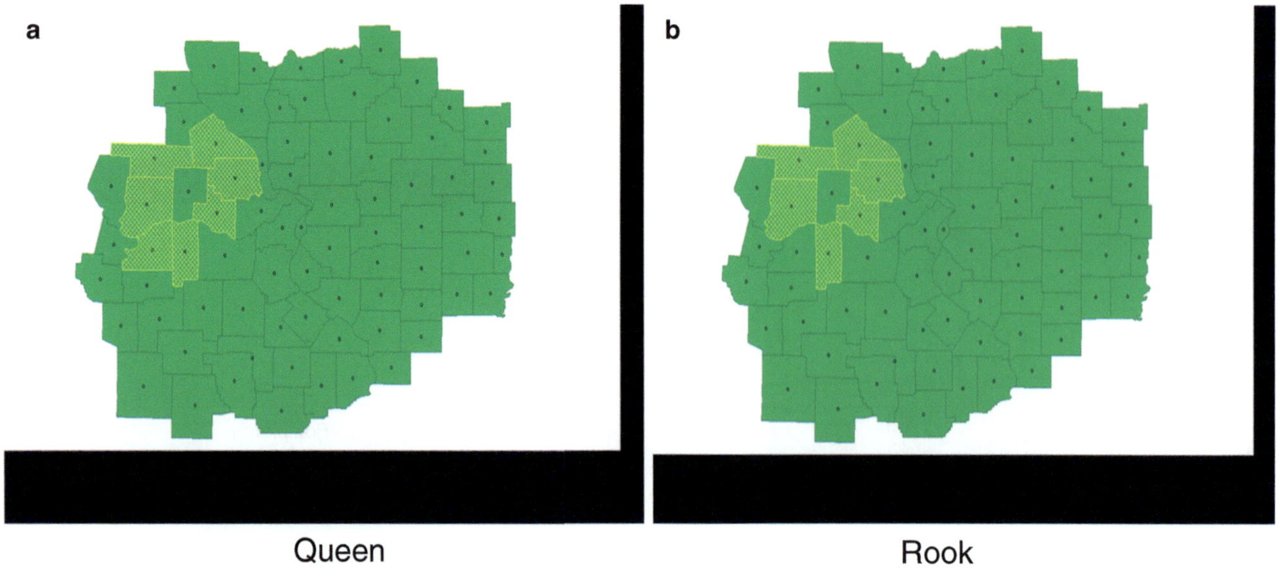

Fig. 13.4 Neighbours according to two different criteria: (**a**) Queen and (**b**) Rook

sample of spatial units, and consequently we could have many islands. In these cases and when spatial units correspond to points, we could define the neighbours according to distances, usually the Euclidean distance, considering the information is appropriately projected. Centroids are commonly used to measure distances between the polygons. There are two types of neighbours for this case:

1. Threshold distance. We define a distance to consider units as neighbours, usually the minimum distance such that every spatial unit has at least one neighbour.
2. k-nearest neighbours. We define a value of k, $k \in N^{+}$ and choose the k closest neighbours for each spatial unit.

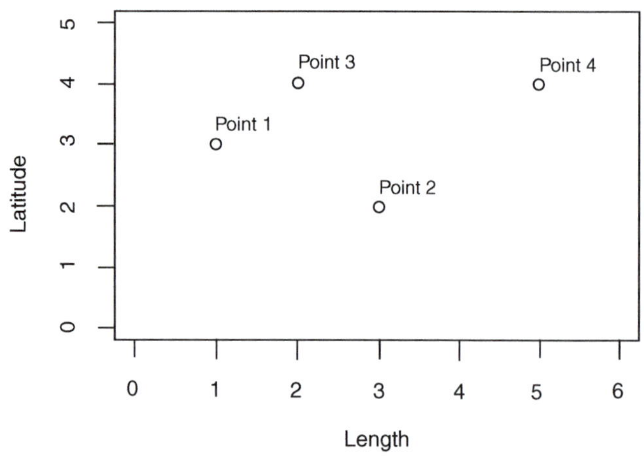

Fig. 13.5 Points representing spatial units in a Cartesian plane

Spatial Weights

According to the neighbours, we can obtain a weight matrix W of dimension $N \times N$, where N is the number of spatial units. Matrix W has entries w_{ij}, where $w_{ij} = 0$ when the spatial units i and j are not neighbours and $w_{ii} = 0$ for all i. The simplest way is assigning a value of 1 to w_{ij} if i y j are neighbours and 0 otherwise, which is called a binary method. Another possibility is calculating a row-standardized matrix W^{s} with entries w_{ij}^{s}, that is, using weights such that their sum for every spatial unit is one.

$$\sum_{j=1}^{N} w_{ij} = 1, \text{ for all } i. \qquad (13.1)$$

There are other types of weights, which vary according to the software.

To illustrate these concepts, consider data corresponding to four spatial units (points) with coordinates (1,3), (3,2), (2,4), and (5,4) as in Fig. 13.5. We can calculate the Euclidean distance between each pair of points, for instance, considering Point 1, we have distances $\sqrt{5} = 2.236$, $\sqrt{2} = 1.414$, and $\sqrt{17} = 4.123$ to Points 2, 3, and 4, respectively. Consider that a threshold distance of 3 is used, that is, two points are neighbours when their distance is between 0 and 3. Hence, the neighbours to Point 1 are Points 2 and 3. According to this process, we have neighbours for each point as follows:

Point	Neighbours		
1	2	3	
2	1	3	4
3	1	2	4
4	2	3	

Hence, the row-standardized weight matrix corresponds to:

$$W^s = \begin{pmatrix} 0 & 1/2 & 1/2 & 0 \\ 1/3 & 0 & 1/3 & 1/3 \\ 1/3 & 1/3 & 0 & 1/3 \\ 0 & 1/2 & 1/2 & 0 \end{pmatrix}$$

When this matrix is applied over one variable, we obtain the average of a variable over all the neighbours. For instance, if $\boldsymbol{Y}$ corresponds to the total income for each of the four Points defined above, with values 150, 155, 200, and 185, for the Points 1,2, 3, and 4, respectively, or

$$Y = \begin{pmatrix} 150 \\ 155 \\ 200 \\ 185 \end{pmatrix},$$

then W^sY includes averages, for instance, for Point 1, we have (155 + 200)/2 = 177.5. A variable calculated in this way, applying a weight matrix, is known as a spatially lagged variable, and it is a measure of how the values of a variable are modified according to its spatial position. Algebraically, the spatially lagged variable associated with a spatial unit i, y_i^s, $i = 1, \ldots, N$, corresponds to:

$$y_i^s = \sum_{j=1}^{N} w_{ij} y_j, \text{ for all } i, \tag{13.2}$$

where y_j is the original value associated to a spatial unit j. The importance of spatially lagged variables is that a measure of spatial autocorrelation, Moran's I [8, 9], is associated with these variables.

Standardizing variable Y, that is, obtaining

$$z_i = \frac{y_i - \bar{y}}{\sqrt{\Sigma\left(y_i - \bar{y}\right)^2 / N}}, \text{ for all } i, \tag{13.3}$$

Moran's I corresponds to

$$I = \frac{\sum_{i=1}^{N} \sum_{j=1}^{N} w_{ij} z_i z_j}{\sum_{i=1}^{N} \sum_{j=1}^{N} w_{ij}}, \tag{13.4}$$

and when the row-standardized weight matrix is used, it simply corresponds to:

$$I = \frac{\sum_{i=1}^{N} \sum_{j=1}^{N} w_{ij}^s z_i z_j}{N}. \tag{13.5}$$

Thus, Moran's I corresponds to a type of correlation, spatially weighted according to the weight matrix W. It has values between −1 and 1, with values near to zero indicating that there is not spatial autocorrelation, and positive values indicating positive autocorrelation, that is, large values (small) of the variable in a spatial unit are associated with large (small) values in their neighbours, whereas negative values indicate dispersion.

Geometrically, Moran's I also corresponds to the slope of a line, the best linear predictor, between a variable and its corresponding spatially lagged version. If we divide the diagram in four segments or quadrants as in the scatter plot shown in Fig. 13.6a, the first (positive values for both variables or high-high) and third quadrants (negative values for both variables or low-low) correspond to spatial units in which there could be possible clustering of high or low values, whereas the other two quadrants (high-low and low-high) correspond to outliers, units with large (small) values surrounded by unit with small (large) values.

We can calculate the contribution of each unit over Moran's I as:

$$I_i = z_i \sum_{j=1}^{N} w_{ij} z_j, \text{ for all } i = 1, \ldots, N$$

which clearly depends on spatially lagged variables, and it is known as the *local indicator of spatial association* or LISA [10]. When the row-standardized matrix W^s is used, Moran's I is the average of these LISA values:

$$I = \frac{\Sigma I_i}{N}.$$

Using similar ideas, a multivariate spatial autocorrelation measure can be defined between vectors corresponding to the values associated with two different vectors z_k $z_l z_l$, as:

$$I_{kl} = \frac{z_k{}' W^s z_l}{N}.$$

Denoting z_k^i as the value associated with spatial unit i, the corresponding multivariate LISA is:

$$I_{kl}^i = z_k^i \sum_{j=1}^{N} w_{ij} z_l^i, \text{ for all } i = 1, \ldots, N.$$

After calculating all these indicators, inference can be obtained to define whether spatial autocorrelation is significant through a hypothesis of the form.

- H_0: No spatial autocorrelation vs H_1: Spatial autocorrelation

Inference is performed through normal approximations or simulation processes.

In terms of the LISA, we test

- H_0: LISA for spatial unit i is not similar as that for the neighbours
- vs
- H_1: LISA for spatial unit i is similar as that for the neighbours

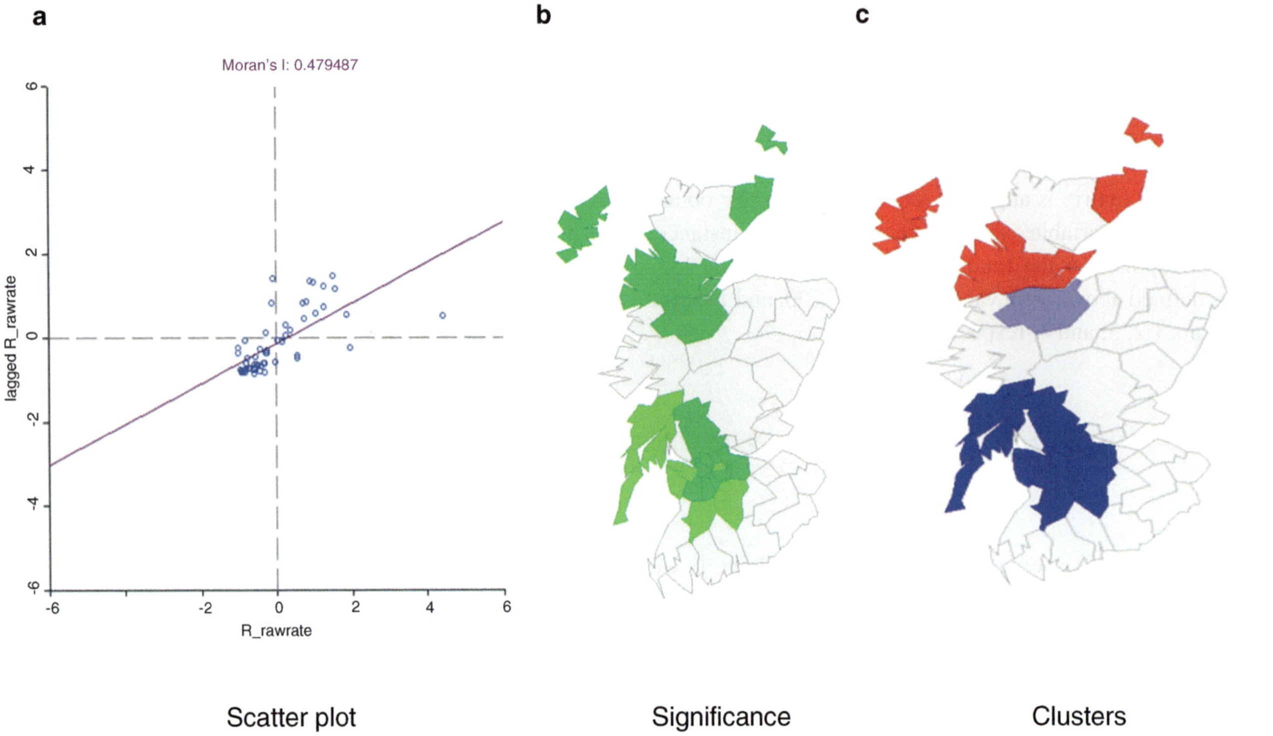

Fig. 13.6 Moran's I and LISA maps associated with male lip cancer incidence in Scotland: (**a**) Scatter plot associated with Moran's I, representing incidence (raw values, R_RAWRATE) and the associated spatially lagged variable, (**b**) significant spatial units at a 0.05 (green) and 0.01 (light green) significance level and not significant units (grey), and (**c**) LISA spatial clustering associated with spatial units with significant clustering according to a 0.05 significance level. Clusters are of the form high-high (red), low-low (blue), and low-high (purple)

LISA maps, as the one shown in Fig. 13.6b, can be obtained to determine where the null hypothesis is rejected at different significance levels. We can also obtain heat maps concerning significant spatial clustering formation, as in Fig. 13.6c, which allows us to see whether a spatial unit is surrounded by spatial units with similar values. These clusters are of the type *high-high* in red, *low-low* in blue, indicating that in a spatial location and their neighbours there are significant similar high or low values, respectively, or high-low and low-high corresponding to significant spatial outliers.

Interpolation and Geostatistics

Another important task in spatial statistics consists of interpolating spatial information, part of the branch of statistics called geostatistics, see for example [11], concerning random variables associated with spatial information. In other words, given information concerning a variable corresponding to a set of points, or centroids in the case of polygons, we want to predict which values take another set of points based on this information. To be able to perform this, we first require to find a rule of how a variable is associated according to the location of the points, at least in terms of the distance between them. The concept of variogram, for example [12], is useful in this process.

Variogram

Mathematically, we have a spatial random process $\mathbf{Z}(\mathbf{s})$, where $\mathbf{s}$ corresponds to a spatial unit in geographical coordinates, usually projected, or in other words a random variable associated with different locations. Additionally, we assume this process is stationary, which means that the association between the values that a variable takes in two locations only depends on the distance or spatial lag between them. Thus, a measure of the association between the values of the variable in two locations separated by a distance h can be given by the variogram:

$$\gamma(h) = \frac{1}{2} E\left[\left(Z(s) - Z(s+h) \right)^2 \right]. \tag{13.6}$$

Another measure of association is the covariogram, which assuming that $s_2 = s_1 + h$, the location moved in h units, corresponds to the covariance $C(h) = Cov(Z(s_1), Z(s_2))$. Both measures are linked between them considering that:

$$\gamma(h) = \sigma^2 - C(h) \tag{13.7}$$

where σ^2 is the variance associated with the process. Both are measures of spatial correlation between spatial units according to the distance separating them.

The variogram and covariogram can be represented in a graph with distances h in the X axis and $\gamma(h)$ or $C(h)$ in the Y axis, respectively. The variogram is a non-decreasing function asymptotic to σ^2, whereas the covariogram is a non-crescent function starting in the value σ^2. These properties make sense, and in the case of the variogram, the form of the function indicates that values of a variable are first similar since the distance between the places is small, and then, they differ more and more once the distance increases until the variogram is close or equal to the full dispersion σ^2, or still, of the process. Meanwhile, in the case of the covariogram, the shape of the function indicates that when there is a distance of zero (in the same location), the covariance corresponds to the variance, and as the distance increases, it is expected that the values in two locations are not or not closely related, and the covariance is then close to zero.

In particular, in data without spatial correlation, we would expect a constant value in the variogram $\gamma(h)$, corresponding to σ^2, since the association between variables does not depend on the distances, whereas $C(h) = 0$, for all distance h.

Both functions are relevant in terms of obtaining spatial interpolation, but also in terms of fitting some spatial linear models.

In practice, we have to estimate these functions using our data. For this process, we define sets of distance intervals or bins, to calculate through moment estimators the sampling variogram considering different separations between points. Since in each bin there could be several points separated by the distances included in an interval, average values are considered. In the end, we can obtain a sampling variogram as in Fig. 13.7a.

Once we obtain the sampling variogram, we replace it with a model [13]. There are several options and we choose the one with the best fit for our data. The models include as parameters some or all of the following terms. A nugget is the variation at a small scale plus a measure error or, in other words, the value corresponding to $\gamma(h)$ for $h = 0$. The sill is the asymptotic value of the variogram, that is, $\gamma(h)$ when $h \to \infty$, and corresponds to the variance of the process. The range is the distance such that the values of the process are no longer associated when distances greater to this value are considered or, alternatively, the value in which the sill is reached. In Fig. 13.7b each of these parts is shown.

Examples of models associated with a variogram are the nugget model, which considering c as a constant term corresponds to:

$$g(h) = \begin{cases} 0 & \text{if } h = 0 \\ c & \text{otherwise} \end{cases}$$

Other possible models are the spherical,

$$g(h) = \begin{cases} c\left(1.5\left(\frac{h}{a}\right) - 0.5\left(\frac{h}{a}\right)^3\right) & \text{if } h \le a \\ c & \text{otherwise} \end{cases}$$

exponential,

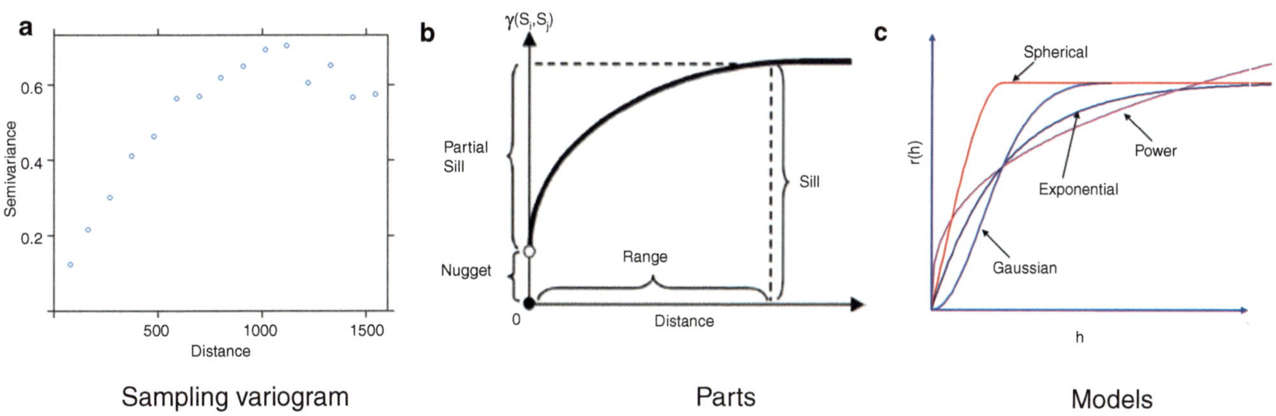

Fig. 13.7 Variogram: (**a**) An example of a sampling variogram, (**b**) parts of a variogram, and (**c**) models associated with a variogram

$$g(h)=c\left(1-\exp\left(\frac{-3h}{a}\right)\right)$$

and Gaussian

$$g(h)=c\left(1-\exp\left(\frac{-3h^2}{a^2}\right)\right).$$

In all these models, a is the range and c is the sill. In Fig. 13.7c, we show the graph associated with instances of these and other models.

The variogram is fitted according to weighted least squares, and once the model with a best fit to our data is obtained, we also derive the correlogram. In simple terms, we obtain from our data the best rules of association of a variable according to the distance separating the points, which allow us to understand the spatial correlation between a variable.

Assuming that our variable depends on a set of explanatory variables through a linear model, we can perform the same process using the residuals. This will be used for interpolation processes and for fitting linear models that consider the spatial relationships between data.

Spatial Interpolation

There are different options to obtain a spatial interpolation; the first one corresponds to a spatial average and the second one is based on the ideas previously presented to define the variogram.

Inverse Distance Weight

We predict the unknown value corresponding to a spatial location using a spatial weighted average, which is called the inverse distance weight (IDW) [14]. The weights correspond to a metric identifying the distance between the point in which the interpolation is required, s_0, and the other points. Mathematically, the value estimated for the process in s_0, $\hat{Z}(s_0)$, is:

$$\hat{Z}(s_0)=\frac{\sum_{i=1}^{N} w(s_i)Z(s_i)}{\sum_{i=1}^{N} w(s_i)}$$

with

$$w(s_i)=\| s_i-s_0 \|^{-p},$$

where s_i corresponds to the coordinates of the locations in which we have the true value of our variable of interest, for $i = 1, \ldots, N$, p refers to the type of distance used, $p \in R$, usually $p = 2$, and consequently $\| \bullet \|$ is the Euclidean distance. If s_0 is the same as s_i, for some $i = 1, \ldots, N$, that is, the location in which we want to predict is the same as the location of one of the available points, then the predicted value is equal to the observed value, $\hat{Z}(s_0)=Z(s_i)$. The weights decrease as the distance to s_0 decreases, and p indicates the degree in which the values associated with the nearest points are preferred, and a large p indicates assigning a greater importance to those points nearest to s_0. Since $w(s_i)/\sum_{i=1}^{N} w(s_i)$ is between 0 and 1, any interpolated value cannot be outside of the range of the observed values.

This is a very simple method since it is just an average; however, it ignores the spatial association, which the following method considers.

Kriging

In this case, we assume that the values of our variable of interest depending on different locations, or process $Z(s)$, can be linearly modelled according to a set of explanatory variables, considering, as it is usual, a random error in the model. Mathematically, and considering that we have N observed locations in which the values of our variable are known, s_i; for $i = 1, \ldots, N$:

$$Z(s)=X\beta+\epsilon(s), \tag{8}$$

where $Z(s)$ are the values of our variable of interest in each location, X is a matrix including the values associated with the explanatory variables, β are the parameters indicating the degree of association of each explanatory variable with our response variable, and $\epsilon(s)$ is a random error, being each component of dimension $N \times 1$, $N \times p$, $p \times 1$, and $N \times 1$, respectively.

We assume that the error associated with the model is close to zero for each location, or $E[\epsilon(s)] = 0$, and that the variance and covariance associated with the process is $Var(\epsilon(s)) = Var(Z(s)) = V$, a matrix of dimension $N \times N$ formed by variances and measures of the association of our variable of interest in two different locations, $Cov(Z(s_i), Z(s_j))$. In simple terms, we define a model including a set of variables explaining the variable we want to predict in a new location, including also the correlation structure between the observations according to where they are located.

Assuming that we have the values of the explanatory variables over the point s_0 in which we want to interpolate, $x(s_0)$ of dimension $1 \times p$, and that v includes the covariances between the values in the observed points $Z(s)$ and the point in which we want to interpolate $Z(s_0)$, that is, v is a vector of dimension N including $Cov(Z(s_i), Z(s_0))$; for $i = 1, \ldots, N$, we can predict the unknown value of our variable in the point s_0, $\hat{Z}(s_0)$, as:

$$\hat{Z}(s_0) = x(s_0)\hat{\beta} + v'V^{-1}\left(Z(s) - X\hat{\beta}\right) \qquad (13.9)$$

where

$$\hat{\beta} = \left(X'V^{-1}X\right)^{-1} X'V^{-1}Z(s).$$

In other words, we estimate the unknown value of our variable in a new location as the sum of two parts. The first part is a linear combination depending on the values that the explanatory variables take in this location and estimated parameters, the latter depending on the values of our response and explanatory variables and the spatial autocorrelation in the points in which the true values of our variable are known. The second part corresponds to another linear combination consisting of the spatial correlation between the values of our variable in the new location and the other locations and a measure of the error between the true values of our variable and the values estimated considering the explanatory variables.

Since we need the spatial variances and covariances, according to V, or in other words the covariogram $C(h)$ if we assume stationarity, we first obtain the variogram and associated model. This type of interpolation is known as Universal Kriging [12, 13], which was first introduced by Krige [15].

We can perform this process at each point in which the values of our variable are unknown. After that, we can even obtain a smoothed map with values predicted for a variable in different locations.

There are simpler variants of Universal Kriging. Ordinary Kriging corresponds to applying Kriging without considering explanatory variables in the model, only a constant term. On the other hand, simple Kriging corresponds to the case in which β is known, or in other words, we know the true linear association between our explanatory variables and the variable we wish to interpolate. In Fig. 13.8, we present examples of a spatial interpolation, through both IDW and Universal Kriging, which are based on the analyses presented in [5].

Kriging, in all their variants, allows us to predict values of a variable in unknown locations. However, as in any predictive model, it is recommended to perform model validation. For instance, we can separate the data into two, the training and test data sets, fitting the model (applying Kriging) in the former and predicting over the other set. If the model fitted well, we should expect the true and estimated values to be similar in the test set.

Of course, Kriging is not the only possible method to obtain spatial predictions. For instance, a simpler method consists of including functions of the coordinates, latitude and longitude, as explanatory variables in a linear model [16, 17], and after fitting this model, we predict values for new

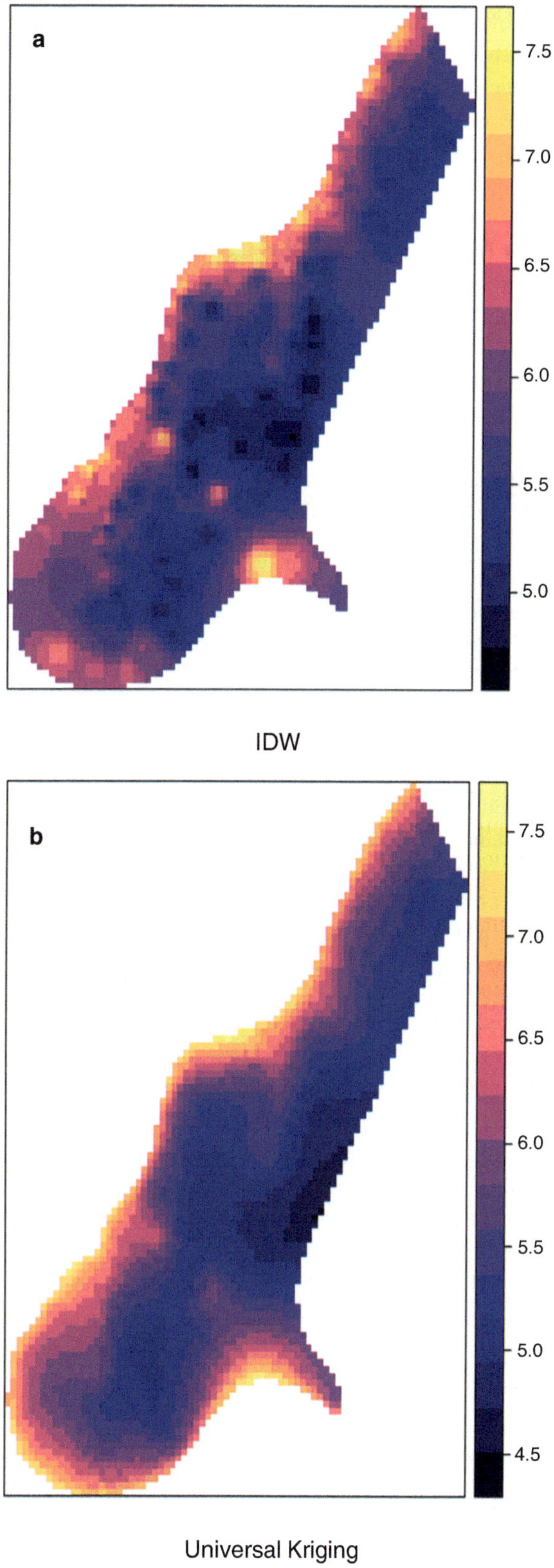

Fig. 13.8 Spatial interpolation for the logarithm of the concentration of a metal: (**a**) Interpolation using inverse distance weight (IDW) with a value of p (inverse distance power) of 2.5 and (**b**) Universal Kriging in which the squared distance to a river is used as explanatory variable and an exponential model is associated with the variogram

points by including their longitude and latitude. However, this method does not consider the spatial association inherent in the phenomenon we are modelling, though it is easier to implement in any available software that fits linear models.

Linear Models for Spatial Data

In this part, our aim is to explain a response variable through a set of explanatory variables, possibly including confounders, considering that each observation corresponds to one of N spatial units. Thus, our main aim is explaining instead of predicting a variable, the latter being the main aim in the geostatistics framework discussed above.

To face this problem, a first possibility is fitting a classical linear model; however, it is possible that there is spatial autocorrelation in the errors, thus violating the independence between observations assumed in these models. The possible spatial autocorrelation could be measured over the residuals using Moran's I, a variogram, or any other available method. If we identify the existence of spatial autocorrelation, we should try to use another type of model.

One first possibility is fitting a linear model considering the heteroscedasticity induced by the spatial information. For instance, we could consider a weighted least squares (WLS) estimator considering the spatial unit size as weight. We could also consider a model fitted by generalized least squares (GLS) by including the variance and covariance matrix V including the spatial correlation [18], for instance that derived from the variogram. Mathematically, we fit a model:

$$Y = X\beta + \epsilon, \text{ with } \epsilon \sim N_n\left(0, \Sigma\right),$$

where the variance and covariance terms in Σ are estimated according to the variogram of the residuals obtained after fitting the classical multivariable linear model:

$$Y = X\beta + \epsilon, \text{ with } \epsilon \sim N_n\left(0, \sigma^2 I\right).$$

Hence, the process consists first of obtaining and estimating the parameters of a model associated with the variogram of the residuals (estimating the sill, nugget, range, etc.) and obtaining the covariogram $C(h)$ and consequently Σ. Hence, considering that Σ is known, the GLS estimators, which are well known, correspond to:

$$\hat{\beta} = \left(X'\Sigma^{-1}X\right)^{-1} X'\Sigma^{-1}Y,$$

with variance and covariance also obtained from the GLS framework as:

$$V\left(\hat{\beta}\right) = \left(X'\Sigma^{-1}X\right)^{-1}.$$

In simple terms, we consider the lack of independence between the spatial units and inference is more precise since we are not violating statistical assumptions associated with the model.

Other possibility is to use variants of linear models that include the spatial correlation by using a spatial weight matrix, W [19].

Spatial Lag Model

A *spatial lag model* or *spatially lagged* y model includes the spatially lagged variable associated with the response as an additional explanatory variable. Mathematically, considering Y as a vector of dimension N associated with the response and X as a matrix corresponding to the regressors or inputs, with β the residuals associated with model is:

$$Y = \rho WY + X\beta + \epsilon, \text{ with } \epsilon \sim N_N\left(0, \sigma^2 I\right), \quad (13.10)$$

where W is a spatial weight matrix, WY is the spatially lagged response variable, and ϵ is a vector of dimension N corresponding to an error, which is assumed to be normally distributed, as in classical linear models.

In other words, we are assuming that the values of the response in the neighbours of a spatial unit are associated with the values in that spatial unit. Parameter ρ, which is an scalar, is such that if $\rho = 0$, then (10) corresponds to a usual regression model; additionally, it is a measure of spatial autocorrelation.

Maximum likelihood estimators associated with the parameters corresponding to the regressors β and spatial association parameter ρ are estimated through a process in which the former are obtained using GLS and the latter by maximizing a function depending on the eigenvalues associated with the weight matrix W. In order to obtain real instead of imaginary eigenvalues, W should be symmetric and consequently neighbours should be obtained with only specific methods. For instance, k-nearest neighbours should not be used to generate W since a spatial unit can be a neighbour of another, the opposite not being true. On the other hand, methods to obtain neighbours as threshold distance, Queen, or Rook using a binary method for constructing the weight matrix could be used. When the number of spatial units is large, a Cholesky decomposition associated with W might be preferred instead of calculating the eigenvalues.

Spatial Autoregressive and Spatial Error Models

A spatial autoregressive (SAR) model is such that the spatial structure, considered through a matrix, is associated with an error term. Mathematically:

$$Y = X\beta + \mathrm{e}, \mathrm{e} = B\mathrm{e} + \epsilon; \text{ with } \epsilon \sim N_N\left(0, \Sigma_\epsilon\right), \quad (13.11)$$

where B is a known squared matrix of dimension N that allows us to identify neighbours and e is a random vector of dimension N, which depends on itself once it is spatially lagged through B, thus the name of the spatial autoregressive model.

Usually, $B = \lambda W$, with W a spatial weight matrix, and

$$Y = X\beta + \lambda W\mathrm{e} + \epsilon; \text{ with } \epsilon \sim N_N\left(0, \sigma^2 I_{N\times N}\right), \quad (13.12)$$

which is known as a *spatial error model*. The model is appropriate when we consider that some unmeasured characteristics are responsible for spatial clustering, having an influence over the response variable in a spatial unit and their neighbours, but that are omitted from the specification of the model.

Once again, the parameters are estimated through maximum likelihood, depending on GLS estimators for the parameters associated with the regressors and depending on the eigenvalues associated with W for the spatial parameter. The same problems as with the spatial lag model arise and λ is a parameter measuring spatial autocorrelation.

Inference with the Models

The two types of spatial models introduced before are not nested between them; however, the regression model (without spatial terms or minimum least squares, LS, model) is nested in both. Hence, to decide whether to use a spatial lagged or a spatial error model, the usual procedure consists of comparing with a linear regression model. Consequently, likelihood ratio tests (LRT) are used to perform tests of the type:

H_0: $\rho = 0$ (usual regression has a good fit) vs
H_1: $\rho \neq 0$ (spatial lag model has a good fit)

and similar for the spatial error model. Other statistics can be used for similar hypothesis tests, for instance, the Wald statistic (normal approximation) or the Lagrange multiplier (LM) and robust LM statistics. Hence, we can decide whether a spatial model has a good fit to the data. For instance, we could first use the LM statistic to decide which of the two spatial models has a good fit, and if both have a good fit, then we could use the robust LM for our final decision. Of course, we can also use information criteria, as for instance the Akaike information criterion (AIC) or similar to compare the models.

Other Spatial Models

Simultaneous Moving Average

A simultaneous moving average (SMA) model is similar to a moving average model in time series, in the sense that the error term associated with the model and its corresponding lagged version, in this case spatially lagged, are included in the linear model. Mathematically, it corresponds to:

$$Y = X\beta + \mathrm{e}, \mathrm{e} = B\epsilon + \epsilon; \text{ with } \epsilon \sim N_N\left(0, \Sigma_\epsilon\right), \quad (13.13)$$

where usually $B = \lambda W$ and $\Sigma_\epsilon = \sigma^2 I_{N\times N}$, similar as with the models in the previous sections. This model is used when there are localized effects, when spatial effects only affect the errors in neighbours according to matrix W.

This and the previous models can be generalized, obtaining the so-called SARMA models, including the spatially lagged response as explanatory plus moving average effects for the error terms. They are analogous to ARMA models in the time series analysis framework. Mathematically, they correspond to:

$$Y = \rho WY + X\beta + \mathrm{e}, \mathrm{e} = \lambda W\epsilon + \epsilon; \\ \text{with } \epsilon \sim N_N\left(0, \sigma^2 I_{N\times N}\right). \quad (13.14)$$

Other type of spatial model corresponds to conditional autoregressive models (CAR) in which the distribution, assumed Gaussian, of the errors associated with each spatial unit depends on the other spatial units, the neighbours of each unit. The parameters in this distribution are restricted as in the previous models.

Geographically Weighted Regression

Geographically weighted regression [20–22] is another type of available model used to understand the relationship between a set of variables and a response. It consists of fitting a linear model for every spatial unit, considering for the estimation process the distance such that the variables have an influence over the response and the degree in which this influence decreases according to different functions of the distance. Mathematically, we have a linear model associating p explanatory variables with the response for each of the N analysed spatial units:

$$Y_i = \beta_0\left(u_i,v_i\right) + \beta_1\left(u_i,v_i\right)x_{1i} + \beta_2\left(u_i,v_i\right)x_{2i} + \ldots \\ + \beta_p\left(u_i,v_i\right)x_{pi} + \epsilon_i, \quad (13.15)$$

where $\beta_j(u_i, v_i)$ is a function of the location for the parameter associated with the explanatory variable j th; $j = 1,..., p$ and $\beta_0(u_i, v_i)$ is a constant term, all depending on the coordinates of a point (u_i, v_i), with $i = 1,..., N$. The error terms correspond to independent random variables with $\epsilon_i \sim N(0,\sigma^2)$.

This method can be applied to point data or to the centroids of polygon data. Since β_j is a function of the coordinates, there are as many parameters for each explanatory variable as spatial units. In other words, we fit a model with p different parameters plus the constant term in each of the N spatial units. Hence, model (15) cannot be estimated since there are more unknown parameters than observations. This problem can be solved by estimating a model using weighted least squares (WLS) for each spatial unit i, where the weights are given by functions that measure how near to i are the other spatial units, that is, we consider heteroscedasticity derived from the closeness to a point i.

Mathematically, considering the vector of parameters associated to a unit i, for $i = 1,...,N$, as.

$$\beta(i) = \begin{pmatrix} \beta_0(u_i,v_i) \\ \beta_1(u_i,v_i) \\ \vdots \\ \beta_p(u_i,v_i) \end{pmatrix},$$

model (15) corresponds to

$$Y_i = (1\, x_{1i} \ldots x_{pi})\beta(i) + \epsilon_i,$$

whose estimator by WLS of $\hat{\beta}(i)$ corresponds to

$$\hat{\beta}(i) = (X'W(i)X)^{-1} X'W(i)Y,$$

where $W(i) = W(u_i, v_i)$ is a diagonal matrix of dimension $N \times N$ associated with the location (u_i, v_i) indicating how much attraction there is to other points. X is as always a matrix corresponding to the explanatory variables for all the observations, and Y is a vector including the values associated with the response. More specifically,

$$W(i) = \begin{pmatrix} w_{i1} & 0 & \ldots & 0 \\ 0 & w_{i2} & \ldots & 0 \\ \vdots & \vdots & & \vdots \\ 0 & 0 & \ldots & w_{iN} \end{pmatrix},$$

with $w_{ij}; j = 1,..., N$, the weight in a calibration from point i to point j, evidently $w_{ii} = 1$, and if $w_{ij} = 1$, for all j, we have the usual regression model. In the model, the values associated with locations closer to the spatial unit i should have more weight and less variability, when compared with those units farther apart, which is reflected in the WLS estimators.

Functions used to generate the spatial weights or kernels have to satisfy that they decrease as the distance from a point increases. Some instances of these functions are:

Bi-square:

$$w_{ij} = \left[1 - \left(\frac{d_{ij}}{b}\right)^2\right]^2.$$

Gaussian:

$$w_{ij} = \exp\left[-(1/2)\left(\frac{d_{ij}}{b}\right)^2\right].$$

Here, d_{ij} corresponds to the Euclidean distance between units i and j and b is the bandwidth, which is the distance such that the explanatory variables still have an important influence over the response variable. For a very large bandwidth, $b \to \infty$, observe that the weights are close to one, $w_{ij} \to 1$, and we have a usual regression model.

The fit depends on the weighting function and the bandwidth, and the latter can be the same in any spatial unit, which is known as a fixed kernel, which in some data can induce certain bias since in some spatial units the bandwidth can be larger than necessary when compared with denser regions. Hence, it can be used as an adaptive kernel, which variates the bandwidth according to the observation.

To select the bandwidth, usually a criterion is chosen such that the bandwidth minimizing it is selected. Examples of criteria correspond to cross validation (CV), generalized CV, Akaike information criterion (AIC), etc. We usually first select the bandwidth (fixed or adaptive) and afterwards we fit the model. Then, we can obtain maps corresponding to the coefficients of the explanatory variables for each spatial unit, that is, the differential effect that an explanatory variable has over the response in each spatial unit, the standard errors, t-statistics, etc. We can even obtain predictions over observations outside of our sample. There are also variants considering generalized linear models. In Fig. 13.9, we present an example of GWR concerning data presented by [23].

Conclusion

In this chapter, we have introduced spatial analyses which can be useful in the epidemiological framework. Of course, there are more types of analyses, for instance, spatiotemporal, which consider spatial information through time. For spatiotemporal analyses, there is direct generalization of the methods shown here; for instance, there is Krigin considering both time and space information or linear models including simultaneously the time and space correlation structures, see for example [24]. There are also spatial models from a

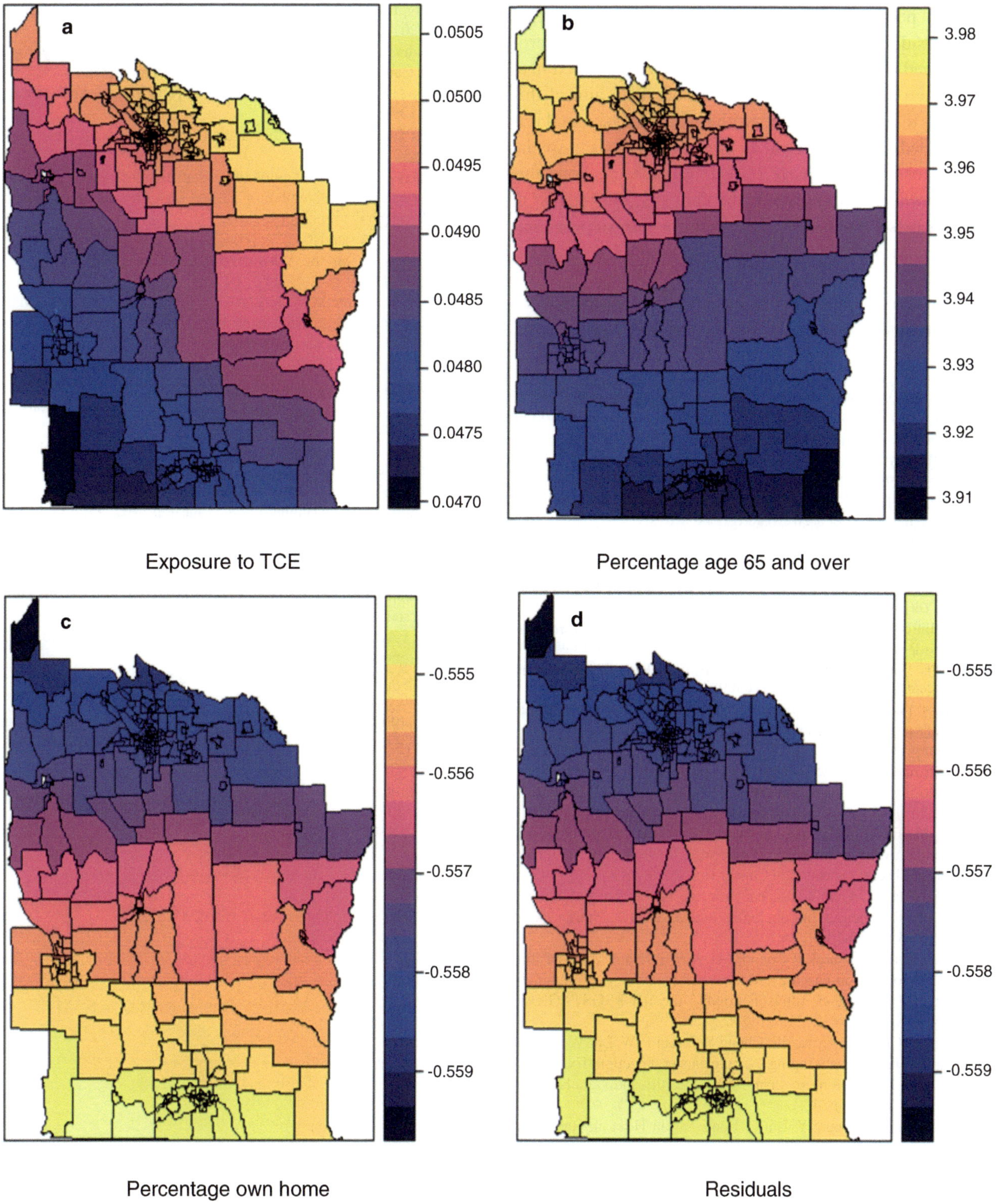

Fig. 13.9 Geographically weighted regression between exposure potential to trichloroethylene (TCE), percentage of people in each spatial unit aged 65 or more (P65), and percentage of people in each spatial unit owning their home (POH) and the logarithmic transformation of the incidence of leukaemia as response, in eight central New York state counties divided into census tracts. A fixed bandwidth obtained through a cross-validation method and Gaussian spatial weighting (kernel) is considered. The estimated coefficients for (**a**) TCE, (**b**) P65, (**c**) POH by spatial unit, and (**d**) the residuals associated with the model are presented

Bayesian perspective [25], or clustering methods parting from considering the spatial information as a random process instead of using weight matrices, as we presented here in the interpolation section. However, the main aim of this chapter is to introduce readers with spatial analyses, and it is not intended as a summary of all available methods in spatial statistics or as an extensive review. As such, we hope this chapter awakens in the reader the curiosity and interest in studying these types of models, being a first step in their learning process.

References

1. Burrough PA, McDonnell RA. Principles of geographical information systems. Oxford: Oxford University Press; 1998.
2. ESRI. ArcGIS Desktop: Release 10.8.1. Redlands, CA.; 2020. Available from: https://www.qgis.org.
3. org Q. QGIS Geographic information System; 2021. Available from: https://www.qgis.org.
4. Anselin L, Ibnu S, Youngihn K. GeoDa: an introduction to spatial data analysis. Geogr Anal. 2006;38(1):5–22.
5. Bivand RS, Pebesma EJ, Gómez-Rubio VG. Applied spatial data analysis with R. New York: Springer-Verlag; 2008.
6. Maling DH. Coordinate systems and map projections. 2nd ed. New York: Pergamon Press; 1992.
7. Anselin L. Spatial econometrics: methods and models. Dordrecht: Springer Science; 1988.
8. Moran PAP. Notes on continuous stochastic phenomena. Biometrika. 1950;37(1/2):17–23.
9. Moran PAP. A test for the serial independence of residuals. Biometrika. 1950;37(1/2):178–81.
10. Anselin L. Local indicators of spatial association-LISA. Geogr Anal. 2010;27(2):93–115.
11. Kitanidis PK. Introduction to geostatistics: applications in hydrogeology. Cambridge: Cambridge University Press; 1997.
12. Cressie NAC. Statistics for spatial data. New York: John Wiley and Sons; 1991.
13. Gaetan C, Guyon X. Spatial statistics and modeling. New York: Springer; 2010.
14. Tomislav H. A practical guide to geostatistical mapping of environmental variables. Luxembourg: Office for Official Publications of the European Communities; 2007.
15. Krige DG. A statistical approach to some basic mine valuation problems on the Witwatersrand. J Chem Metal Min Soc. 1951;52:119–39.
16. Webster R, Oliver MA. Geostatistics for environmental scientists. Statistics in practice. Chichester: Wiley; 2001.
17. Ripley BD. Spatial statistics. Hoboken: Wiley; 1981.
18. Basu S, Reinsel GC. Regression models with spatially correlated errors. J Am Stat Assoc. 1994;89(425):88.
19. Ward MD, Gleditsch KS. Spatial regression models. Thousand Oaks: Sage Publications; 2008.
20. Brunsdon C, Fotheringham AS, Charlton ME. Geographically weighted regression: a method for exploring spatial nonstationarity. Graph Anal. 1996;28(4):281–98.
21. Brunsdon C, Fotheringham AS, Charlton ME. Geographically weighted regression-modelling spatial non-stationarity TheStatistician. 1998;47 Part 3:431–443.
22. Fotheringham AS, Brunsdon C, Charlton ME. Geographically weighted regression: the analysis of spatially varying relationships. Chichester, West Sussex: John Wiley and Sons; 2002.
23. Waller LA, Gotway CA. Applied spatial statistics for public health data. Hoboken: Wiley; 2004.
24. Wikle CK, Zammit-Mangion A, Cressie N. Spatio-temporal statistics with R. Boca Raton: CRC Press; 2019.
25. Blangiardo M, Cameletti M. Spatial and spatio-temporal Bayesian models with R-INLA. Chichester, West Sussex: Wiley; 2015.

Principles of Network Models and Systems Epidemiology

14

Ricardo Ramírez-Aldana, Otto Hahn-Herrera, Ricardo Quiroz-Baez, and Juan Carlos Gomez-Verjan

Abbreviations

ANN	Artificial neural networks
DAGs	Directed acyclic graphs
FI	Frailty index
HIV	Human immunodeficiency virus
MHAS	Mexican Health Aging Study
MLP	Multi-layer perceptron network
NIAID	National Institute of Allergy and Infectious Diseases
NIH	National Institute of Health
PGM	Probabilistic graphical models
PGM	Probabilistic graphical models
SB	Systems Biology
SEp	Systems Epidemiology
SNA	Social Network Analysis
SNA	Social Network Analysis
SNA	Social network analysis
TGF-β	Transforming growth factor beta

Introduction

Classical epidemiology relates to lifestyle and environmental exposure to several diseases. The *one-level paradigm*, a method most used in research, focuses mainly on modeling one risk for a disease, thus limiting epidemiology advances since [1]: most of the diseases depend on the interaction of genetics and environmental variables; there are individual differences among the expression of diseases (*there are no diseases, there are patients*); there is a high variability on the pharmacological responses (*each individual has different reactions to treatment*); there are no clinical studies or approximations to evaluate the complex dynamics on the exposition to multiple risk factors during the whole life of an individual (*we change habits, locations, and customs during life*); phenotypes of diseases are codified at different levels of complexity, which diminish average statistical power to associate the disease with only one variable (*association is not causality*). Thus, the classical epidemiology paradigm should be reconsidered since it becomes clear that the health state is more complicated than previously known since it involves the interaction of multiple variables. Consequently, a real approximation of causality and association among diseases requires a holistic approach, as epidemiology is the first field to understand relationships of diseases, risk, and exposure factors [2].

In this sense, one of the best ways to analyze complex relationships over several amounts of variables and several members of a system is those based on networks that have proven to be useful. By using them, one can have a big picture of the whole relationship between different pieces of information and identify whether there are members of a system that are more important than others. These methods have been introduced recently, mainly because of the computational power they require, which is nowadays available. Interestingly, most of these have been implemented with relative success by experts in computational data sciences, engineering, and mathematics. Hence, the terms used to define them and the way they are presented and described is generally aimed for experts in these areas. Therefore, in the present chapter, we attempt to present in a simple format the most used network-based models that could be useful for epidemiological and biomedical problems that could not be approached with standard linear models.

R. Ramírez-Aldana · O. Hahn-Herrera · R. Quiroz-Baez
Dirección de Investigación, Instituto Nacional de Geriatría (INGER), Mexico City, Mexico

J. C. Gomez-Verjan (✉)
Dirección de Investigación, Instituto Nacional de Geriatría (INGER), Ciudad de México, Mexico
e-mail: jverjan@inger.gob.mx

J. C. Gomez-Verjan, N. A. Rivero-Segura (eds.), *Principles of Genetics and Molecular Epidemiology*,
https://doi.org/10.1007/978-3-030-89601-0_14

Systems Epidemiology

Systems Biology (SB) represents an attempt to understand how biological systems work at different levels of complexity and multidimensions. One definition of SB is that of Serrano in 2007. It defines it as computational modeling of molecular systems, which, due to its interdisciplinary nature, is presented in a wide variety of fields [3]. In this sense, as defined by the National Institute of Health (NIH), SB could be defined as an attempt to understand the larger picture (depending on the level of complexity, i.e., organism, tissue, or cell) by putting its pieces together, an exciting contrast to the standard approach used by reductionist biology. SB has taken an essential momentum due to the advantage in multi-omics technologies (genomics, transcriptomics, proteomics, epigenomics, metabolomics, and health informatics) and bio-bigdata rapidly generated by them.

SB has been applied to different fields in medicine, including epidemiology and public health, to integrate all the different levels of complexity in such disciplines, including epidemiological data, physiology, environment, genetics, socioeconomic variables, and others. Thus, a combination of complex mathematical models could help understand the causality among different states of health and risk factors, leading to a deeper understanding of black-boxes in epidemiology by relating genetics to the environment and making possible approaches to different types of interventions [4]. In this sense, Systems Epidemiology (SEp) has been defined by Lund and Dumeaux [5] as the observational side of SB that "*seeks to integrate pathway analyses on different observational studies, to improve our understanding of biological processes in the human organism.*" Moreover, methods in networks could help to create causality diagrams, where different variables at different levels can be grouped as a daily method to inferring the causality of diseases [6].

SEp could be helpful to quickly implement Bradford Hill criteria to assess causality covering strength and specific association, biological plausibility, dose–response relationship, rationality, and experimental evidence [7]. In the field of causality directed acyclic graphs (DAGs), diagrams have been applied to clarify ideas to visualize hypotheses. DAG's per se could be considered as SEp, since they attempt to visualize critical concepts that correlate the clinical variables with the biomedical results at different levels, due to exposure (Fig. 14.1).

Several examples of the multidisciplinary approach of SEp have already been published, especially on infectious diseases. For instance, the National Institute of Allergy and Infectious Diseases (NIAID) and the Broad Institute presented a couple of initiatives that allow having an in-depth knowledge of the type of interaction that exists in the epidemiological triad (Koch theory) of tuberculosis, supported by

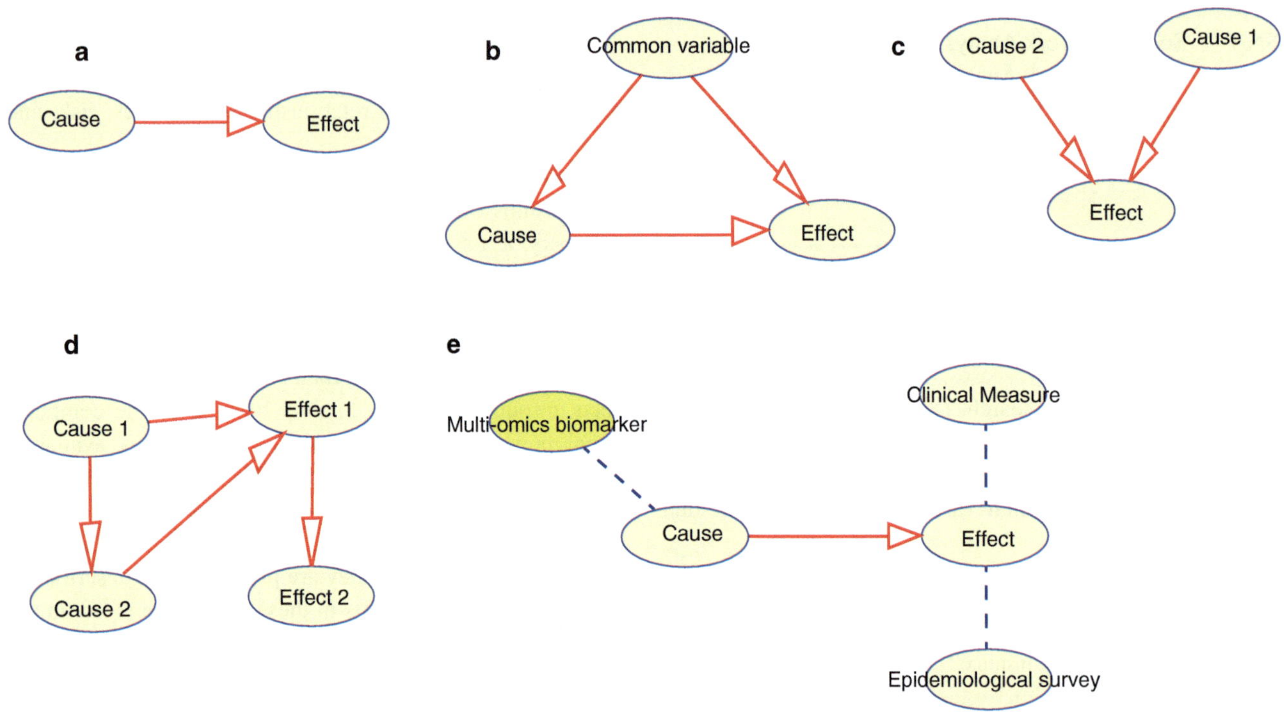

Fig. 14.1 Directed acyclic graphs for causality as tools for SEp. (**a**) General network for a simple DAG. (**b**) Network with common variables related to cause and effect. (**c**) Multiple independent causes for a single effect. (**d**) Dependent causes for multiple effects related with other effects. (**e**) Application of technologies for DAG's causality improvements. Red arrows are causal relations

computer sequencing and modeling technologies [8]. Results show a strong relationship between the different genotypes of *Mycobacterium tuberculosis* strains and human genotypes, which impacts the susceptibility of a population. They explained the different geographical distribution of diverse strains, as a result of some environmental factors, like diet and socioeconomic variables [9]. Another interesting example was with Ebola outbreaks in West Africa, where models in epidemiology have been successful in efficiently monitoring not only at the public health level but from the capability to sequence samples quickly. Epidemiological fences cover a different perspective since phylogenetic divergence could help to establish the origin of the outbreak or even where it could be mutating [10]. In this context, the importance of spatial and ecological variables, the so-called geoepidemiology, have proven to be quite a useful tool for epidemiological studies as, for instance, Pybus O.G. et al. demonstrated by using phylogenetics reconstructions with spatial autocorrelations that it was possible to calculate the diffusion coefficient, a robust nonparametric estimator of the spreading of the West Nile virus across North America [11]. Moreover, the use of remote sensing data from global satellites demonstrated an evident seasonal variation in the vulnerability in the scattered areas of urban and mixed horticulture land due to the climate factor [12].

Interestingly, the introduction of SB network models has been successfully implemented to different chronic diseases (that per se are quite complex) and aging to simplify the way we approach them and its possible interventions [13]. For instance, Drenos et al., in 2015, demonstrated a loss of complexity of association network maps before a cardiovascular event [14]. On the other side, the introduction of a new chi-squared model network to detect differences between groups, derived from biomarkers and exposure indicators, proved a significant association between smoking and insulin signaling pathways with DNA methylation [15]. On the other hand, Ji et al. developed a statistical methodology based on a bootstrap on different samples from acute myeloid leukemia, showing that the transforming growth factor beta (TGF-®) route could be the most important signaling pathway and a potential therapeutic target for this disease [16]. In this sense, Mitnitski et al. implemented a complex theoretical network model of health deficits to better understand the changes in health captured by the frailty index (FI) [17]. Moreover, Garcia-Peña et al. demonstrated using Bayesian networks with data from the Mexican Health Aging Study (MHAS) that not all deficits are equally related in the construction of the FI [18].

Another exciting topic where network models have been successfully implemented is on network meta-analysis. A novel statistical method for comparing multiple treatments in a single analysis by combining direct (evidence from randomized clinical trials) and indirect (evidence obtained through one or more common comparators) evidence within a network of randomized controlled trials and is quite useful for pharmacology treatments among others [19]. These types of analyses have been beneficial over the last years for the pharmaceutical industry; nodes represent treatments connected by standard comparators that represent evidence. Therefore, the most connected two nodes are with the most evidence (direct or indirect) exists. Since these models offer a unique opportunity to obtain information from clinical trials, statistical heterogeneity and incoherence, as well as conceptual coherence, must be seriously considered in the network construction before implementing these models [20].

In this context, considering previous reports, in the present chapter, we will define SEp as a novel tool where holistic approaches (network models) and sophisticated mathematical models (particularly non-linear models) are implemented to understand complex problems in epidemiology and public health. Such approaches must be based on analysis at multiple levels with the implementation of novel tools from multi-omics approaches to remote sensing data with conventional tools from epidemiology such as clinical trials, longitudinal studies, and surveys. Interestingly, the core of most SEp approximation is the implementation of network methodologies that will allow us to understand the association, causality relationships, classification, importance, and hierarchy, to understand variables as measures associated with individuals or even as individuals per se. In the following sections, we will present the most used types of methodologies at date based on networks: probabilistic graphical models (PGM), social network analyses (SNA), and artificial neural networks (ANN). We show examples of their use, as well as their classification, advantages and disadvantages, and examples of software in which they can be implemented.

Network Models

In Table 14.1, we present the main three types of networks (social, probabilistic, and neural), how the user should provide the data, the types of variables according to the type of analysis, and the model names identifying each combination of a model with the data type.

In Table 14.2, we present each type of model defined in the previous table (Table 14.1), the goals or utility of each model, and the limitations or problems a researcher could find when using each model inherent to any technique under development. We also present examples of software or libraries (e.g., in R) that can be implemented for each analysis.

Table 14.1 Type of network analysis. Social network analysis (SNA), probabilistic graphical models (PGM), and artificial neural networks (ANN), data provided by the user, type of admitted variables, and models associated with each type

Model	Data	Data type	Models
SNA	Through an adjacency, similarity, distance, or social matrix or through expert's knowledge. Matrices can be obtained from a bibliometric analysis. Weighted networks (ties weighted) are available	The importance is the relationships between nodes and their cause, not the variable type	Exponential (statistical) random graph (ERG), random, and dynamic network models
PGM	From data through structural learning (using algorithms, e.g., hc and PC) or using experts' knowledge to build the network, associated probabilities, or both	Discrete	Loglinear graphical (undirected) Discrete Bayesian networks (directed) Discrete chain models (both)
		Continuous	Undirected Gaussian graphical models Directed Gaussian graphical models Gaussian chain graph models (directed and undirected)
		Mixed	Mixed interaction models (undirected) Mixed chain graph models (directed and/or undirected)
ANN	From data: specify an input, output, or another kind of node, number of hidden layers and nodes in each layer	Discrete and continuous	Feed-forward networks Recurrent neural networks (RNN) or feedback network (use of loops allowed) Pattern recognition (mainly unsupervised, i.e., the number of clusters and clusters obtained from data): Kohonen network (self-organized map), Hopfield network, Boltzmann machines, generative adversarial networks, and deep belief networks

Table 14.2 Network models, goals, disadvantages, and examples of software available for each model

Types	Models	Goals	Disadvantages	Software examples
SNA	ERG, random, and dynamic network models	Nodes relevance, social cohesion and community detection; understand the nature and causes of the relationships between nodes and simulate networks	It is time-consuming; convergence problems can be difficult, particularly for dynamic network models	R (igraph, statnet, intergraph, UserNetR, ergm, tnet, etc.), Cytoscape, Python (networkX)
PGM	Loglinear graphical	Understand marginal and conditional independence between variables	The computational time directly proportional to the number of nodes, particularly for structural learning; thus, other techniques and/or graph restrictions (e.g., use of trees, approximations, etc.) can be necessary. Arcs direction or forbidding of certain direction must be validated by experts. Not all types of graphical models (particularly the mixed type) are as well developed, particularly in the same software. Gaussian distribution assumed in the continuous and mixed network types	R (gRbase, gRain, gRim, bnlearn, ggm, pcalg, etc.), Hugin
	Discrete Bayesian networks Discrete chain models	Understand marginal and conditional independence between variables, understand causality between all variables (not just one), and evidence propagation (prediction and classification)		
	Undirected Gaussian graphical models	Understand marginal and conditional independence between variables		
	Directed Gaussian graphical models Gaussian chain graph models	Understand marginal and conditional independence between variables, understand causality between all variables, and evidence propagation (prediction and classification)		
	Mixed interaction models	Understand marginal and conditional independence between variables		
	Mixed chain graph models	Understand marginal and conditional independence between variables, understand causality between variables, and evidence propagation (prediction and classification)		
ANN	Feed-forward networks	Prediction and classification. Some models allow parameter interpretation (causal explanation)	There is not a standard of how many hidden layers and associated nodes should be used. Risk of overfitting, for example, a model could correctly classify all data in the train data but misclassify data in the test data	R (neuralnet, nnet, RSNNS, rnn, mxnet, h2o, etc.), Matlab, SPSS, Python (scikit-learn, theano, keras, tensor-flow, pyTorch)
	RNN	They have memory, and feedback used to improve model		
	More complex pattern recognition networks	Clustering and classification (images, voice recognition, etc.)		

Probabilistic Graphical Models (PGM)

Probabilistic graphical models, including Bayesian networks, are multidimensional models, in which, according to a graph, the joint probability associated with all variables is factorized, representing a set of marginal and conditional independencies, also known as Markov properties. For instance, in an undirected network, the absence of an edge between two variables indicates that such variables are conditionally independent, given the remaining variables. The edges associated with the graph can be undirected or directed, where in both cases independence is represented, but in the latter (Bayesian networks), the dependence between variables is represented through conditional probabilities; technical details can be found in Lauritzen, 1996 [21] and Sucar, 2015 [22]. For instance, considering that two nodes u and v point to a node w, u and v are said to be the parent nodes of w, and we have the probability density associated with w given specific values to u and v. For a better understanding, in Fig. 14.2, we show a practical example of a Bayesian network concerning arterial damage. We include the associated conditional (depending on the parent nodes) and marginal probabilities (when there are no parent nodes), and this is a directed network for discrete data. When continuous variables are involved, conditional Gaussian distributions are applied.

Experts can provide the network structure and parameters (e.g., probabilities) associated with the models according to their knowledge. They can also be learned from the data through algorithms and statistical methods or a combination of both processes. We must identify coherent relationships, forbidding those illogical or forced, using different algorithms, comparing the networks obtained, or randomly simulating several networks to identify the most repeated relationships. Additionally, when directed networks are used, appropriate directions must be validated.

In Bayesian networks, we can also assign values to a set of nodes A (evidence) and see how these values affect another set of variables B, getting the conditional probability of those variables in B given specific values to nodes in A [23] [24]. Consequently, these networks can be used to understand dependence and causality or even to establish a classification. PGM and Bayesian networks represent dependence between all variables at the same time and not associations as in other commonly used models. A recent work shows the application of this type of analysis in epidemiology, for instance, Haddawy et al. who obtained a good prediction model through the use of Bayesian networks to obtain a spatiotemporal malaria prediction at the village level in Thailand [25]. On the other side, Bui et al. analyze the spread routes of avian influenza subtypes H5N1 and H7N9 through a phylogeographical analysis. They found differences in transmission dynamics between both subtypes, suggesting the need for discrete control strategies for each one [26]. These two examples represent accurate applications of the use of Bayesian networks to public health, particularly in the contention of diseases.

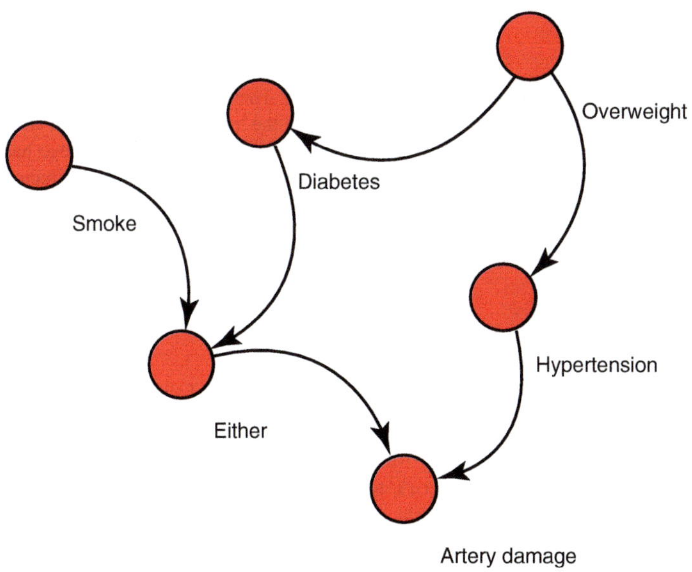

<table>
<tr><td>P(smoke)</td><td>P(overweight)</td></tr>
<tr><td>yes no</td><td>yes no</td></tr>
<tr><td>0.5 0.5</td><td>0.11 0.89</td></tr>
<tr><td>P(diabetes | overweight)</td><td>P(either | diabetes, smoke = no)</td></tr>
<tr><td>Overweight</td><td>diabetes</td></tr>
<tr><td>diabetes yes no</td><td>either yes no</td></tr>
<tr><td>yes 0.1 0.01</td><td>yes 10</td></tr>
<tr><td>no 0.9 0.99</td><td>no 01</td></tr>
<tr><td>P(hypertension | overweight)</td><td>P(artery_damage | hypertension, either = yes)</td></tr>
<tr><td>overweight</td><td>hypertension</td></tr>
<tr><td>hypertesnsion yes no</td><td>artery_damage yes no</td></tr>
<tr><td>yes 0.6 0.3</td><td>yes 0.9 0.7</td></tr>
<tr><td>no 0.4 0.7</td><td>no 0.1 0.3</td></tr>
<tr><td>P(either | diabetes, smoke = yes)</td><td>P(artery_damage | hypertension, either = no)</td></tr>
<tr><td>diabetes</td><td>hypertension</td></tr>
<tr><td>either yes no</td><td>artery_damage yes no</td></tr>
<tr><td>yes 11</td><td>yes 0.8 0.1</td></tr>
<tr><td>no 00</td><td>no 0.2 0.9</td></tr>
</table>

Fig. 14.2 Network representing a model concerning artery damage. Being overweight may be considered as a cause of both diabetes and hypertension. Smoking and diabetes are linked into a node (either) indicating whether a subject smokes or has diabetes. The presence of any of these two problems, represented in the node named as either, and the possible presence of hypertension are possible causes of artery damage. Using data, or according to expert knowledge, the conditional and marginal probabilities associated with each node can be obtained, and possible values are shown beside the network

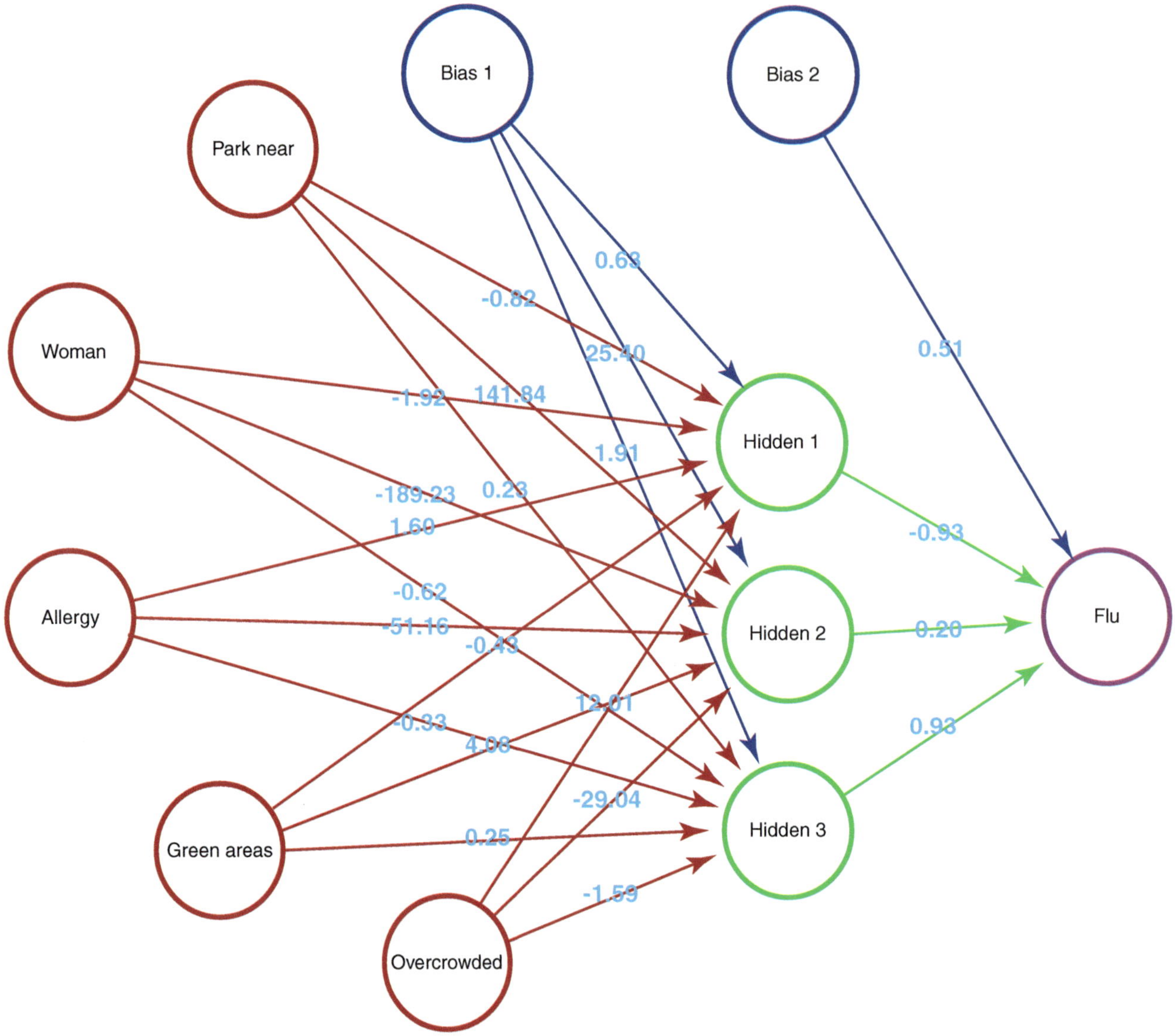

Fig. 14.3 Neural network built to predict the number of cases of flu in each district; all districts are of a specific geographical area. The inputs (red) correspond to proportion of people that reports having a park near home, the proportion of women, the number of people with respiratory allergies, the approximate proportion of a district corresponding to green areas, and the proportion of people that, according to the last census, live in an overcrowded home. The output (purple) corresponds to the number of cases of flu in each district. A hidden layer (green) formed by three hidden nodes is used. There are directed edges from each of the input nodes to each node in the hidden layer, and from each node of the hidden layer to the output variable, each directed edge has an associated weight. A bias term (blue) and its associated directed edges are associated with each node in the hidden layer (like a constant term from the input layer), and another bias term is associated with the output node, and they also have associated weights

Neural Networks

Neural networks are another new tool that has raised the baseline for a significant number of tasks, including visual pattern recognition, natural language translation, and mathematical modeling. The most basic model of a neural network could be appreciated in Fig. 14.3. In this sense, there is a multitude of network variations (topologies) and training schemes suited to different tasks.

All neural systems try to incorporate common elements observed in nature, such as the connectivity of the network, the response function, the learning rate, and the memory (the ability of the network to incorporate information seen in a past sample to predict a response). Neural networks receive this name since they mimic the brain; the neurons are a function of their inputs and outputs that lead to other functions. The most basic and oldest kind of neural network is the perceptron [27]. These neurons are usually arranged in a multi-layer perceptron network (MLP) and proved to be a universal function approximator [28]. MLP relies on supervised learning for either classification or prediction tasks. However, neural networks are not only used for supervised learning, but there are also variations used for unsupervised learning (cluster identification) [29].

MLPs are arranged in layers, where the input layer is a vector formed of the input variable's set, and the desired result set forms the output layer. A set of neurons can be specified as hidden layers to give more discernment power to the network. All neurons are characterized by their inputs and the connection weights assigned to them; sometimes an additional neuron is added in each input and hidden layer, not receiving information, in a single-layer perceptron; this neuron acts as a constant term as in a linear regression model and is referred to as the bias neuron. The weights adjust the signal strength of the different inputs, and the non-linear part of the approximation is achieved by applying a non-linear activation function to the output [30]. A weight is associated with each edge either joining an input variable with each variable in a hidden layer or joining an input variable with each variable in the output layer when hidden layers are not present. The weights of a neural network are estimated with algorithms; one of the most popular is the error backpropagation algorithm [31]. The same occurs with those edges joining variables in a hidden layer (when two or more hidden layers are present) or the output layer (Fig. 14.3). There are network variations that have edges linking a variable with itself; this is called a recurrent neural network (RNN), which can be useful in time series forecasting.

Networks are mostly used as predictive instead of explanatory ones, to avoid overfitting a neural network; it is recommended to use training and test sets to validate our model and avoid using too many hidden layers. There are different kinds of learning algorithms used to train a neural network. These methods are optimization algorithms that optimize the network weights and the network topology [32]. This second class of algorithms is pruning algorithms that eliminate little-used nodes (nodes with small weights), making it smaller and faster. There exist different kinds of network topologies and neuron types, allowing for different kinds of uses. If an output corresponds to a quantitative variable, we can predict values associated with the output (as in a regression model); on the contrary, if the output is a qualitative variable, a classifier can be obtained. Application of neural networks to epidemiology has been made regarding the generation of risk models [33, 34] and for survival prediction [35] in pediatric medicine. In Fig. 14.3, we show a neural network with an output variable, the number of cases of flu in each geographical district, five input variables, and one hidden layer with the corresponding estimated weights.

Social Network Analysis (SNA)

SNA is the most common network model performed at the date in several studies on epidemiology, and it is a quantitative and qualitative analysis using graph theory. SNA maps and measures the flow and changes of social relationships, and usually it is represented with points (nodes) and lines (edges), being the graphs representing either symmetric or asymmetric relations between discrete pairs of objects. The networks can be built from experts' knowledge, bibliometric studies, using an association or correlation measure between pairs of actors, etc. The nodes can represent objects, people, variables, diseases, etc. The edges usually represent interaction among nodes. They can contain associated values or weights, can be direct or undirected, and can define the strength of the interaction or relationship, among others.

SNA can be used to determine the relevance of nodes according to several properties and centrality measures, most of which are expressed in Table 14.3. This model could be used to obtain whether there is social cohesion and detect communities of nodes (clusters or modules), compare the relation of the network structure with well-known models based on simulation, or explain such structure through statistical models cross-sectional or longitudinal. Applications of social network methodology in biology include logistical networks, gene regulatory networks, metabolic networks, and, more recently, the interactomics or connectomics (Fig. 14.4). Mainly, in molecular biology, this approach has been used since there are signaling pathways that behave as a social network with quite connected nodes (hubs) and nodes connecting different clusters of vertices (bottlenecks). The number of publications with SNA and molecular networks to approach epidemiological studies has increased over the last years due to the development and improvement of genomic and proteomic techniques, as well as the improvement of computer programs capable of processing these data. For instance, the study by Gardy et al. in 2011 uses SNA and whole genome sequencing of *Mycobacterium tuberculosis* in order to describe the outbreak dynamics at a higher resolution in a medium-sized community in British Columbia [36]. This methodology has also been used in longitudinal studies of cognitive performance and depression [37], studies of human immunodeficiency virus (HIV) epidemiology [38, 39], studies of inclusion of marginalized women into government support programs, [40] or genome-wide association studies in bipolar disorder [41], just to mention few examples of its applications. A large-scale meta-analysis from epigenome-wide association studies of 24 birth cohorts shows that birth weight is associated with widespread differences in DNA methylation in neonates. The difference persisted only minimally across childhood and into adulthood [42]. Hence,

Table 14.3 Main network properties, which can be calculated to analyze the properties and main characteristics of networks

Property	Characteristics or definition
Size	The number of nodes in a network.
Density	It is defined as a relation between the number of edges to the number of possible edges in a network. It corresponds to the proportion of potential edges in a network.
Connectedness and connected components	An undirected graph is connected when it has at least one vertex, and there is a path between every pair of vertices. A graph that is not connected is said to be disconnected, and the number of connected components (maximal set of nodes such that a path connects each pair of nodes) can be calculated.
Clustering coefficient	It is a measure of the extent to which nodes in a graph tend to cluster together. It is defined as the proportion of closed triangles to the total number of open or closed triangles.
Diameter	It corresponds to the longest of all the calculated shortest paths in a network. When the graph is disconnected, it is calculated using the largest connected component.
Average shortest path length	It is a measure defined as the average number of steps along the shortest paths for all possible pairs of network nodes.
Node centrality	It depends on the type of relations and interactions among nodes, but in general, it measures the most important nodes within a graph. Three commonly used measures of centrality are degree, betweenness, and closeness.
Degree	The number of connections associated with each node in an undirected graph.
Closeness	A measure of the extent a node is near or far to all other nodes in the network. It is calculated as the reciprocal of the sum of the length of the shortest paths between the node and all other nodes in the graph.
Betweenness	It is a measure that represents how much a node stands between other nodes.
Average degree	Characteristic of an undirected graph used to measure the number of edges compared to the number of nodes. To calculate it, we divide the summation of all nodes' degree by the total number of nodes.
Number of hubs	A hub is a node with a number of degrees greatly than the average. When a hub is connected to several hubs, it is called "meta-hub."
Bottlenecks	A bottleneck is a hot spot, central nodes that provide the only connection between different parts of the network; usually, they are nodes that connect clusters. Sometimes they are hubs, but not necessarily.

the applications of these methodologies allow an in-depth and detailed analysis, having the ability to find new associations that facilitate the study of complex pathologies and the finding of new prognostic marker determinants in the transition process between physiologically healthy states toward a pathological one.

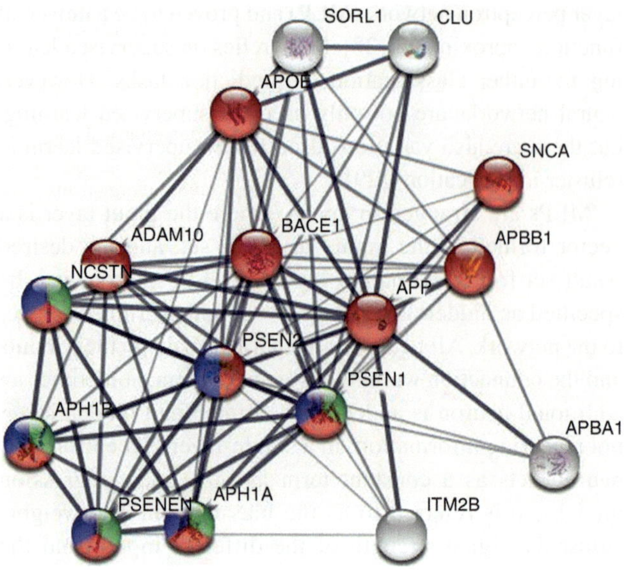

Fig. 14.4 Network representing APP interacting proteins. The image shows a set of proteins able to interact with amyloid precursor protein (APP), a key element in Alzheimer's disease (AD) pathology. Nodes including red color represent relevant proteins involved in AD pathways. Meanwhile, the nodes including blue represent those involved with NOTCH (membrane receptor) signaling pathway. Components of the gamma-secretase complex are nodes including green color, and those nodes including white represent other pathways. The data demonstrate the complexity of finding new drugs for gamma-secretase due to the overlapping of pathways involved in AD and NOTCH (neuronal survival). Line thickness represents the strength of data support, which depends on an index (edge confidence) obtained from experimental proof, available data sets, and text mining: low (>0.15), medium (0.16–0.40), high (0.41–0.70), and highest (<0.70). When the tridimensional protein structure is known, the node is included inside the corresponding 3D representation. The network analysis was performed in the STRING consortium website (https://string-db.org)

Perspectives

In this chapter, we have seen that network and graphical models are useful mathematical approaches that study complex systems in epidemiological research, having in common a graphical representation that considers multidimensional relationships between sets of variables, giving results to a different branch of SB the so-called SEp. Additionally, we briefly review the leading techniques and software available according to the data type.

Mainly, PGM and Bayesian networks are applied to analyze causality relationships to see how some variables affect others in descriptive analyses or even for classification purposes. On the other hand, neural networks are machine learning models that can be used for prediction and classification. There are several types of neural networks; the simpler ones include a set of inputs and a set of outputs related through non-linear functions, and even some include additional hidden nodes to improve the discernment power. One of the main disadvantages of neural networks is the opacity of the generated model, as the network deepens, and more nodes are added, except for the single-layer feed-forward perceptron where we can refer directly from the network. Finally, SNA are, at the date, the most used models since they can be easily obtained and interpreted. Networks can be built from any pairwise relationship: people, genes, socioeconomic measures, among others, which can or cannot have a direction or even a weight indicating the strength of the relationship. Several results could be obtained when this model of the network is implemented such as relationships, the most relevant nodes, clustering, structure, dyad-level predictors (e.g., friendships could be more likely between people of the same sex), or using the information of a network representing another type of relationship with the same actors.

Conclusions

Lastly, SEp models represent a challenge since they are mathematically more complex than those traditionally used. Nevertheless, they allow us to combine and generate models of diseases considering several levels of complexity. In this context, these models could help us to understand causality and open black-boxes in public health, since we could get to know how infectious or chronic diseases spread, how environmental changes interact with disease, and how this will later impact over whole human populations, meaning *taking every variable and everyone into account*, as a novel paradigm in epidemiology.

Acknowledgments This review is part of a project registered at Instituto Nacional de Geriatría as DI-PI-003/2018.

Conflict of Interest Authors declare NO conflict of interest.

References

1. Haring R, Wallaschofski H. Diving through the "-omics": the case for deep phenotyping and systems epidemiology. Omi A J Integr Biol. 2012;16:231–4. https://doi.org/10.1089/omi.2011.0108.
2. Lund, Eiliv. "Between Epidemiology and Basic Genetic Research – Systems Epidemiology." (2012). IntechOpen. London, UK.
3. Serrano L. Synthetic biology: promises and challenges. Mol Syst Biol. 2007;3:158. https://doi.org/10.1038/msb4100202.
4. Galea S, Riddle M, Kaplan GA. Causal thinking and complex system approaches in epidemiology. Int J Epidemiol. 2010;39:97–106. https://doi.org/10.1093/ije/dyp296.
5. Lund E, Dumeaux V. Systems epidemiology in cancer. Cancer Epidemiol Biomark Prev. 2008;17:2954–7. https://doi.org/10.1158/1055-9965.EPI-08-0519.
6. Bartocci E, Lió P. Computational modeling, formal analysis, and tools for systems biology. PLoS Comput Biol. 2016:1–22. https://doi.org/10.1371/journal.pcbi.1004591.
7. Hill AB. The environment and disease: association or causation? Proc R Soc Med. 1965;58:295–300.
8. Comas I, Gagneux S. A role for systems epidemiology in tuberculosis research. Trends Microbiol. 2011;19:492–500. https://doi.org/10.1016/j.tim.2011.07.002.
9. Gutacker MM, Mathema B, Soini H, Shashkina E, Kreiswirth BN, Graviss EA, et al. Single-nucleotide polymorphism–based population genetic analysis of mycobacterium tuberculosis strains from 4 geographic sites. J Infect Dis. 2006;193:121–8. https://doi.org/10.1086/498574.
10. Gire SK, Goba A, Andersen KG, Sealfon RSG, Park DJ, Kanneh L, et al. Genomic surveillance elucidates Ebola virus origin and transmission during the 2014 outbreak. Science (80-). 2014;345:1369–72. https://doi.org/10.1126/science.1259657.
11. Pybus OG, Suchard MA, Lemey P, Bernardin FJ, Rambaut A, Crawford FW, et al. Unifying the spatial epidemiology and molecular evolution of emerging epidemics. Proc Natl Acad Sci U S A. 2012;109:15066–71. https://doi.org/10.1073/pnas.1206598109.
12. Pham NTT, Nguyen CT, Vu HH. Assessing and modelling vulnerability to dengue in the Mekong Delta of Vietnam by geospatial and time-series approaches. Environ Res. 2020;186:109545. https://doi.org/10.1016/j.envres.2020.109545.
13. Rasmussen AL, Katze MG. Review genomic signatures of emerging viruses : a new era of systems epidemiology. Cell Host Microbe. 2016;19:611–8. https://doi.org/10.1016/j.chom.2016.04.016.
14. Drenos F, Grossi E, Buscema M, Humphries SE. Networks in coronary heart disease genetics as a step towards systems. Epidemiology. 2015:1–16. https://doi.org/10.1371/journal.pone.0125876.
15. Yuan Z, Ji J, Zhang T, Liu Y, Zhang X, Chen W, et al. A novel chi-square statistic for detecting group differences between pathways in systems epidemiology. Stat Med. 2016;35:5512–24. https://doi.org/10.1002/sim.7094.
16. Ji J, Yuan Z, Zhang X, Li F, Xu J, Liu Y, et al. Detection for pathway effect contributing to disease in systems epidemiology with a case-control design. BMJ Open. 2015;5:e006721. https://doi.org/10.1136/bmjopen-2014-006721.
17. Mitnitski AB, Rutenberg AD, Farrell S, Rockwood K. Aging, frailty and complex networks. Biogerontology. 2017;18:433–46. https://doi.org/10.1007/s10522-017-9684-x.
18. García-Peña C, Ramírez-Aldana R, Parra-Rodriguez L, Gomez-Verjan JC, Pérez-Zepeda MU, Gutiérrez-Robledo LM. Network analysis of frailty and aging: empirical data from the Mexican health and aging study. Exp Gerontol. 2019;128:110747. https://doi.org/10.1016/j.exger.2019.110747.
19. Rouse B, Chaimani A, Li T. Network meta-analysis: an introduction for clinicians. Intern Emerg Med. 2017;12:103–11. https://doi.org/10.1007/s11739-016-1583-7.
20. Mills EJ, Thorlund K, Ioannidis JPA. Demystifying trial networks and network meta-analysis. BMJ. 2013;346:f2914. https://doi.org/10.1136/bmj.f2914.
21. Lauritzen SL. Graphical models 17. Clerendon Press; 1996. Oxford, UK.
22. Sucar LE. Probabilistic graphical models principles and applications. 2015; https://doi.org/10.1007/978-1-4471-4929-3.
23. Højsgaard S, Edwards D, Lauritzen S. Graphical models with R. Springer; 2012. https://doi.org/10.1007/978-1-4614-2299-0.
24. Cowell RG. Probabilistic networks and expert systems : exact computational methods for Bayesian networks. Springer; 2007. New York, USA.
25. Haddawy P, Hasan AHMI, Kasantikul R, Lawpoolsri S, Sa-angchai P, Kaewkungwal J, et al. Spatiotemporal Bayesian networks for malaria prediction. Artif Intell Med. 2018;84:127–38. https://doi.org/10.1016/j.artmed.2017.12.002.
26. Bui CM, Adam DC, Njoto E, Scotch M, MacIntyre CR. Characterising routes of H5N1 and H7N9 spread in China

using Bayesian phylogeographical analysis. Emerg. Microbes Infect. 2018;7 https://doi.org/10.1038/s41426-018-0185-z.
27. McCulloch WS, Pitts WH. A logical calculus of the ideas immanent in nervous activity. Syst Res Behav Sci A Sourceb. 2017:93–6. https://doi.org/10.1016/s0092-8240(05)80006-0.
28. Cybenko G. Approximation by superpositions of a sigmoidal function. Math Control Signals Syst. 1989;2:303–14. https://doi.org/10.1007/BF02551274.
29. Ciaburro G, Venkateswaran B. Neural networks with R : smart models using CNN, RNN, deep learning, and artificial intelligence principles. (2017). Packt Publishing. Birmingham, UK.
30. Hastie T, Friedman J, Tibshirani R. Model assessment and selection BT – the elements of statistical learning: data mining, inference, and Prediction. New York: Springer; 2001. https://doi.org/10.1007/978-0-387-21606-5_7.
31. Rumelhart DE, Hinton GE, Williams GJ. Learning representations by back-propagating errors. Cogn Model. 1988; https://doi.org/10.7551/mitpress/1888.003.0013.
32. Pelillo M, Fanelli AM. A method of pruning layered feed-forward neural networks. Lect. Notes Comput. Sci. (including Subser. Lect. Notes Artif. Intell. Lect. Notes Bioinformatics), vol. 686, 1993, p. 278–83. https://doi.org/10.1007/3-540-56798-4_160.
33. Sherriff A, Ott J. Artificial neural networks as statistical tools in epidemiological studies: analysis of risk factors for early infant wheeze. Paediatr Perinat Epidemiol. 2004;18:456–63. https://doi.org/10.1111/j.1365-3016.2004.00592.x.
34. Chan CH, Chan EY, Ng DK, Chow PY, Kwok KL. Application of artificial neural networks to establish a predictive mortality risk model in children admitted to a paediatric intensive care unit. Singap Med J. 2006;47:928–34.
35. Sato F, Shimada Y, Selaru FM, Shibata D, Maeda M, Watanabe G, et al. Prediction of survival in patients with esophageal carcinoma using artificial neural networks. Cancer. 2005;103:1596–605. https://doi.org/10.1002/cncr.20938.
36. Gardy JL, Johnston JC, Sui SJH, Cook VJ, Shah L, Brodkin E, et al. Whole-genome sequencing and social-network analysis of a tuberculosis outbreak. N Engl J Med. 2011;364:730–9. https://doi.org/10.1056/NEJMoa1003176.
37. Kuiper JS, Smidt N, Zuidema SU, Comijs HC, Oude Voshaar RC, Zuidersma M. A longitudinal study of the impact of social network size and loneliness on cognitive performance in depressed older adults. Aging Ment Heal. 2019; https://doi.org/10.1080/13607863.2019.1571012.
38. Auerbach DM, Darrow WW, Jaffe HW, Curran JW. Cluster of cases of the acquired immune deficiency syndrome. Patients linked by sexual contact. J Urol. 1984;132:421. https://doi.org/10.1016/s0022-5347(17)49667-4.
39. Brown A, Leigh Brown A, Lycett S, Weinert L, Hughes G, Fearnhill E, et al. Transmission network parameters estimated from HIV sequences for a Nationwide epidemic. J Infect Dis. 2011:1463–9.
40. Loutfi D, Andersson N, Law S, Salsberg J, Haggerty J, Kgakole L, et al. Can social network analysis help to include marginalised young women in structural support programmes in Botswana? A mixed methods study. Int J Equity Health. 2019;18:12. https://doi.org/10.1186/s12939-019-0911-8.
41. Stahl EA, Breen G, Forstner AJ, McQuillin A, Ripke S, Trubetskoy V, et al. Genome-wide association study identifies 30 loci associated with bipolar disorder. Nat Genet. 2019;51:793–803. https://doi.org/10.1038/s41588-019-0397-8.
42. Küpers LK, Monnereau C, Sharp GC, Yousefi P, Salas LA, Ghantous A, et al. Meta-analysis of epigenome-wide association studies in neonates reveals widespread differential DNA methylation associated with birthweight. Nat Commun. 2019;10:1893. https://doi.org/10.1038/s41467-019-09671-3.

Molecular Pharmacological Tools Applied to Epidemiology

15

Oscar Salvador Barrera-Vázquez, Edgar Flores-Soto, and Juan Carlos Gomez-Verjan

Abbreviations

AABPP	Amino acid–based protein or peptide prediction
ADME	Absorption, distribution, metabolism and excretion
CoMFA	Comparative molecular field analysis
DFT	Density functional theory
EMA	European Medicines Agency
FB-QSAR	Fragment-based two-dimensional QSAR
FDA	Food and Drug Administration
HIV/AIDS	Human immunodeficiency virus/acquired immunodeficiency syndrome
ICH	International Council for Harmonisation of Technical Requirements for Pharmaceuticals for Human Use
IDLS	Iterative double least squares
MF-3D-QSAR	Multiple field three-dimensional QSAR
NDA	New Drug Application
PD	Pharmacodynamics
PDB	Protein Data Bank
PI3K-AKT	Phosphatidylinositol 3-kinase/serine/threonine-specific protein kinase
PK	Pharmacokinetics
QSAR	Quantitative structure relationship
QSP	Quantitative systems pharmacology
QSPR	Quantity structure–property relationship SAR structure–activity relationship
SBDD	Structure-based drug design
US	United States of America

O. S. Barrera-Vázquez · E. Flores-Soto
Departamento de Farmacología, Facultad de Medicina, Universidad Nacional Autónoma de México (UNAM), Mexico City, Mexico

J. C. Gomez-Verjan (✉)
Dirección de Investigación, Instituto Nacional de Geriatría (INGER), Ciudad de México, Mexico
e-mail: jverjan@inger.gob.mx

Introduction

Drug development is essential to understand modern medicine because of the appearance of novel infectious diseases and the increase of the incidence of chronic degenerative conditions related to the human lifestyle. In this sense, the introduction of new drugs has reduced the various complications such as peptic ulcers [1] or HIV/AIDS, which has improved seriously from a deadly disease to a chronic condition [2]. Drug discovery involves rational design through medicinal chemistry for the efficient identification and optimization of active compounds. For this reason, it is crucial that clinicians deeply understand the drug discovery process since several observations could only be identified from the clinic and epidemiological perspective. Such an understanding will drive innovation for patients to understand the importance of participating in clinical trials, report any adverse events to improve pharmacovigilance, improve personalized medicine, and improve the communication between pharmacists and clinicians [3, 4].

Drug Discovery

A drug discovery program is a process through which potential new medicines are identified. It involves a wide range of scientific disciplines, including biology, chemistry and pharmacology. It started because there is a disease or clinical condition with epidemiological relevance and no pharmacological or inefficient treatment. Initial research often occurs in the academic world [5]. Nevertheless, drug development is a multifaceted process involving public and private institutions

J. C. Gomez-Verjan, N. A. Rivero-Segura (eds.), *Principles of Genetics and Molecular Epidemiology*,
https://doi.org/10.1007/978-3-030-89601-0_15

that generate enough data to develop a relevant hypothesis (inhibition or activation of a target from a specific pathway), which will eventually lead to the identification of a potentially active compound biotechnological product [5, 6].

Finding and progressing a new drug from the original idea to the launch of a finished product is a complex process that can take between 12 and 15 years and requires an investment of approximately $ 900 million to 2 billion dollars [7]. The stages of drug discovery and development focused on identifying a possible target or pathway and validating such a target against different compounds. These processes are traditionally carried out in preclinical, experimental models, which lead to the identification of compounds that could modulate somehow the target, and then through a process of optimization (*hit-to-lead*), choosing the best of the hits based not only on their pharmacodynamic potential but also on their pharmacokinetic and pharmaceutical properties wherein such a compound could not only be easily formulated and administered but also produced so that it could reach the patient as rapidly as possible (Fig. 15.1).

Target Identification

The first step in drug discovery is identifying the biological origin of a disease and the possible *targets* for intervention. A *target* could be defined as a molecule (nucleic acid, metabolite, or protein) involved in gene regulation or intracellular signaling disease processes [9]. The ideal *target* must be effective and safe and meet clinical and commercial requirements. Target identification techniques may be based on principles from molecular biology, biochemistry, genetics, biophysics, and other disciplines [10].

Validation of the Target

Once we identify a potential target (metabolite, gene, protein or nucleic acid), it must be demonstrated that such target is involved in the progression of a given disease and that its activity can be modulated [5]. Validation experiments range from in silico and in vitro to in vivo models. Validation of efficacy and toxicity, including mutagenicity, must be careful and precise for successful drug development in the following stages [11].

Lead Identification

The hit-to-lead process involves a procedure to identify from the complete set of active molecules (hits) a compound that could be considered as the leader due to its unique interaction with the selected target (potency and selectivity) and to its toxicological and pharmaceutical properties (absorption, distribution, metab-

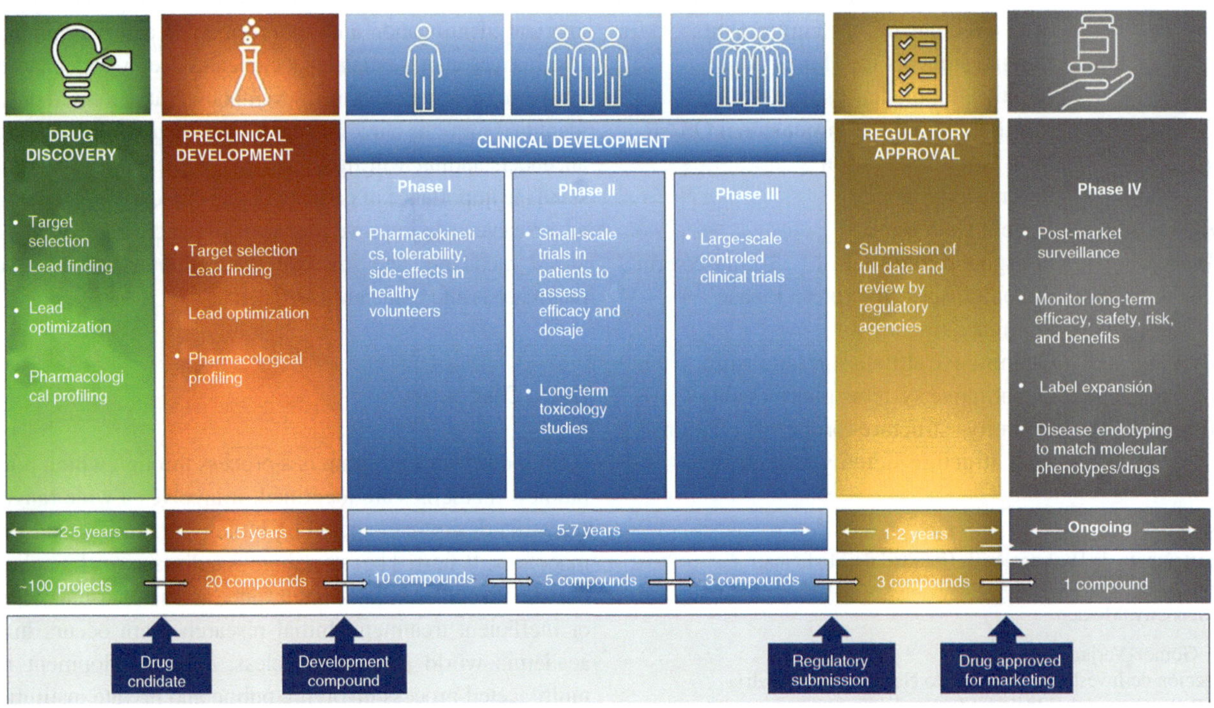

Fig. 15.1 Schematic representation of the drug discovery process. For example, of all the phases that cover drug discovery and the approximate amount of compounds and time of each process, only one ultimately gets approval [8]. Success requires immense resources, from scientific to highly sophisticated laboratories and technologies

olism, and excretion) [5]. It is worth mentioning that this evaluation reveals compounds that could fail during the drug development process. Additionally, it is important to remember that screening experiments could be performed with already known drugs with already proven pharmaceutical properties to identify novel activities. Alternatively, synthetic compounds can be specifically designed to target without interfering with other cellular processes.

Lead Optimization

After a lead compound is identified, the goal is to move into clinical trials; therefore, possible deficiencies in its structures may need to be improved to produce a *drug candidate*. Skipping this stage could lead to a lack of efficacy, pharmacokinetics (PK), and compounds' safety issues. This stage can determine how chemical structure and biological activity are related to interacting with the target and its metabolism [12]. For this process, a combination of specialists in computational chemistry, medical chemistry, drug metabolism, and other areas is recommended to provide insight into this last stage. Additionally, it is crucial to consider the manufacture of new drugs since you could have a lead compound with unusual pharmacological activity but challenging to meet the dosage needs of the population affected by the disease; let us think, for example, of a drug that will be used for the treatment of diabetes type 2 worldwide and the amount of compound you have to obtain to meet the need. After this stage, the next phase is the clinical trials, which serve as the gold standard to evaluate tested drugs' efficacy and safety before marketing authorization.

Clinical Trials

Phase I

In general, during this phase, compounds are tested in 20 to 100 healthy volunteers or people with the disease/condition depending on the study's design; the purpose of this phase is mainly focused on determining safety and dosage. The duration of the study, on average, is several months [13].

Phase II

Phase II is mainly focused on characterizing the compound's efficacy and the possible side effects; in such a phase, the number of participants increases to several hundred people with the disease/condition. In this phase, novel information is obtained, such as the optimal dose, frequency of intake, and the disease's effect, and the duration of this phase could be between several months and 2 years [13].

Phase III

In Phase III trials, researchers study the drug in participant groups of 300 to 3000 volunteers who have the disease or condition, generate statistically significant data, and study the changes of doses and efficacy over different populations. The duration of this phase is around 1–4 years [13].

Phase IV

In Phase IV, participants increase to several thousands who have the disease/condition. The purpose is to validate its safety and efficacy on significant populations and perform a new drug's follow-up, including a pharmacovigilance phase [13] (Fig. 15.2).

In most cases, a drug research agency (such as the FDA in the USA or the EMA in the European Union) must approve the sale's effectiveness. Phase 4 trials: Here, you do ongoing monitoring after an investigational drug agency approves a drug. The purpose of this phase is to monitor and find more information about the risks, benefits, and optimal use of the approved drug.

In Silico vs In vitro

Scientific knowledge arises from applying the scientific method, which is based on the observation of a natural phenomenon, the formulation of a hypothesis, and its verification through experiments [14]. The experiment is a generally controlled situation in which an observed phenomenon is reproduced. In this sense, biological sciences rely on different experimental models to test hypotheses such as in vivo, in vitro, and/or in silico [15]. In vivo tests are defined as those carried out in the conditions closest to the observed phenomenon of incomplete living organisms [16]. Meanwhile, in vitro methods are used as preliminary tests, carried out on experimental isolated models such as cell cultures, microbiological cultures, and organoids, among others, reducing the number of animals [17]. There are several opinions that in vitro tests may be better than in vivo tests because there is better control of the conditions, ethical considerations, and a lower cost, among others [17]. It is important to carry out in vivo studies since results are easily extrapolated to other species, and you could analyze how the drug behaves in a complete living organism. Nevertheless, the best model to study a complete phenomenon does not exist because each of them will play a key role in explaining a natural phenomenon. Complementing models are recommended with the appropriate experimental designs and statistical tests.

With the advancement of technology, the possibility of modeling natural phenomena, also known as in silico

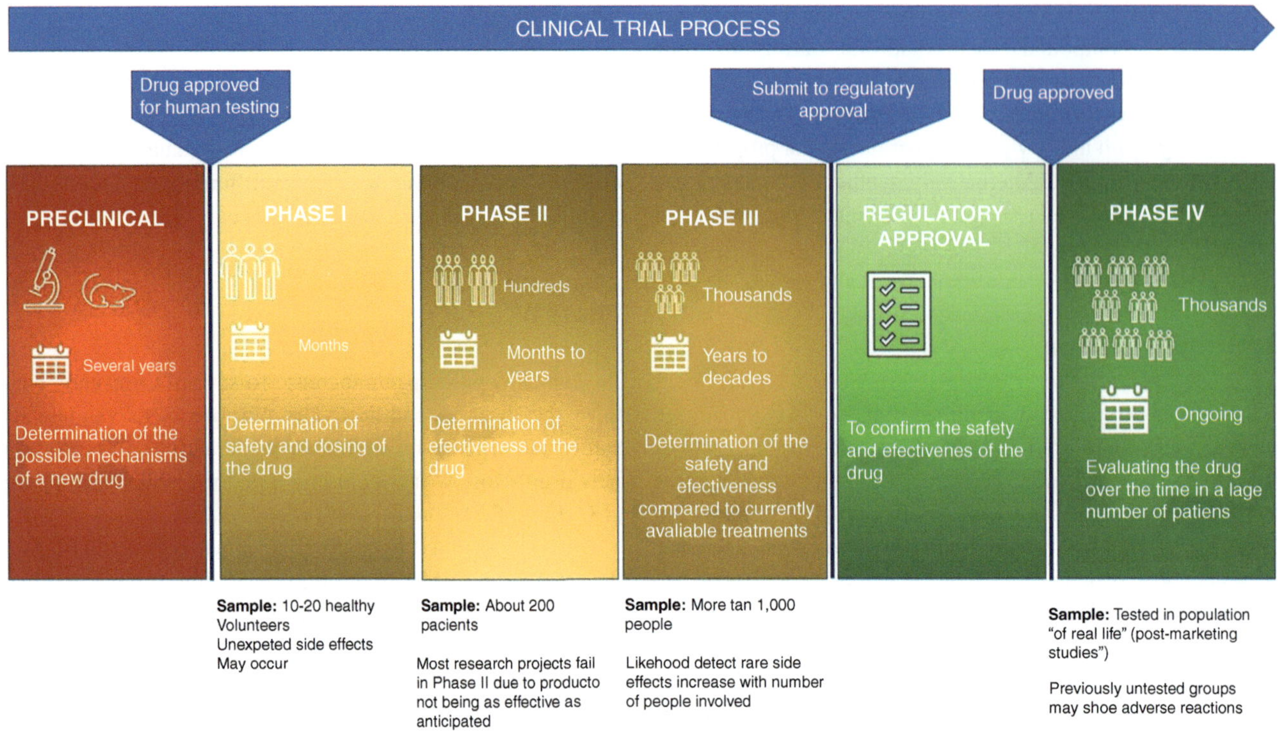

Fig. 15.2 Clinical trial phases. Phase 1 trials: in this phase, the drug's safety is evaluated, and side effects are identified; these tests are performed in a small group of healthy people (10–20). Phase 2 trials: here, the drug's efficacy is determined, and its safety further evaluated; these trials are performed in a larger group of patients (100–300). Phase 3 trials: In this phase, different aspects of the new drug are corroborated, such as the drug's efficacy, monitoring of side effects, and its comparison with standard or equivalent treatments, as well as collecting information to ensure that it is new. The drug can be used safely. Large groups of people (1000–3000) are used in this phase

modeling, has become more common due to the speed at which models can be solved with computer science [18, 19]. The interrelation of biological sciences with computer science and engineering has made it possible to build and solve mathematical models that allow a specific phenomenon to be incorporated by a computer [20]. These computational mathematical models or in silico models allow us to simulate real situations [21, 22], modifying the values of the variables involved in a wide range of values and contrasting them with the actual values. In this way, they will show us behaviors that might take decades to obtain from observation or experimentation.

In silico models at the biological level are emerging slowly but steadily; the number of periodical publications in this area supports it. However, the current deficiency of these models is based on the scarce necessary knowledge in the phenomena at the biological level and their interaction with other phenomena. The benefits of in silico models are evident in countless studies on pharmacodynamics (PD) and pharmacokinetics [23]. A recent development that deserves mention is a simulator for preclinical studies in diabetes, which the Food and Drug Administration (FDA) approved as a substitute for animal studies [24]. Such a simulator uses algorithms that model the human metabolic system based on data from 300 diabetic patients of different ages and has 26 parameters that allow modeling individual states of each patient. It allows you to enter variables such as physical exercise, diet, and insulin injection and study the effectiveness of new products or compare them with existing ones.

Sources of Pharmacologically Active Compounds

Natural Products

The nature of a pharmacological compound can be synthetic or derived from a natural source (natural products) either in its unaltered state or with several chemical modifications [25, 26]. A natural product is understood as a chemical compound or substance produced by a biological organism (it could be a primary or secondary metabolite) with pharmacological activity [27]. Historically, at the core of traditional medicine, we have turned to natural products as a source of treatment for many diseases [28, 29]. Over 35% of available drugs are currently derived either directly or indirectly from a natural product source [30]. Moreover, in certain areas, the discovery of natural products has been particularly prolific, such as antibiotics and antineoplastic agents, where 60–80% of them are derived from natural sources [31]. In this sense, many notable discoveries revolutionized medicine and encouraged the continuation of research involving natural products in drug discovery, such as penicillin, derived from a species of *Penicillium notatum*, by Alexander Fleming [30, 32]. Other notable discoveries are that of ivermectin and artemisinin, derived from *Streptomyces avermitilis* and *Artemisia annua*, respectively, which were discoveries awarded with Nobel prizes [33].

The study of these natural products has long been explored, especially with the rise of metabolomics. The study of these products' intricate molecular structures serves as a framework and inspiration for drug discovery [26, 34]. Metabolomics is defined as the study of the entirety of the molecular profile found within a biological system, facilitated with the improved high throughput screening technologies in chromatography and spectroscopy [34, 35]. Through this process, we can elucidate the standout compounds comprising the natural products that produce the therapeutic effects, often consisting of simultaneous synergistic action from several of the chemicals within [26, 34]; please refer to Chap. 7, focused only on metabolomics.

The innovative development and the implementation of novel technologies in informatics and analytical technologies such as quantum computing, profiling technologies, computational biology techniques, big data, and microfluidics will enable a multidisciplinary approach to the study of the metabolite profile of natural products; it serves for a quick and efficient analysis of large data sets of metabolite libraries, providing great insight into the potential of these specimens in the medical field and their potential in drug discovery to meet the evergrowing needs for novel treatments of the health challenges of today [26, 34]. The focus of metabolomics in the identification of metabolites and the study of the biological mechanisms activated following the effects of a pharmacological compound [26, 34]. A growing interest in this field came along the necessity for databases, both open and commercial, dedicated to natural products that optimizes their analysis, some examples being: *Super Natural II*, *Universal Natural Product Database*, *Chinese Natural Product Database*, *Drug Discovery Portal*, *iSmart*, *NuBBE*, and many more [34].

Quantitative Structure Relationship (QSAR)

According to the rational drug design approach's primary hypothesis, the essential effects of drugs are generated from the molecular recognition and the binding of ligands to the active site of targets (nucleic acids, receptors, and enzymes). These drugs affect both inhibiting or activating signal transduction through the binding of enzymatic activity or molecular transport. Medicinal chemistry used small chemical compounds that already existed in nature, and their activity was revealed by empiricism [36].

The quantitative structure–activity relationship (QSAR) is used to build computational or mathematical models whose purpose is to find a statistically significant correlation technique between structure and function using a chemometric technique. The aims of QSAR include correlating the relationship between trends in chemical structure alterations and changes in the biological endpoint of their biological activities. Optimize the available leads to improve your biological activities and predict the biological activities of untested compounds; Table 15.1 shows the traditional methodology of QSAR [37].

Quantitative structure–activity relationships (QSAR) and quantitative structure–property relationships (QSPR) are of

Table 15.1 Examples of classification of QSAR methodologies

QSAR models	Concept	Explanation of each model	References
1D-QSAR	Molecular representations and molecular fragments, such as H-donors, H-acceptors, pKa, log P with biological activity.	Such models reflect only the composition of the molecule. It is impossible to solve "structure–activity" tasks using such approaches adequately. These models have an auxiliary role.	[43]
2D-QSAR	Representation that contains topological information as physicochemical properties with biological activity.	These models are trendy and reflect only the molecule's topology, which contains information about possible conformations of the compound.	[43]
3D-QSAR	Correlation of various properties generates a 3D representation of the molecule.	These models are the most widespread. However, the choice of the analyzed conformer is primarily accidental.	[43]
4D-QSAR	Representation of ligand receptor interactions of the drug molecule with the 3D properties.	These models have similarity to 3D models; however, when 4D models are compared to them, the structural information is considered for a set of conformers (conditionally, the fourth dimension) instead of one fixed conformation.	[43]
5D-QSAR	Representing different induced fit models in 4D-QSAR.	This fifth dimension helps to identify the active site's molecular fragments that are responsible for the biological activity of the molecule.	[44]
6D-QSAR	Incorporating different solvation models in 5D-QSAR.	It considers the solvation function in QSAR analysis, which is an expansion of the QUASAR (5D-QSAR), employing simulations for different solvation models.	[45]
7D-QSAR	One more dimension has been added to the 6D-QSAR to introduce another higher dimension (7D-QSAR).	This analysis comprises real receptor or target-based receptor model data.	[45]

great importance in current chemistry and biochemistry. QSAR was conceived as a comprehensive and competent method widely used in pharmaceutical problems; it uses molecular descriptors to predict the relationships between the target's molecular properties and its biological context [38]. These molecular descriptors are universal variables used for QSAR-based activity prediction modeling [39].

For a model drug candidate to be considered, it needs to possess specific properties, such as chemical properties, solubility, enzymatic stability, permeation through biological membranes, low clearance from the liver or kidneys, potency, and safety [36]. Although there are countless molecular descriptors, selecting these is the most critical challenge in a QSAR. Therefore, to understand the QSAR model, decrease overfitting, speed up training, and improve the model's overall predictability, choosing appropriate and interpretable descriptors to configure QSAR models is a highly crucial, challenging, and complicated step. The primary assumption of the QSAR methodology is that the observed discrepancy in biological activity is correlated with molecular structure [36].

The construction of QSAR models for drug discovery consists of a general, systematized protocol and several modular steps that involve cheminformatics and machine learning techniques. The first step of the protocol is "molecular coding," where chemical characteristics and properties are obtained from chemical structures or the search for experimental results. The next step is to perform feature selection using unsupervised learning techniques to identify the most relevant properties and reduce the feature vector's dimensionality. During the final phase, a supervised machine learning model is applied to discover an empirical function (it can be explicitly or implicitly) to achieve an optimal mapping between the input feature vectors and the biological responses. An accurate QSAR model's construction requires careful consideration and selection of the SAR data sets used for training and model validation [40].

Besides, conventional QSAR methods have recently changed, due to the introduction of sophisticated mathematical tools and well-designed theoretical models, thus leading to three modern QSAR methods [41, 42]. The first model is fragment-based two-dimensional QSAR (FB-QSAR). A series of drug candidates' molecular structures are segregated into several fragments depending on the surrogate being examined. The physicochemical characteristics of the molecular fragments are compared with the drug candidate's bioactivities with the help of two sets of coefficients, of which one is for the molecular fragments while the other is for the physicochemical characteristics. The second QSAR type is the multiple field three-dimensional QSAR (MF-3D-QSAR). In this category, the additional molecular potential field (thermodynamic and non-thermodynamic) is incorporated into a comparative molecular field analysis (CoMFA), using two sets of coefficients, one for the position of the three-dimensional Cartesian space and the other for the field of potential, and this was used for the first time for the analysis of the three-dimensional structure of the ligands and to describe the structure–activity relationships. This three-dimensional plane or network corresponds to a surrogate for the actual biological receptor binding site. The third category of the QSAR comprises the amino acid–based protein or peptide prediction (AABPP), which is used for the analysis of peptide and protein activity using two sets of coefficients, one for the physicochemical properties of amino acids and the other for residues in the peptide chain [36]. These three recent QSAR approaches are characterized by simultaneous three-dimensional equations that enclose two sets of indeterminate coefficients. Furthermore, these new QSAR approaches, compared to traditional QSAR approaches, can increase the predictive power of QSAR and offer more information on the molecular structure.

Docking

In recent years, the number of proteins with known three-dimensional structure has shown an increase in their quantity, thanks to technological advances and the fact that these structures are available to the public through servers present on the network [46, 47]. On the other hand, improvements in techniques for determining structures, such as high-performance X-ray crystallography, have led to an increase in the number of structural targets [47].

Molecular docking is a computational method used to predict two interactions of molecules, generating a binding model. The docking method has proven helpful in small molecule drug discovery and design; in this context, the docking method is performed between a small molecule and a macromolecule creating a protein–ligand docking. However, for the molecular modeling of significantly more flexible and larger peptides, these coupling methods designed for small molecule interactions are not entirely suitable [48]. On the other hand, interest in peptide therapy [49] led to new techniques focused on protein–peptide coupling [50, 51]. Protein–peptide coupling methods fall into three categories: template-based coupling, local docking, and global docking. These different approaches provide different prediction precision levels, which are focused on the amount of interaction information provided as input.

Comparative coupling methods use known structures (templates) as scaffolds to generate a model of the complex. One of the most common practices is receptor threads and/or peptide sequences via a template structure. This method can be efficient if the template is like the complex investigated [49]. This type of template-based coupling is commonly performed manually or semi-automatically using a set of tools for sequence–structure comparison and analysis. Complex models are subsequently built using energy-based optimization and refinement that allow for structural flexibility.

Template-based docking of highly homologous complexes is provided by protocols focused on predicting and designing peptide binding specificity [52]. On the other hand, template-based modeling methods can also use monomeric protein fragments and protein–protein complex interfaces, which are used to build modeling structures—the structures of the interaction of protein–protein interfaces help design peptide inhibitors of protein–protein interactions [49].

Global Docking

Global docking methods are used to perform a coupled search for the peptide binding site and pose. This global protein–peptide coupling is focused on treating the input protein and peptide conformations as rigid and, subsequently, performing an exhaustive body-rigid docking. The more sophisticated methods automatically predict peptide conformation from a user-supplied sequence, which is divided into three steps: (i) generation of input peptide conformations, (ii) rigid docking model, and (iii) scoring of the models and/or refinement. Alternatively, the overall coupling can be combined with binding site predictions. Three main challenges have been encountered for protein–peptide docking: (i) flexibility problems, which are generated by modeling significant conformational changes of both peptide and protein molecules, (ii) scoring problems, which are caused by the selection of the most precise structure among many generated models, and (iii) integrative modeling or integration of experimental data and computational predictions in the protein–peptide docking scheme.

The difficulty of docking and the accuracy of prediction lie in the number of flexible bonds in a peptide, size, and defined secondary structure. Small molecule docking programs are usually limited to very short peptides, down to a few residues [53, 54]. Longer peptide patterning can be overcome by docking peptide fragments followed by their fusion [55, 56]. The receptor's flexibility upon binding can range from minor side-chain rearrangement to large-scale spinal rearrangements [57, 58]. The difficulty of coupling increases with increasing receptor conformational changes, and explicitly addressing spinal flexibility can become a significant challenge [57, 59]. The most straightforward approach is to perform a rigid body coupling ignoring the flexibility of the receiver. This method's main advantage is the low computational cost, which allows a comprehensive sampling of the receptor surface in search of a binding site. Rigid body coupling is often used as the main or one of the main components of global coupling protocols. However, those protocols allow at least side-chain flexibility in other modeling steps. Finally, coarse-grained protein models can be used to model large-scale rearrangements of the spine, such as disordered regions of significant length [60] or a loop region near the binding site [61].

Among the various challenges that docking presents, it is the most successful selection of the most accurate model to use. In most cases, the highest-ranked models are of lower quality than the most accurate models present in the docking results. Most docking tools use energy-based scoring methods for model classification. Except for energy-based scoring, some of the protein–peptide docking tools use additional methods to improve model selection; these methods include structural clustering and selecting the most significant clusters, incorporating co-evolutionary technical information, mutagenesis data comparison with template structures, or sequence-based predictions [49].

Network Pharmacology and Drug Repurposing

Among the most recent tools implemented for the acceleration of drug discovery is the so-called *network pharmacology*, which takes advantage of the enormous amount of information generated by the recent advances in several omics international projects, as well as the development of novel tools in bioinformatics in a combination of systems biology and polypharmacology knowledge [62]. To understand such an approach, it is essential to define some of the primary systems biology principles. Systems biology could be defined as an effort to understand from a holistic point of view the complexity of biological systems (biomolecules, cells, tissues, human body, groups of humans with a particular characteristic) as a whole (*"the whole of living organisms are more than the simple sum of their parts"*). To get closer to such knowledge, one integral approach is the development of mathematical models that capture the complexity of biological systems and their emergent properties [63], which in general obey nonlinear dynamics; in this context, computational approaches are needed to solve such models and their equations. There are several options for choosing an adequate model depending on the system and the complexity you are looking to approach, for instance, time-dependent phenomena, biological processes, concentrations, sampling, and the possibility to access experimental data. It is not the same to study the molecular changes associated with gout than to study the changes in social behavior that having gout represents, although both are closely associated. In this sense, some of the most fundamental and valuable approaches to understanding how several variables on a biological system interact with each other under specific parameters and at different levels are the network approaches. These models have shown to be quite helpful and relatively easy to implement and understand the *"big picture"* of the relationships (edges) among variables (nodes) [28]. Network approaches are often built with experimental knowledge and sometimes are data hungry; the more accurate you want your model, the more information is needed [64]. At the same time, these models are helpful; you could use a network to describe from our genome to the cells that make up the organs in our bodies to ourselves in our world; as stated by the Institute of Systems Biology, *"we are fundamentally a network of networks*." An interaction network could be more than a simple binary relationship between individuals and could be probabilistic relationships or confidential information at different; for a complete description of network models, please refer to Chap. 14.

On the other hand, the traditional pharmacology paradigm is "*one disease-one target-one drug*" nevertheless; this approach oversimplifies therapeutics for diseases. It makes it extremely difficult to discover novel drugs since you have to focus on one target; this may be pretty selective; however, diseases result from several mechanisms from environmental to genetic. Moreover, diseases share clinical symptoms making them even more complex [65]. In this context, to increase the therapeutic tools, the network paradigm from systems biology was applied to pharmacology, giving birth to the so-called *systems pharmacology*, also known as *network pharmacology*, which is considered the next paradigm in drug discovery [66]. Network pharmacology could be defined as a systems biology–based methodology based on biological networks to search for multitarget drugs [67]. It uses molecular pathway knowledge to integrate several concepts, including polypharmacology, toxicology, and drug repurposing, based on evaluating the selectivity among shared targets. In this sense, this tool could improve the potency of drugs and diminish the adverse effects [62].

There are several ways to approach *network pharmacology* analyses; for instance, you can reconstruct and predict a network based on the drug targets, or you can base your analysis on the chemical characteristics of the drug itself. Additionally, you can base your analysis on the network equilibrium model based on the structural network analysis rather than pure nodes properties. In general, there is a pipeline that can vary depending on the analysis but that describes well how to perform a network pharmacology analysis (Fig. 15.3).

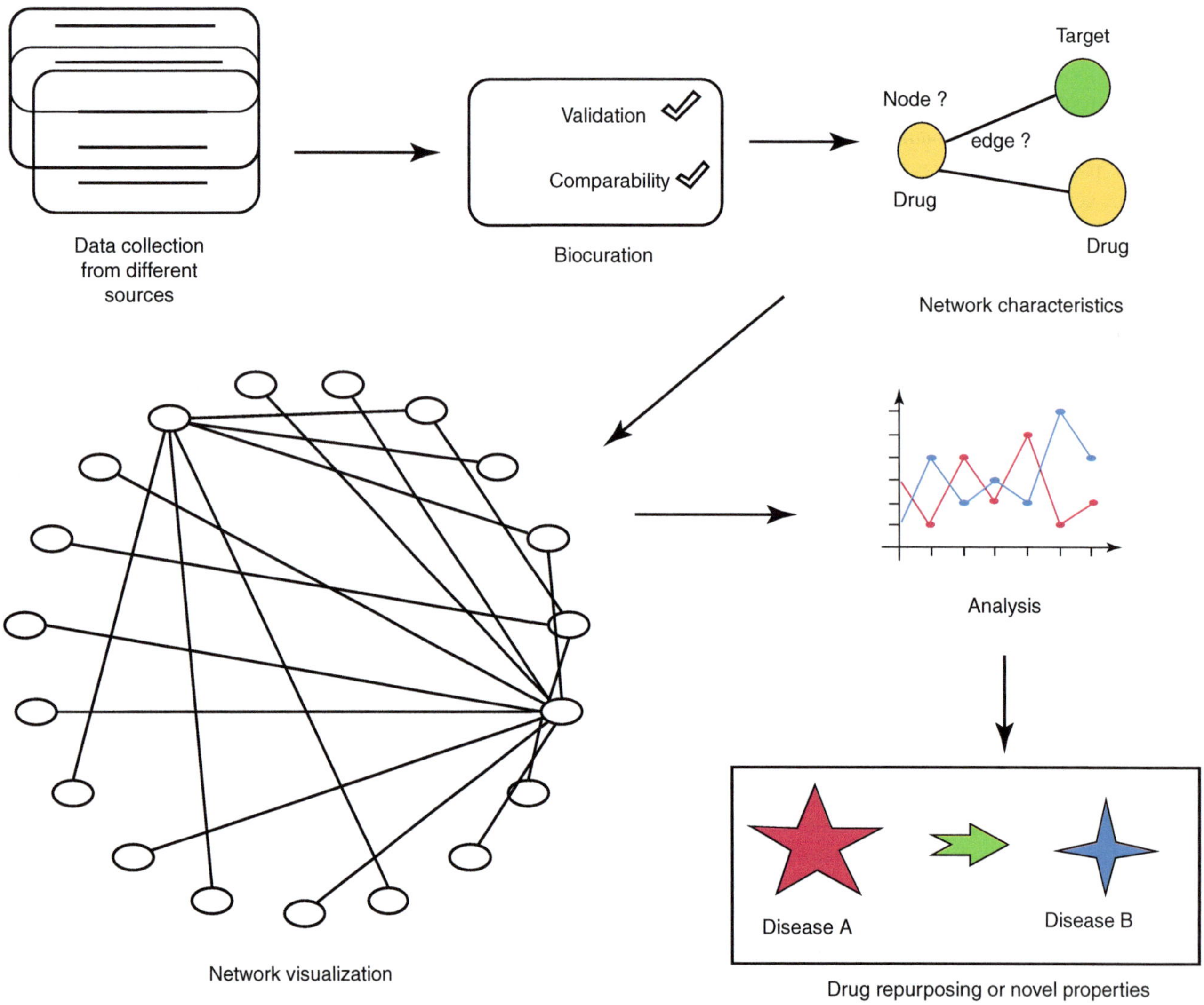

Fig. 15.3 General pipeline for a network pharmacology analysis. Data collection and the biocuration of the data are among the first steps in any bioinformatic analysis. To establish nodes and edge properties and characteristics, it is pretty important so we could pass to the network visualization and consequent analysis. Finally, once we have the results, we could establish novel properties for drug repurposing or similar

Table 15.2 Most widely network pharmacology tools and software freely available

Tool	Description	Link
Cytoscape	It is one of the most widely used open-source software for visualizing molecular interaction networks and biological pathways and integrating them with several data form annotations to gene expression profiles. It has several apps and tools that could be used for enrichment analysis and genetic expression or in general to draw and analyze network properties [71].	https://cytoscape.org/
GUESS	It is an exploratory data analysis and visualization tool for graphs and networks. It offers a visualization front end that supports the export of static images and dynamic movies.	http://graphexploration.cond.org/index.html
KNIME	It is a commercial tool for data-driven innovation, designed for discovering the potential hidden in data. It possesses several extensions for chemoinformatic analysis and easy to use. It has two versions, one for simple data scientist analysis and a commercial one for companies.	https://www.knime.com/software-overview
SmartGraph	It is a predictive web-based platform that supports complex cheminformatics workflows. It allows integration of additional biomedical data layers, such as pharmacological action of drugs, non-small molecule drugs, disease information, and target categories [72].	https://smartgraph.ncats.io/
FangNet	This freely available tool ranks several herbs used in traditional Chinese medicine based on their relative topological importance using a PageRank algorithm and the constructed symptom–herb network from a collection of empirical clinical prescriptions [73].	http://fangnet.herb.ac.cn.
YaTCM	It is a free web-based toolkit, which provides comprehensive information about traditional Chinese medicine and correlates several active compounds with different molecular pathways [74].	http://cadd.pharmacy.nankai.edu.cn/yatcm/home
Comparative Toxicogenomics database	It is an innovative web tool that relates toxicological information from different chemicals, genes, phenotypes, diseases, and exposures and is literature-based and manually curated [75].	http://ctdbase.org/
Swiss target prediction	This web tool aims to predict the most probable protein targets of small molecules. Its predictions are mainly based on the similarity principle. It is possible to use a set of targets from *Mus musculus*, *Homo sapiens*, and *Rattus norvegicus* [76].	www.swisstargetprediction.ch
STITCH (search tool for interactions of chemicals)	This tool integrates information about interactions from metabolic pathways, crystal structures, binding experiments, and drug–target relationships. It uses information from metabolic pathways, crystal structures, binding experiments, and drug–target relationships to predict relations between chemicals [77].	http://stitch.embl.de/
Drug set enrichment analysis	Drug set enrichment analysis works with the same principles of a gene set enrichment analysis to use a set of drugs that are tested against a database of pathways [78].	http://dsea.tigem.it

There are several successful examples of network pharmacology analysis already implemented in several fields; for instance, Wang et al. used a network approach to identify the possible mechanisms of *Zingiberis rhizoma*, and *Coptidis rhizoma* reported as antitumoral herbal medicine on Chinese medicine and identified that phosphatidylinositol 3-kinase/serine/threonine-specific protein kinase (PI3K-AKT) pathway might be targeted by different active compounds previously isolated on such plants [68]. Interestingly, Qin-Qin et al. identify that carvedilol may help treat ischemic cerebrovascular disease through the use of a disease–disease association network-assisted model [69]. Moreover, several studies have demonstrated that network pharmacology analysis is quite reproducible at the practical level, such as Casas et al. [65] and Gomez-Verjan et al. [70]. Additionally, there are several tools developed by different international groups freely available online to be used in this pharmacology novel (Table 15.2).

Quantitative Systems Pharmacology

We obtain a very recent subdiscipline called quantitative systems pharmacology (QSP) if we combine systems pharmacology with data from pharmacometrics analyses. Such subdiscipline has taken an interest over recent years in the pharmaceutical industry [79] since it bears the promise to support and improve the drug development process mainly to accelerate the knowledge on drug absorption, distribution, metabolism, and excretion (ADME). In this context, QSP uses systems biology tools to generate quantitative models that involve both pharmacokinetics (PK) and pharmacodynamics (PD). The ultimate goal of QSP is to generate approaches that involve toxicology quantitatively, are biological, and involve diseases in response to different therapeutic regimes, giving the idea of a "spatiotemporal" mechanism of action that could be used to improve the preclinical data [79] (Fig. 15.4).

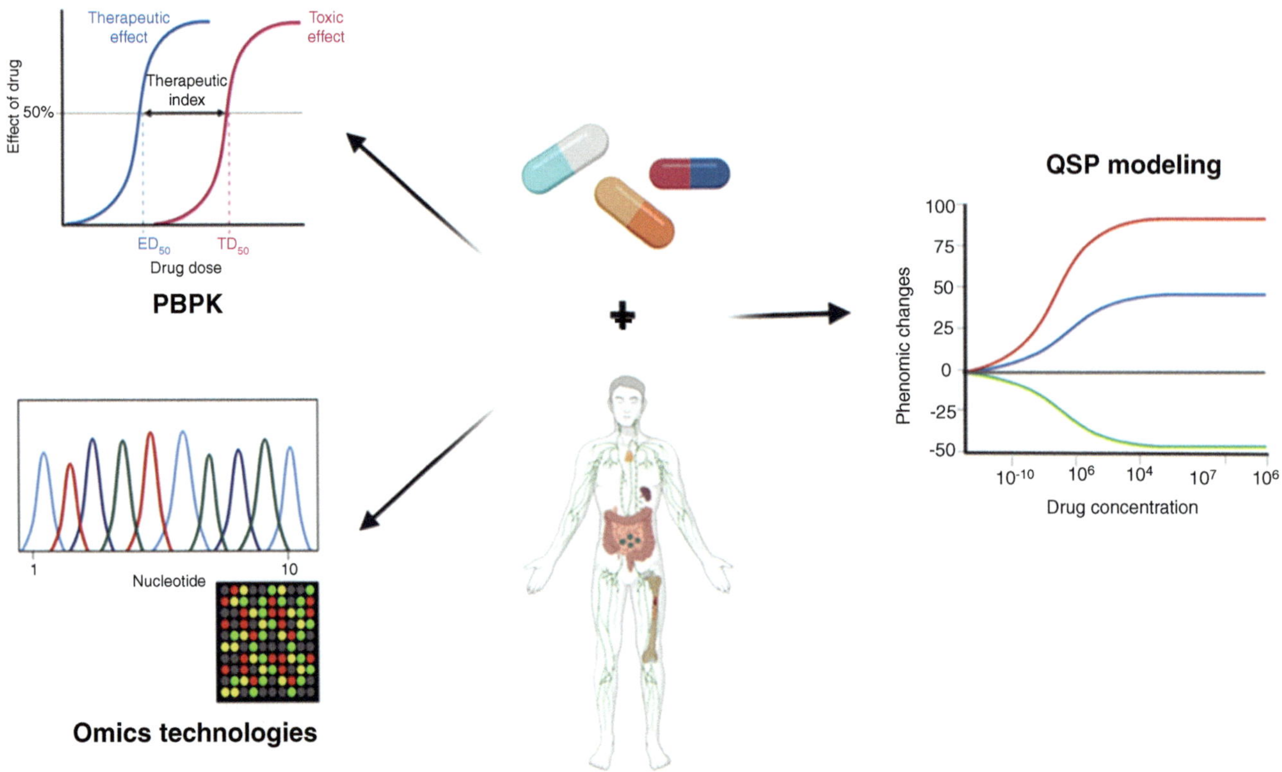

Fig. 15.4 QSP modeling. QSP approaches could encompass and use different tools such as omics technologies and PBPK models

In part, much of the QSP approaches' success has been linked to the advances in physiological-based pharmacokinetics (PBPK) models. Such models have been used and refined for several decades but have recently been successful, thanks to advances in computing power [80]. In general, most PBPK models are based on the assumption that the body's physiological organs are compartments linked by different tubes (circulating blood system); in this sense, each compartment has a particular volume and a specific blood rate [81]. Moreover, each body compartment is defined by both perfusion rate and permeability rate, and kinetics governing such phenomena is delimited by pharmaceutical properties of the molecules, including lipophilicity, metabolic rate, molecular size, hydrogen bond acceptors or donors, albumin, and p-glycoprotein interaction, just to mention a few examples [82]. In this context, ordinary differential equations have proven to be quite valuable to perform mass balance analysis for QSP models. Such models are algae composed of two main sides; the first side always contains the physiological parameters (humans or experimental animal models) independent of the drug itself. The second side consists of the previously mentioned pharmaceutical properties of the drug itself. Interestingly, PBPK modeling has gained attention in regulatory bodies such as the FDA and the European Medicines Agency (EMA) [81, 82]. Interestingly, there has been an increase in the modeling work in the documentation for novel drug submission over recent years, helping pharmaceutical companies perform clinical development decisions for novel candidate selection.

Conclusions

Pharmacology is a discipline that has taken advantage of the recent innovations in bio- and cheminformatics, leading to the development of novel tools as web servers, network pharmacology, QSAR, and docking analysis, just to mention a few examples. In this sense, in the present chapter, we try to resume some of the most exemplary methods, emphasizing their importance in epidemiology. In this sense, it is essential to mention that both branches are more closely related than expected. The more we learn from epidemiological models, the more we will learn about the necessities to develop drugs epidemiologically based on the needs of the population priorities in public health. Similarly, the more we learn to develop novel molecules, the more tools we will have to counter the effects of diseases in the population.

Acknowledgments This chapter is part of a project registered at the Instituto Nacional de Geriatría (INGER) about geroprotectors (DI-PI-008/2021). Oscar Salvador Barrera-Vázquez receives a postdoctoral fellowship from the DGAPA-UNAM.

References

1. Kuna L, Jakab J, Smolic R, Raguz-Lucic N, Vcev A, Smolic M. Peptic ulcer disease: a brief review of conventional therapy and herbal treatment options. J Clin Med Res. 2019; https://doi.org/10.3390/jcm8020179.
2. Antiretroviral Therapy Cohort Collaboration. Survival of HIV-positive patients starting antiretroviral therapy between 1996 and 2013: a collaborative analysis of cohort studies. Lancet HIV. 2017;4:e349–56.
3. Singh DB. Success, limitation and future of computer aided drug designing translational medicein. 2014. https://doi.org/10.4172/2161-1025.1000e127.
4. Macalino SJY, Gosu V, Hong S, Choi S. Role of computer-aided drug design in modern drug discovery. Arch Pharm Res. 2015;38:1686–701.
5. Hughes JP, Rees S, Kalindjian SB, Philpott KL. Principles of early drug discovery. Br J Pharmacol. 2011;162:1239–49.
6. Gad SC. Drug discovery handbook. Wiley New Jersey, USA; 2005.
7. Website. The drug development process. Available at: https://www.fda.gov/ForPatients/Approvals/Drugs/default.htm. Accessed 26 Feb 2021.
8. Paul SM, Mytelka DS, Dunwiddie CT, Persinger CC, Munos BH, Lindborg SR, Schacht AL. How to improve R&D productivity: the pharmaceutical industry's grand challenge. Nat Rev Drug Discov. 2010;9:203–14.
9. Lindsay MA. Target discovery. Nat Rev Drug Discov. 2003;2:831–8.
10. Imming P, Sinning C, Meyer A. Drugs, their targets and the nature and number of drug targets. Nat Rev Drug Discov. 2006;5:821–34.
11. Deore AB, Dhumane JR, Wagh R, Sonawane R. The stages of drug discovery and development process. Asian J Pharm Res Develop. 2019;7:62–7.
12. Chen J, Luo X, Qiu H, Mackey V, Sun L, Ouyang X. Drug discovery and drug marketing with the critical roles of modern administration. Am J Transl Res. 2018;10:4302–12.
13. Website. US Food and Drug Adminitration, FDA. https://www.fda.gov/patients/drug-development-process/step-3-clinical-research. 2018. Accessed 26 Feb 2021.
14. Blystone RV, Blodgett K. WWW: the scientific method. CBE Life Sci Educ. 2006;5:7–11.
15. Semple JL, Woolridge N, Lumsden CJ. In vitro, in vivo, in silico: computational systems in tissue engineering and regenerative medicine. Tissue Eng. 2005;11:341–56.
16. Jafari SM. Biopolymer nanostructures for food encapsulation purposes: volume 1 in the Nanoencapsulation in the food industry series. Academic Press Massachusetts, USA; 2019.
17. Fini M, Giardino R. In vitro and in vivo tests for the biological evaluation of candidate orthopedic materials: benefits and limits. J Appl Biomater Biomech. 2003;1:155–63.
18. Ionescu C, Lopes A, Copot D, Machado JAT, Bates JHT. The role of fractional calculus in modeling biological phenomena: a review. Commun Nonlinear Sci Num Simul. 2017;51:141.
19. Metzcar J, Wang Y, Heiland R, Macklin P. A review of cell-based computational modeling in cancer biology. JCO Clin Cancer Inform. 2019;3:1–13.
20. Malik-Sheriff RS, Glont M, Nguyen TVN, et al. BioModels—15 years of sharing computational models in life science. Nucleic Acids Res. 2019;48:D407–15.
21. Viceconti M, Pappalardo F, Rodriguez B, Horner M, Bischoff J, Musuamba Tshinanu F. In silico trials: verification, validation and uncertainty quantification of predictive models used in the regulatory evaluation of biomedical products. Methods. 2021;185:120–7.
22. Bartocci E, Lió P. Computational modeling, formal analysis, and tools for systems biology. PLoS Comput Biol. 2016;12:e1004591.
23. Gharaghani S, Khayamian T, Ebrahimi M. Molecular dynamics simulation study and molecular docking descriptors in structure-based QSAR on acetylcholinesterase (AChE) inhibitors. SAR QSAR Environ Res. 2013;24:773–94.
24. Kovatchev BP, Breton M, Dalla Man C, Cobelli C. In silico preclinical trials: a proof of concept in closed-loop control of type 1 diabetes. J Diabetes Sci Technol. 2009;3:44–55.
25. Newman DJ, Cragg GM. Natural products as sources of new drugs from 1981 to 2014. J Nat Prod. 2016;79:629–61.
26. Lahlou M. The success of natural products in drug discovery. m.scirp.org › papersm.scirp.org › papers. 2013.
27. Medina-Franco JL. New approaches for the discovery of pharmacologically-active natural compounds. Biomol Ther. 2019; https://doi.org/10.3390/biom9030115.
28. Newman M, Barabási A-L, Watts DJ. The structure and dynamics of networks. Princeton University Press Maryland, USA; 2006.
29. Newman DJ, Cragg GM. Natural products as sources of new drugs over the last 25 years. J Nat Prod. 2007;70:461–77.
30. Calixto JB. The role of natural products in modern drug discovery. An Acad Bras Cienc. 2019;91(Suppl 3):e20190105.
31. Khazir J, Riley DL, Pilcher LA, De-Maayer P, Mir BA. Anticancer agents from diverse natural sources. Nat Prod Commun. 2014;9:1655.
32. Sand M. Did Alexander Fleming deserve the Nobel Prize? Sci Eng Ethics. 2020;26:899–919.
33. Molyneux DH, Ward SA. Reflections on the Nobel Prize for Medicine 2015–the public health legacy and impact of avermectin and artemisinin. Trends Parasitol. 2015;31:605–7.
34. Harvey AL, Edrada-Ebel R, Quinn RJ. The re-emergence of natural products for drug discovery in the genomics era. Nat Rev Drug Discov. 2015;14:111–29.
35. Issaq HJ, Veenstra TD. Proteomic and Metabolomic approaches to biomarker discovery. Academic Press Massachusetts, USA; 2019.
36. Tandon H, Chakraborty T, Suhag V. A concise review on the significance of QSAR in drug design. Chem Biomol Eng. 2019;4:45.
37. Mahalakshmi PS, Sree Mahalakshmi P, Jahnavi Y. A review on QSAR studies. Int J Adv Pharm Biotechnol. 2020;6:19–23.
38. Tandon H, Chakraborty T, Suhag V. A new model of atomic nucleophilicity index and its application in the field of QSAR. IJQSPR. 2019;4:99–117.
39. Yang H, Sun L, Li W, Liu G, Tang Y. In silico prediction of chemical toxicity for drug design using machine learning methods and structural alerts. Front Chem. 2018;6:30.
40. Tropsha A. Best practices for QSAR model development, validation, and exploitation. Mol Inform. 2010;29:476–88.
41. Du Q-S, Huang R-B, Wei Y-T, Du L-Q, Chou K-C. Multiple field three dimensional quantitative structure-activity relationship (MF-3D-QSAR). J Comput Chem. 2008;29:211–9.
42. Du Q-S, Wei Y-T, Pang Z-W, Chou K-C, Huang R-B. Predicting the affinity of epitope-peptides with class I MHC molecule HLA-A*0201: an application of amino acid-based peptide prediction. Protein Eng Des Sel. 2007;20:417–23.
43. Kuz'min VE, Artemenko AG, Polischuk PG, Muratov EN, Hromov AI, Liahovskiy AV, Andronati SA, Makan SY. Hierarchic system of QSAR models (1D–4D) on the base of simplex representation of molecular structure. J Mol Model. 2005;11:457–67.
44. Gupta PP, Bastikar VA, Bastikar A, Chhajed SS, Pathade PA. Computational screening techniques for Lead design and development. Comput Aided Drug Des. 2020:187–222.
45. Roy K, Kar S, Das RN. Newer directions in QSAR/QSPR. SpringerBriefs in Molecular Science. New York, USA; 2015. p. 105–121.
46. Blundell TL, Jhoti H, Abell C. High-throughput crystallography for lead discovery in drug design. Nat Rev Drug Discov. 2002;1:45–54.

47. Westbrook J, Feng Z, Chen L, Yang H, Berman HM. The protein data bank and structural genomics. Nucleic Acids Res. 2003;31:489–91.
48. London N, Raveh B, Schueler-Furman O. Druggable protein-protein interactions--from hot spots to hot segments. Curr Opin Chem Biol. 2013;17:952–9.
49. Ciemny M, Kurcinski M, Kamel K, Kolinski A, Alam N, Schueler-Furman O, Kmiecik S. Protein-peptide docking: opportunities and challenges. Drug Discov Today. 2018;23:1530–7.
50. Diller DJ, Swanson J, Bayden AS, Jarosinski M, Audie J. Rational, computer-enabled peptide drug design: principles, methods, applications and future directions. Future Med Chem. 2015;7:2173–93.
51. Schueler-Furman O, London N. Modeling peptide-protein interactions: methods and protocols. Humana Press New Jersey, USA; 2017.
52. Alam N, Schueler-Furman O. Modeling peptide-protein structure and binding using Monte Carlo sampling approaches: Rosetta FlexPepDock and FlexPepBind. Methods Mol Biol. 2017;1561:139–69.
53. Rentzsch R, Renard BY. Docking small peptides remains a great challenge: an assessment using AutoDock Vina. Brief Bioinform. 2015;16:1045–56.
54. Martín FJG. El fin del mito masculino: la entrada en el siglo de la mujer. Erasmus Ediciones. 2007.
55. Antunes DA, Moll M, Devaurs D, Jackson KR, Lizée G, Kavraki LE. DINC 2.0: a new protein-peptide docking webserver using an incremental approach. Cancer Res. 2017;77:e55–7.
56. Peterson LX, Roy A, Christoffer C, Terashi G, Kihara D. Modeling disordered protein interactions from biophysical principles. PLoS Comput Biol. 2017;13:e1005485.
57. Antunes DA, Devaurs D, Kavraki LE. Understanding the challenges of protein flexibility in drug design. Expert Opin Drug Discov. 2015;10:1301–13.
58. Buonfiglio R, Recanatini M, Masetti M. Protein flexibility in drug discovery: from theory to computation. ChemMedChem. 2015;10:1141–8.
59. Zacharias M. Protein-protein complexes: analysis, modeling and drug design. World Scientific Singapore; 2010.
60. Ciemny MP, Debinski A, Paczkowska M, Kolinski A, Kurcinski M, Kmiecik S. Protein-peptide molecular docking with large-scale conformational changes: the p53-MDM2 interaction. Sci Rep. 2016;6:37532.
61. Kmiecik S, Gront D, Kolinski M, Wieteska L, Dawid AE, Kolinski A. Coarse-grained protein models and their applications. Chem Rev. 2016;116:7898–936.
62. Zhang R, Zhu X, Bai H, Ning K. Network pharmacology databases for traditional Chinese medicine: review and assessment. Front Pharmacol. 2019;10:123.
63. Bellouquid A, Delitala M. Mathematical modeling of complex biological systems: a kinetic theory approach. Springer Science & Business Media New York, USA; 2007.
64. Nookaew I. Network biology. Springer New York, USA; 2017.
65. Casas AI, Hassan AA, Larsen SJ, et al. From single drug targets to synergistic network pharmacology in ischemic stroke. Proc Natl Acad Sci U S A. 2019;116:7129–36.
66. Hopkins AL. Network pharmacology: the next paradigm in drug discovery. Nat Chem Biol. New York, USA 2008;4:682–90.
67. Luo T-T, Lu Y, Yan S-K, Xiao X, Rong X-L, Guo J. Network pharmacology in research of Chinese medicine formula: methodology, application and prospective. Chin J Integr Med. 2020;26:72–80.
68. Wang M, Qi Y, Sun Y. Exploring the antitumor mechanisms of Zingiberis Rhizoma combined with Coptidis Rhizoma using a network pharmacology approach. Biomed Res Int. 2020;2020:8887982.
69. Zhao Q-Q, Li X, Luo L-P, Qian Y, Liu Y-L, Wu H-T. Repurposing of approved cardiovascular drugs against ischemic cerebrovascular disease by disease-disease associated network-assisted prediction. Chem Pharm Bull. 2019;67:32–40.
70. Gómez-Verjan JC, Rivero-Segura NA, Estrella-Parra E, Rincón-Heredia R, Madariaga-Mazón A, Flores-Soto E, González-Meljem M, Cerbón M, Reyes-Chilpa R. Network pharmacology uncovers anticancer activity of Mammea-type Coumarins from Calophyllum brasiliense. Planta Med. 2019;85:14–23.
71. Shannon P, Markiel A, Ozier O, Baliga NS, Wang JT, Ramage D, Amin N, Schwikowski B, Ideker T. Cytoscape: a software environment for integrated models of biomolecular interaction networks. Genome Res. 2003;13:2498–504.
72. Zahoránszky-Kőhalmi G, Sheils T, Oprea TI. SmartGraph: a network pharmacology investigation platform. J Cheminform. 2020;12:5.
73. Bu D, Xia Y, Zhang J, et al. FangNet: mining herb hidden knowledge from TCM clinical effective formulas using structure network algorithm. Comput Struct Biotechnol J. 2021;19:62–71.
74. Li B, Ma C, Zhao X, Hu Z, Du T, Xu X, Wang Z, Lin J. YaTCM: yet another traditional Chinese medicine database for drug discovery. Comput Struct Biotechnol J. 2018;16:600–10.
75. Davis AP, Grondin CJ, Johnson RJ, Sciaky D, Wiegers J, Wiegers TC, Mattingly CJ. Comparative Toxicogenomics Database (CTD): update 2021. Nucleic Acids Res. 2021;49:D1138–43.
76. Daina A, Michielin O, Zoete V. SwissTargetPrediction: updated data and new features for efficient prediction of protein targets of small molecules. Nucleic Acids Res. 2019;47:W357–64.
77. Kuhn M, von Mering C, Campillos M, Jensen LJ, Bork P. STITCH: interaction networks of chemicals and proteins. Nucleic Acids Res. 2008;36:D684–8.
78. Napolitano F, Sirci F, Carrella D, di Bernardo D. Drug-set enrichment analysis: a novel tool to investigate drug mode of action. Bioinformatics. 2016;32:235–41.
79. Manca D. Quantitative systems pharmacology: models and model-based systems with applications. Elsevier Amsterdam, Netherlands; 2018.
80. El-Khateeb E, Burkhill S, Murby S, Amirat H, Rostami-Hodjegan A, Ahmad A. Physiological-based pharmacokinetic modeling trends in pharmaceutical drug development over the last 20-years; in-depth analysis of applications, organizations, and platforms. Biopharm Drug Dispos. 2020; https://doi.org/10.1002/bdd.2257.
81. Zhang X, Yang Y, Grimstein M, Fan J, Grillo JA, Huang S-M, Zhu H, Wang Y. Application of PBPK modeling and simulation for regulatory decision making and its impact on US prescribing information: an update on the 2018–2019 submissions to the US FDA's Office of Clinical Pharmacology. J Clin Pharmacol. 2020;60(Suppl 1):S160–78.
82. Wu F, Zhou Y, Li L, Shen X, Chen G, Wang X, Liang X, Tan M, Huang Z. Computational approaches in preclinical studies on drug discovery and development. Front Chem. 2020;8:726.

Systems Medicine Applied to Epidemiology

16

Juan Carlos Yustis-Rubio and Juan Carlos Gomez-Verjan

Abbreviations

ODEs Ordinary differential equations
GRNs Gene regulatory networks

Bioinformatic Tools for Medicine

The development of new computational methods and the ever-growing decrease in sequencing costs have generated a massive amount of biological data. Basic research applications aside, the clinical field has also greatly benefited from this information, as novel insights into a wide array of pathological conditions and diseases have been discovered through data analysis. Digital web repositories and international consortiums have been created from the data generated by novel high-throughput technologies. This data has been instrumental in a wide variety of clinical applications. In Table 16.1, we briefly summarize some of these online information repositories and their clinical applications.

These massive amounts of data have opened the way to new research methodologies and approaches in a broader sense. Among these novel fields, systems biology is an exciting new area that allows us to study biological phenomena in a new light.

J. C. Yustis-Rubio
Departamento de Ecología Funcional, Instituto de Ecología, UNAM, Mexico City, Mexico
e-mail: jcyustis@ciencias.unam.mx

J. C. Gomez-Verjan (✉)
Dirección de Investigación, Instituto Nacional de Geriatría (INGER), Ciudad de México, Mexico
e-mail: jverjan@inger.gob.mx

What Is Systems Biology?

Talking about systems biology is a complex issue. Having been established as a novel discipline relatively recently and named only two decades ago [1, 2], attempting to find a consensus definition of systems biology is a futile task. Its definition seems to change depending on its source you ask. However, a consensus exists on what systems biology does and how it does it. Driven by François Jacob's statement that "every object that biology studies are a system of systems [3]," scientists began to use tools from seemingly unrelated fields to start dissecting biological systems as a whole instead of independently analyzing the components that make up the said system. Firstly cells, tissues, and organs were studied as complex biological systems [4], then the accelerated development of omics technologies (genomics, transcriptomics, proteomics, metabolomics, etc.) and decreasing costs of sequencing technologies generated enormous data sets that uncovered the essential components that formed these complex systems [5]. Finally, novel computational methods and technologies allowed scientists to study how the interactions between these essential components gave rise to behaviors that could not be predicted or explained if you only studied the isolated components of the system [6, 7].

Breitling (2010) argues that systems biology's foundations and what makes it a genuinely independent discipline rests upon the three main aesthetic principles: diversity, simplicity, and complexity. Breitling states that any research that claims to be systems biology must have all three of these principles as a theoretical and philosophical background [8]. Diversity in a systems biology context refers to the myriad of molecules that form a complex system, such as a cell, genes, transcripts, proteins, and metabolites, which all can be components of this complex biological system. The said diversity comes from the development of all the omics technologies currently available and the massive amounts of data that have been generated from them. However, in the world of systems biology, all these molecules do not come alone; in this

J. C. Gomez-Verjan, N. A. Rivero-Segura (eds.), *Principles of Genetics and Molecular Epidemiology*,
https://doi.org/10.1007/978-3-030-89601-0_16

Table 16.1 Summary of some existing clinically relevant biological information databases

Database	Objective	Information	Website
The Cancer Genome Atlas (TCGA)	Molecular characterization of different types of cancer with applications toward diagnosis, treatment, and prevention	Genomic, epigenomic, transcriptomic, and proteomic data	https://www.cancer.gov/about-nci/organization/ccg/research/structural-genomics/tcga
Database of Genotypes and Phenotypes (dbGaP)	Study of the interaction between genotype and phenotype in humans	Genome-wide association studies, medical sequencing, molecular diagnostic assays, and association between genotype and non-clinical traits	https://www.ncbi.nlm.nih.gov/gap/
Human Cell Atlas (HCA)	"To create comprehensive reference maps of all human cells—The fundamental units of life—As a basis for both understanding human health and diagnosing, monitoring, and treating disease"	High-throughput data generated from single-cell technologies	https://www.humancellatlas.org/
UK Biobank	"Improving the prevention, diagnosis, and treatment of a wide range of serious and life-threatening illnesses—Including cancer, heart diseases, stroke, diabetes, arthritis, osteoporosis, eye disorders, depression, and forms of dementia"	Imaging, genetic, health-related records, biomarkers, physical activity data, data from online questionnaires	https://www.ukbiobank.ac.uk/
Roadmap Epigenomics Mapping Consortium	Production of a public resource of epigenomic maps for stem cells and primary ex vivo tissues selected to represent the regular counterparts of tissues and organ systems frequently involved in human disease	Genome-wide, histone modification, DNase, DNA methylation, and RNA-Seq data sets	http://www.roadmapepigenomics.org/
BioModels	"Provide the systems modelling community with reproducible, high-quality, freely-accessible models published in the scientific literature"	Literature-based physiologically and pharmaceutically relevant mechanistic models	https://www.ebi.ac.uk/biomodels/

context, they all come with their peculiarities and molecular characteristics and their interactions with other molecules and their place in a more general context (or network). Simplicity is a fundamental principle of systems biology that may seem contradictory as it was previously stated that systems biology studied complex phenomena and because diversity is anything but simple. However, simplicity in systems biology stems from its roots in physical sciences, as systems biology attempts to identify general laws and principles applicable beyond the specific object of study. Even then, the relevance of searching for general laws in biology is debatable, although some patterns and structural arrangements seem to be recurrent and follow some general laws. However, it is essential to understand that theoretical principles that derive from these observations might be nothing more than useful predictive tools [8]. According to Breitling, complexity is the final and most fundamental principle of systems biology [8]. As mentioned earlier, the interactions of individual components of a complex system give rise to complex behaviors that cannot be observed when studying these components independently. Defining complexity, especially in biological systems, can be a daunting endeavor because when studying a complex system, one must define basic features of the said system while excluding irrelevant or random features. Defining which features of a complex system are essential can fall to a certain degree into an arbitrary task. The fact that complex biological systems have evolved from non-adaptive processes makes it even more difficult [9, 10]. Nevertheless, data integration is the most critical aspect of systems biology. A complex system is the sum of all the interactions of every single isolated component that makes up the said system [11].

In terms of systems biology, complexity may seem to be antithetical to simplicity, taking into account that the latter aims to discover and establish general and, in some cases, more specific laws. In that sense, systems biology may seem to be incompatible with traditional scientific research methodology. Individual components of a more extensive network are examined, isolated from the rest of the system or delimited in specific network modules. However, this apparent controversy may be overrated. Although systems biology may seem incompatible with the reductionist approach that permeates classical research methods and hypothesis testing, systems biology must also be predictive [12], and this is best done by testing (and/or refuting) hypotheses by introducing perturbations in individual components of the system, therefore doing so in a reductionist approach. Also, and maybe most importantly, systems biology is dependent on the experimental data that comes from such "reductionist" research.

Principles of Network Biology

The complexity aspect of systems biology is deeply rooted in the research areas of systems and network theory. As mentioned earlier, biological systems are the sum of all the different and intricate interactions of all their different components. Therefore biological systems can be studied as a network generated by the interactions of different molecular components (genes, transcripts, proteins, metabolites, etc.) [11]. Five biological networks exist: genetic interaction, transcription factor binding, protein–protein interactions, protein phosphorylation, and metabolic interaction networks [13]. Each one of the components of a network is termed a node, and the interactions between each one of these nodes are called edges for networks without directionality (such as protein–protein and genetic interaction networks) or arcs for networks that have directionality (transcription factor binding, phosphorylation, and metabolic networks). Undirected networks represent the interaction between nodes without any hierarchical specifications, and the data used to construct the network usually comes from sizable high-throughput data sets. Directed networks specify how a signal is propagated and the hierarchy of the said propagation, from one upstream node to a downstream node, for example [11].

Networks can be mathematically defined as graphs and studied as computational units and systems [14, 15]. Computation is one of the defining features of systems biology [11]. As with any graph, networks possess certain topological features that are important to analyze and to understand the said network (graph) (Fig. 16.1) [13]. The degree of a node is the number of links connected to a defined node or, simply said, the number of interactions of a node in a network (Fig. 16.1a). A node with a degree higher than the average of all nodes in the network is known as a hub [16]. There are cases in which the interactions between nodes are not fixed but conditional. The networks where edges are probabilistic are called Bayesian networks [17]. This type of networks allows for the discovery of probabilistic relationships among nodes of a network and aids in defining the conditions that increase or decrease the probability of the said relationships [18–20]. The degree distribution is the probability distribution of all degrees of nodes in a network. The assembly of biological data into networks shows that these "real" networks behave differently than random networks and have a specific degree distribution [21].

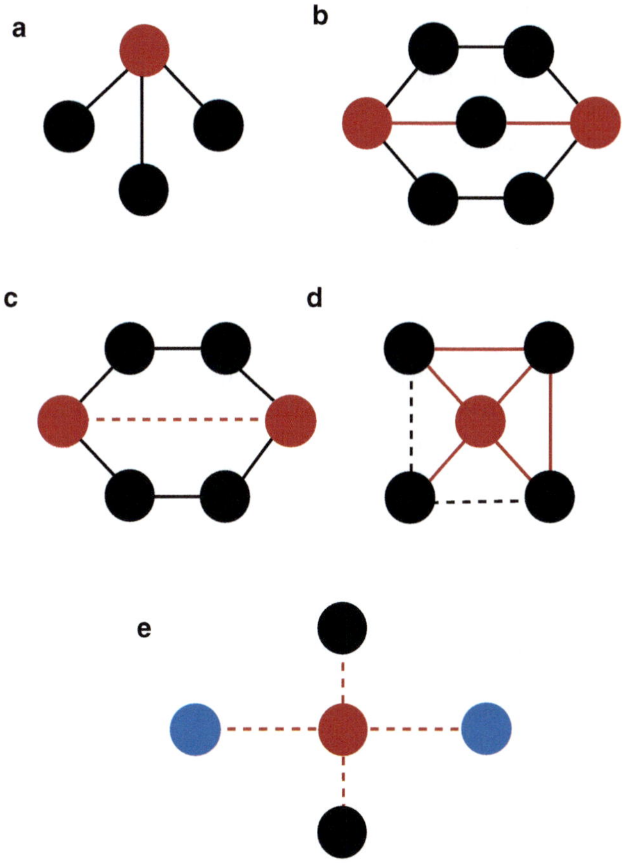

Fig. 16.1 Graph representation of commonly used topological parameters of a network. (**a**) A degree is the number of connections a specific node has. (**b**) The distance is the shortest path between nodes. (**c**) Diameter represents the maximum distance between two nodes. (**d**) The clustering coefficient is the number of connections that are present in the neighborhood of a single node. (**e**) Betweenness is the number of shortest paths from all nodes to all others that pass through that node

Other important properties of a network are its robustness and its sensitivity to perturbations. Perturbations can be introduced into a network by removing specific nodes and analyzing the network's resistance to change. Biological networks are highly robust and remain primarily unaffected by random removal of a node; however, removing a hub can severely alter the function of these types of networks [14, 22, 23]. The shortest path between two nodes is called distance (Fig. 16.1b), and the maximum distance between two nodes in a network is called (graph) diameter (Fig. 16.1c). Next, the clustering coefficient is the percentage of existing interactions among the neighborhood of one node (Fig. 16.1d). A network with a high clustering coefficient indicates that the said network is a small-world network; the said type of network is that any two network nodes can be connected through relatively short paths [24]. Finally, betweenness is the fraction of the shortest paths between all pairs of nodes that pass through one node (Fig. 16.1e). Betweenness estimates the traffic load that goes through one specific node, assuming that the flow of information follows the shortest paths available [13].

Networks present subunits called modules formed by groups of densely associated components (nodes) that are loosely connected, interact to receive a signal, process it, and transduce it to other modules [25, 26]. Integration and analyses of a large amount of data in a network can give rise to smaller common functional patterns or motifs used with a relatively higher frequency relative to randomized networks [13, 27, 28]. These network motifs present a specific dynamical behavior, and a group of motifs with a particular function is called a functional module [28, 29].

The interactions of a biological network exhibit complex temporal dynamics regarding signal propagation and processing [11, 30]. Network-based models present limitations when trying to account for the temporal aspects of the system. However, this complex dynamic behavior allows the components of a biological network (e.g., cells) to react to various conditions or states [13]. Therefore, network analysis needs to be combined with dynamic quantitative mathematical models. Studying the temporal dynamics of a network can be done by translating its components into ordinary differential equations (ODEs) [31, 32]. Another way of translating the components of a network into a mathematical workspace is by using a modelling approach named Boolean logic. This approach assigns each component of the network either an active or "on" state (actual state) or a deactivated or "off" state (false state) [33–35]. These types of networks are used to approximate the dynamics of gene regulatory networks (GRNs). This modelling approach may introduce rough approximations as the binary states that the components are assigned neglect possible intermediate states; however, it is useful when analyzing the robustness and stability of GRNs [36]. Qualitative networks enhance the possibilities of working with Boolean networks by allowing its components to assume a finite number of possible values and increasing the variety of states possible to model through this approach [37]. Merging both ODEs-based modelling and Boolean/qualitative state-based networks, hybrid models use a combination of Boolean logic and functions with differential equations, representing discrete nodal values with continuous dynamics in each state [38]. This type of modelling is suitable when combining qualitative and quantitative information [39].

Choosing which type of modelling approach best depends entirely on the objectives, scope, and system under study, as every type of model comes with its strengths and limitations. More frequently than not, multiple modelling approaches are necessary to generate an accurate predictive system model. Statistical models generate probabilistic relationships built upon correlations. These types of models are helpful in a clinical environment because most complex diseases and pathological conditions are associated with molecular markers (genes) in a probabilistic manner. However, these models cannot explain the underlying mechanisms of the said conditions [40, 41]. Network and dynamic models allow us to understand the nature and direction of interactions of the components of the system (cell). Even then, experimental evidence is needed to build these types of models [11].

Systems Medicine, a Novel Branch

As mentioned earlier, systems biology can be defined as the analysis of interactions with different biological systems at different complex levels (molecules, cells, tissues, organs, individuals, societies, and ecosystems) through different network approaches [42]. In this sense, systems biology has permeated to other fields such as epidemiology (see the chapter of Systems Epidemiology) and the health sciences known as *systems medicine*, which uses novel advanced omics technologies to impact personalized medicine. *Systems medicine* could interpret and understand the pathogenesis and pathophysiology of different diseases with different perspectives through the use of complex computational models from the molecular search biomarkers to discover novel therapeutic targets [43]. In this context, since considerable amounts of information are needed, most of these models are "data-hungry" to understand a system and its dynamics. Systems medicine workflows need the use of omics technologies and the free availability of databases with clinically relevant information from patients. The correct use and implementation of such technologies with clinical perspective allow for the so-called *interactome* to be defined as a complex representation of functional interactions between molecules either within a cell or within the organism as a whole [44] could be implemented into a novel concept or perspective with the final object of diagnosis or therapeutics. However, since there are no specific tools for pathologies, a higher degree of understanding and integrating all this novel information will be needed. Clinicians need to have an open mind for its implementation and potential applications to change the predominant years of reductionism. Systems medicine could be defined as a holistic approach where human health is integrated from different perspectives, from biomedical to environmental and social.

Applications of Systems Medicine to Clinical Research

Systems medicine has begun to be implemented in several fields; for instance, in novel drug research, one of the main challenges is searching for novel drug adverse reactions. In this context, a clear example is the flux balance analysis of genome-scale metabolic network of human hepatocyte, which can qualitatively link gene activity perturbation with bile acid homeostasis, permitting the assessment of the role of genetic polymorphism in toxicity [45], and another exciting example of the systems medicine applications is the development of the so-called physiologically based pharmacokinetic (PBPK) models, which through the use of complex mathematical modelling based on ODEs could predict the concentration of a xenobiotic in the body and with the use of omics technologies could help to develop pharmacodynamic prediction models [46]. Other exciting applications of systems medicine are cardiovascular disease, where the CARDIoGRAMplusC4D consortium has identified 152 loci related to such disease, shading light on novel pathways and phases of clinical phenotypes [47, 48]. Moreover, the concept of genetic risk score has been implemented to assess predictive values to single-nucleotide polymorphisms [49, 50]. Also, on the health-care approach, a new paradigm is being depicted with the so-called "*systems healthcare*" [50]. Based on a model of clinical care that compresses conventional clinical information, imaging data, biological (omics data) and linked information from social media it was easy to provide a precise risk stratification and a more precise diagnoses and prognoses for patients. In this sense, social media (interpersonal communication, information sharing, crowd-sourcing, mobility information, among others) are helping public health surveillance to replace traditional information. For instance, in 2015, during the Middle East respiratory syndrome coronavirus (MERS-CoV) outbreak in South Korea, it was demonstrated that social media could significantly increase preventive behaviors via the self-relevant emotion and the public's risk perception [51]. In this context, the correct synchronization of such technologies with the novel omics tools and biomarkers could help develop a novel public health approach (Fig. 16.2).

Future Directions and Conclusion

Studying biological phenomena as complex systems allows clinicians to integrate various qualitative and quantitative information, temporal and spatial. In this sense, applying systems biology methodologies such as statistical, dynamical, and network models to health leads to the development of the so-called *systems medicine*, which has impacted the discovery of emerging functions in therapeutics, at different scales, from the cell level to whole organisms including those of social behavior. Such advances applied to clinical settings will lead to precision and individualized medicine treating each patient and their needs as a specific complex system. Current and future developments in computational and experimental methodologies will be fundamental to expand the applications of systems medicine further to advance diagnoses and therapeutics.

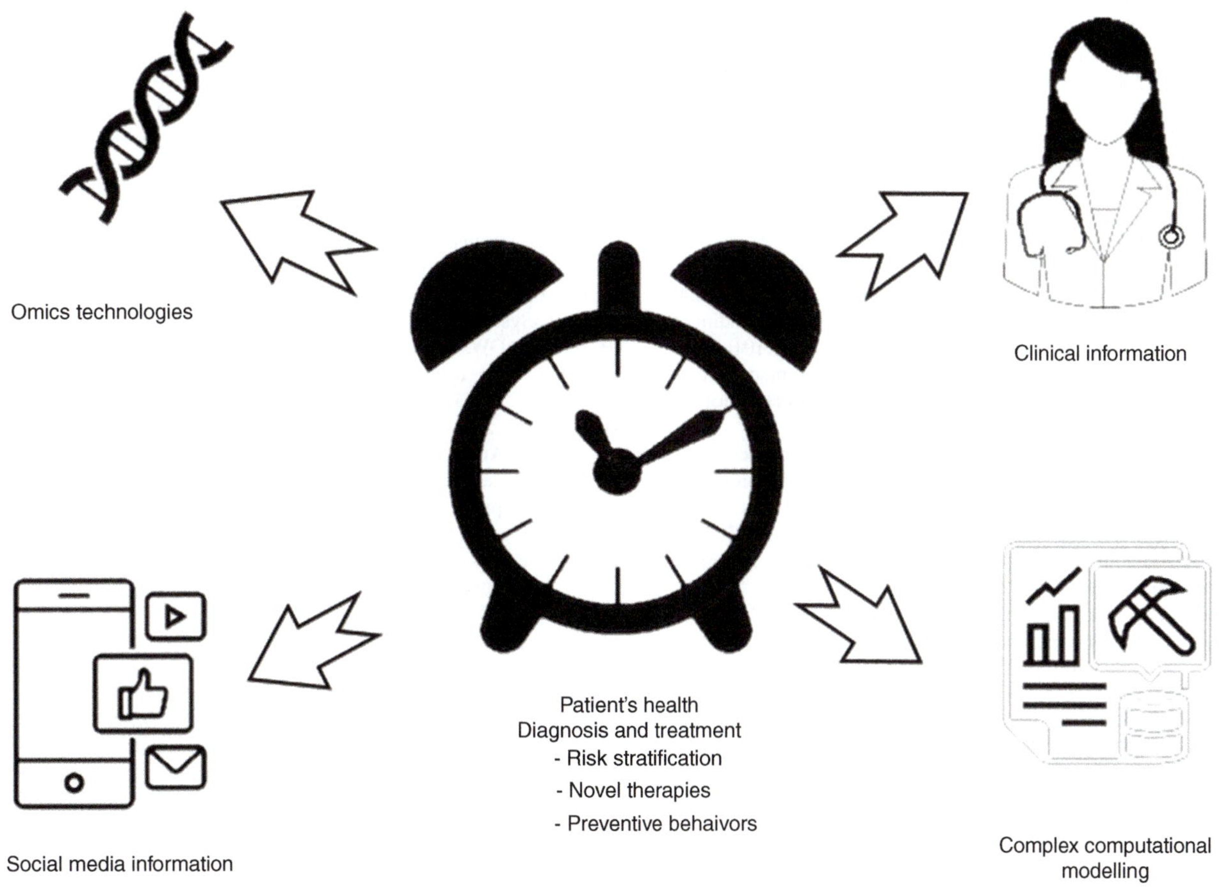

Fig. 16.2 Systems healthcare. Novel approaches could help to have a novel perspective on public health to develop novel strategies for diagnostics, treatments, therapeutics, and preventive behaviors with the correct timing for patients

Acknowledgments This chapter is part of a registered project at the Instituto Nacional de Geriatría with the number DI-PI-003/2018.

References

1. Ideker T, Galitski T, Hood L. A new approach to decoding life: systems biology. Annu Rev Genomics Hum Genet. 2001;2:343–72.
2. Kitano H. Systems biology: a brief overview. Science. 2002;295:1662–4.
3. Trewavas A. A brief history of systems biology: "Every object that biology studies is a system of systems." Francois Jacob (1974). Plant Cell. 2006;18:2420–30.
4. Jensen HJ. Self-organized criticality: emergent complex behavior in physical and biological systems. Cambridge University Press, Cambridge, UK; 1998.
5. Ideker T, Thorsson V, Ranish JA, Christmas R, Buhler J, Eng JK, Bumgarner R, Goodlett DR, Aebersold R, Hood L. Integrated genomic and proteomic analyses of a systematically perturbed metabolic network. Science. 2001;292:929–34.
6. Bhalla US, Iyengar R. Emergent properties of networks of biological signaling pathways. Science. 1999;283:381–7.
7. Tyson JJ, Chen KC, Novak B. Sniffers, buzzers, toggles and blinkers: dynamics of regulatory and signaling pathways in the cell. Curr Opin Cell Biol. 2003;15(2):221–31. https://doi.org/10.1016/s0955-0674(03)00017-6. PMID: 12648679.
8. Breitling R. What is systems biology? Front Physiol. 2010;1:9.
9. Lynch M. The evolution of genetic networks by non-adaptive processes. Nat Rev Genet. 2007;8:803–13.
10. Lynch M. The frailty of adaptive hypotheses for the origins of organismal complexity. Proceedings of the National Academy of Sciences. (2007), 104 (suppl 1). 8597–8604; https://doi.org/10.1073/pnas.0702207104.
11. Tavassoly I, Goldfarb J, Iyengar R. Systems biology primer: the basic methods and approaches. Essays Biochem. 2018;62:487–500.
12. Chuang H-Y, Hofree M, Ideker T. A decade of systems biology. Annu Rev Cell Dev Biol. 2010;26:721–44.
13. Zhu X, Gerstein M, Snyder M. Getting connected: analysis and principles of biological networks. Genes Dev. 2007;21:1010–24.
14. Barabási A-L, Oltvai ZN. Network biology: understanding the cell's functional organization. Nat Rev Genet. 2004;5:101–13.
15. Ryan CJ, Roguev A, Patrick K, et al. Hierarchical modularity and the evolution of genetic interactomes across species. Mol Cell. 2012;46:691–704.
16. Han J-DJ, Bertin N, Hao T, et al. Evidence for dynamically organized modularity in the yeast protein–protein interaction network. Nature. 2004;430:88–93.
17. Sachs K, Perez O, Pe'er D, Lauffenburger DA, Nolan GP. Causal protein-signaling networks derived from multiparameter single-cell data. Science. 2005;308:523–9.
18. Sachs K, Gifford D, Jaakkola T, Sorger P, Lauffenburger DA. Bayesian network approach to cell signaling pathway modeling. Sci STKE. 2002;2002:e38.
19. Needham CJ, Bradford JR, Bulpitt AJ, Westhead DR. A primer on learning in Bayesian networks for computational biology. PLoS Comput Biol. 2007;3:e129.
20. Jha SK, Clarke EM, Langmead CJ, Legay A, Platzer A, Zuliani P. A Bayesian approach to model checking biological systems. Comput Meth Syst Biol. 2009;5688:218–34.
21. Albert R. Scale-free networks in cell biology. J Cell Sci. 2005;118:4947–57.
22. Barabási A-L. Scale-free networks: a decade ``and beyond. Science. 2009;325:412–3.
23. Hornung G, Barkai N. Noise propagation and signaling sensitivity in biological networks: a role for positive feedback. PLoS Comput Biol. 2008;4:e8.
24. Milgram S. The small-world problem. PsycEXTRA Dataset. 1967; https://doi.org/10.1037/e400002009-005.
25. Rives AW, Galitski T. Modular organization of cellular networks. Proc Natl Acad Sci U S A. 2003;100:1128–33.
26. Newman MEJ. Modularity and community structure in networks. Proc Natl Acad Sci. 2006;103:8577–82.
27. Milo R, Shen-Orr S, Itzkovitz S, Kashtan N, Chklovskii D, Alon U. Network motifs: simple building blocks of complex networks. Science. 2002;298:824–7.
28. Tyson JJ, Novák B. Functional motifs in biochemical reaction networks. Annu Rev Phys Chem. 2010;61:219–40.
29. Alon U. Network motifs: theory and experimental approaches. Nat Rev Genet. 2007;8:450–61.
30. Tyson JJ, Chen K, Novak B. Network dynamics and cell physiology. Nat Rev Mol Cell Biol. 2001;2:908–16.
31. Jones PA, Baylin SB. The epigenomics of cancer. Cell. 2007;128:683–92.
32. Rual J-F, Venkatesan K, Hao T, et al. Towards a proteome-scale map of the human protein-protein interaction network. Nature. 2005;437:1173–8.
33. Kauffman SA. Metabolic stability and epigenesis in randomly constructed genetic nets. J Theor Biol. 1969;22:437–67.
34. Thomas R, Kaufman M. Multistationarity, the basis of cell differentiation and memory. I. Structural conditions of multistationarity and other nontrivial behavior. Chaos. 2001;11:170–9.
35. Thomas R, Kaufman M. Multistationarity, the basis of cell differentiation and memory. II. Logical analysis of regulatory networks in terms of feedback circuits. Chaos. 2001;11:180–95.
36. Bartocci E, Lió P. Computational modeling, formal analysis, and tools for systems biology. PLoS Comput Biol. 2016;12:e1004591.
37. Schaub MA, Henzinger TA, Fisher J. Qualitative networks: a symbolic approach to analyze biological signaling networks. BMC Syst Biol. 2007;1:4.
38. Alur R, Courcoubetis C, Henzinger TA, Ho P-H. Hybrid automata: an algorithmic approach to the specification and verification of hybrid systems. Hybrid Syst. 1993;736:209–29.
39. Fromentin J, Eveillard D, Roux O. Hybrid modeling of biological networks: mixing temporal and qualitative biological properties. BMC Syst Biol. 2010; https://doi.org/10.1186/1752-0509-4-79.
40. Boran ADW, Iyengar R. Systems approaches to polypharmacology and drug discovery. Curr Opin Drug Discov Devel. 2010;13:297–309.
41. Hansen J, Zhao S, Iyengar R. Systems pharmacology of complex diseases. Ann N Y Acad Sci. 2011;1245:E1–5.
42. Wolkenhauer O. Systems Medicine: Integrative, Qualitative and Computational Approaches. 1st Ed. 2020 Academic Press. Cambridge, Massachusetts US.
43. Kanodia AK, Kim I, Sturmberg JP. A personalized systems medicine approach to refractory rumination. J Eval Clin Pract. 2011;17:515–9.
44. Wang L, Eftekhari P, Schachner D, et al. Novel interactomics approach identifies ABCA1 as direct target of evodiamine, which increases macrophage cholesterol efflux. Sci Rep. 2018;8:11061.
45. Fisher CP, Plant NJ, Moore JB, Kierzek AM. QSSPN: dynamic simulation of molecular interaction networks describing gene regulation, signalling and whole-cell metabolism in human cells. Bioinformatics. 2013;29:3181–90.
46. Hartung T, FitzGerald RE, Jennings P, Mirams GR, Peitsch MC, Rostami-Hodjegan A, Shah I, Wilks MF, Sturla SJ. Systems toxicology: real world applications and opportunities. Chem Res Toxicol. 2017;30:870–82.

47. Nikpay M, Goel A, Won H-H, et al. A comprehensive 1,000 genomes-based genome-wide association meta-analysis of coronary artery disease. Nat Genet. 2015;47:1121–30.
48. Mäkinen V-P, Civelek M, Meng Q, et al. Integrative genomics reveals novel molecular pathways and gene networks for coronary artery disease. PLoS Genet. 2014;10:e1004502.
49. Smith JA, Ware EB, Middha P, Beacher L, Kardia SLR. Current applications of genetic risk scores to cardiovascular outcomes and subclinical phenotypes. Curr Epidemiol Rep. 2015;2:180–90.
50. Fiandaca MS, Mapstone M, Connors E, Jacobson M, Monuki ES, Malik S, Macciardi F, Federoff HJ. Systems healthcare: a holistic paradigm for tomorrow. BMC Syst Biol. 2017;11:142.
51. Oh S-H, Lee SY, Han C. The effects of social media use on preventive behaviors during infectious disease outbreaks: the mediating role of self-relevant emotions and public risk perception. Health Commun. 2020:1–10.

Genomic Surveillance in Public Health

17

Oscar Salvador Barrera-Vázquez, Elizabeth Sulvaran-Guel, Gibrán Pedraza-Vázquez, and Juan Carlos Gomez-Verjan

Abbreviations

AD	Alzheimer's disease
ApoA-1	Apolipoprotein A1
Apoc1	Apolipoprotein C1
Apoc2	Apolipoprotein C2
Apoe	Apolipoprotein E
APP	Amyloid precursor protein
ATP	Adenosine triphosphate
CD4	Cluster of differentiation 4
CDKAL1	Cdk5 regulatory associated protein 1-like 1
CDKN2A/B	Cyclin-dependent kinase inhibitor 2A/B
CHRNA3	Cholinergic receptor nicotinic alpha 3 subunit
CHRNA5	Cholinergic receptor nicotinic alpha 5 subunit
COPD	Chronic obstructive pulmonary disease
COVID-19	Coronavirus disease 19
DNA	Deoxyribonucleic acid
EBV	Epstein–Barr virus
ERBB3	Erb-B2 receptor tyrosine kinase 3
ESR1	Estrogen receptor 1
FAM13A	Family with sequence similarity 13 member A
FEV1	Forced expiratory volume in 1 second
FVC	Forced ventilatory capacity
GABRP	Gamma-aminobutyric acid type A receptor subunit pi
GIS	Geographic information system
GISAID	Global Initiative on Sharing All Influenza Data
GPS	Global Positioning System
GS	Genome sequencing
GSK3B	Glycogen synthase kinase 3 beta
GWAS	Genome-wide association study
HA	Hemagglutinin
HDAC9	Histone deacetylase 9
HDL-C	High-density lipoprotein cholesterol
HELLS	Helicase, lymphoid specific
HEY2	Hes related family BHLH transcription factor with YRPW motif 2
HHIP	Hedgehog interacting protein
HIV	Human immunodeficiency virus
HLA-DQ2/8	Human leukocyte antigen DQ2/8
HLA-DR	Human leukocyte antigen-death receptor
HLA-DR3/4	Human leukocyte antigen-death receptor 3/4
HSP60	Heat shock protein 60
HSP70	Heat shock protein 70
IGFBP1	Insulin-like growth factor binding protein 1
il13	Interleukin 13
IREB2	Iron-responsive element-binding protein 2
ITS	Internal transcribed spacer
KCNQ1	Potassium voltage-gated channel subfamily Q member 1
KDM5A	Lysine demethylase 5A
LCK	Lymphocyte-specific protein tyrosine kinase
lncRNAs	Long non-coding RNAs
LRBA	LPS-responsive beige-like anchor protein
M	Matrix
M. leprae	*Mycobacterium leprae*

O. S. Barrera-Vázquez
Departamento de Farmacología, Facultad de Medicina, Universidad Nacional Autónoma de México (UNAM), Mexico City, Mexico

E. Sulvaran-Guel
Dirección de Investigación, Instituto Nacional de Geriatría (INGER), Mexico City, Mexico

Licenciatura en Ciencias Genómicas, Universidad Nacional Autonoma de Mexico, Mexico City, Mexico
e-mail: sulvaran@lcg.unam.mx

G. Pedraza-Vázquez
Posgrado en Biología Experimental, Departamento de Ciencias de la Salud, Universidad Autónoma Metropolitana Unidad Iztapalapa, Mexico City, Mexico
e-mail: gpv@xanum.uam.mx

J. C. Gomez-Verjan (✉)
Dirección de Investigación, Instituto Nacional de Geriatría (INGER), Ciudad de México, Mexico
e-mail: jverjan@inger.gob.mx

J. C. Gomez-Verjan, N. A. Rivero-Segura (eds.), *Principles of Genetics and Molecular Epidemiology*,
https://doi.org/10.1007/978-3-030-89601-0_17

MAPK	Mitogen-activated protein kinase
MDT	Multi-drug therapy
MEF2	Myocyte enhancer factor-2
miRNA	MicroRNA
MMD	Monocyte to macrophage differentiation associated
MTB	*Mycobacterium tuberculosis*
NA	Neuraminidase
NGS	Next generation sequencing
NOLC1	Nucleolar and coiled-body phosphoprotein 1
p53	Protein p53
pCPF5603	Plasmid pCPF4969
PCR	Polymerase chain reaction
PD	Parkinson's disease
PFGE	Pulsed-field gel electrophoresis
PGR	Progesterone receptor
PNH	Non-human primates
qRT-PCR	Quantitative reverse transcription PCR
RNA	Ribonucleic acid
RNA-seq	RNA-sequencing
ROS	Reactive oxygen species
RT-PCR	Reverse transcription PCR
RUNX3	RUNX family transcription factor 3
SARS-CoV-2	Severe acute respiratory syndrome coronavirus 2
scRNA-seq	Single-cell RNA sequencing
SLC39A6	Solute carrier family 39 member 6
SNPs	Single nucleotide polymorphism
Spp.	*Species pluralis*
SRF	Serum response factor
STAT1	Signal transducer and activator of transcription 1
STAT4	Signal transducer and activator of transcription 4
TBC1D9	TBC1 domain family member 9
TIGIT	T cell immunoreceptor with Ig And ITIM domains
US	United States
VMAT2	Vesicular monoamine transporter 2
VNTRs	Variable number of tandem repeats
WGS	Whole-genome sequencing
YFV	Yellow fever virus

Genomics in Public Health

Advances in sequencing techniques and the gradual decrease in costs have made genomics a powerful tool for public health. Since the human genome project, sequencing technologies have made the genome of various pathogenic organisms easy to detect and follow up on diseases and their prognosis. Genomic techniques applied to public health are becoming increasingly crucial for identifying, detecting, and following up on the dynamics of diseases in populations representing an excellent advantage for their correct management [1].

We are used to putting all the patients with certain diseases in the same "*therapeutic box*," getting them the same treatment. However, human genomics can specifically distinguish the best treatments, showing that they can be directed to the best available pharmacological treatment based on their genomic profile, making personalized treatments more effective and less risky (pharmacogenomics) [2]. The advancement of sequencing technology and the prevalence of diseases in various populations is the right niche for applying pharmacogenomics. A clear example is the use of small personalized sequencers known as MinION® adapted to detect *Mycobacterium tuberculosis* (MTB) variants and provide the appropriate treatment based on the drug resistance in a record time [3]. Similarly, several infectious diseases have found genomic techniques invaluable for diagnosis and treatment [4, 5].

However, these technologies are not limited to infectious diseases; it is also beneficial for chronic diseases. Among the most studied are cancer, neurodegenerative, and metabolic diseases (see below). Genomic studies in these diseases can detect variants and establish a prognosis of these diseases using genetic signatures [5]. It is also important to emphasize that since most of these diseases are heavily influenced by external factors, public health authorities could use these technologies to accurately diagnose the condition of each patient to be treated appropriately and develop new health policies [6]. Therefore, in the present chapter, we discuss the importance of genomic tools in the context of epidemiological surveillance of infectious and chronic diseases.

Genomic Tools for Epidemiological Surveillance

Different tools allow us to delve into the expression levels of genes associated with different pathologies or conditions. These include next generation sequencing (NGS) techniques that allow us to know the complete genome sequence (genomic DNA), transcriptome (mRNA), and even part of the epigenome (miRNAs or DNA methylation). NGS allows us to know in detail every aspect of sequence to determine in subsequent analyses all the changes between individuals or conditions. A similar case is that of microarrays that, unlike NGS techniques, serve to determine whether some mRNAs or miRNAs are present in the sample and a relative proportion of these genes. Both NGS and microarrays allow us to identify genetic signatures associated with variants or the prognosis of various disease scenarios. They can even help determine the affinity of available treatments. On the other hand, if the purpose is to know the special status of a gene or a small set of genes, PCR and qRT-PCR allow us to determine the presence and proportion compared to another sample quickly. The latter can be handy as a diagnostic tool or to verify whether genetic signatures of diseases are present, helping determine the patient's prognosis and the most effective treatment [2, 5, 7, 8].

The existence of these tools makes the identification and detection of diseases faster and more accurate. Besides, the availability of information in open-access databases of various health institutions online allows for a broader view of the dynamics of infectious diseases and patterns associated with economic, social, or environmental variables in the development and prevalence of cancer, metabolic, or neurodegenerative diseases [5, 6]. Moreover, groups can access this data and start working on treatments to mitigate the effects of the disease [1, 4, 5].

The information generated from real-time disease surveillance is important, because in the case of an outbreak somewhere in the world, changes in contagion dynamics and the region where it originated, such as environmental factors, demographics, and economic conditions, can be observed [5]. Moreover, such results could synergize with remote perception technologies, for instance, GPS data from the detected cases of an outbreak, allowing its spatial analysis and distribution in a particular region, followed by a geographic information system (GIS) analysis using satellite data that often helps to reveal an environmental factor, such as zoonotic risk caused by deforestation of the nearby area. Then, an alert is triggered. The field response team takes samples of the patients, immediately analyzed using portable DNA sequencers like MinION®, coupled to a phone or computer that allows a report to be generated of the clinical metagenomic results in real time revealing the identity of the pathogen. The sequencing data derived from this intervention is immediately uploaded to a public repository, tagged with metadata about the host, sample type, and location, allowing scientists, specialists, and governments to collaborate against the outbreak [5]. Another exciting tool or novel approach is mobile apps that helped to understand the movement of individuals across neighborhoods; in this sense, such technologies seem to play an interesting novel role in the recent outbreak of COVID-19 [9]. This information allows health authorities to quickly attend to the affected population and determine whether it is a known disease, more susceptible populations or communities to the disease, and the sanitary measures necessary for its containment. One of the clearest examples is the case of SARS-CoV-2, where NGS techniques were used to determine the genome sequence of the virus just a few weeks after the beginning of the pandemic. With the help of different techniques, it was possible to determine a qRT-PCR for correct diagnosis in all countries and the follow-up and development of novel strains all around the globe [5, 10] (Fig. 17.1).

The changing dynamics of emerging viruses have made it increasingly crucial to discover and diagnose viruses in clinical medicine and public health. The globalization of travel and trade in domestic animals and animal products, bushmeat trafficking, political instability, and bioterrorism, as well as climate change and its effects on vector geographical distribution, have facilitated the emergence and re-emergence of zoonosis, as we have seen with the current pandemic caused by SARS-Cov-2 or some of the novel dengue outbreaks [11, 12]. Previously, some of these viruses were restricted to a host species or geographic regions; however, this has changed. Moreover, these alterations have generated confusion among clinicians, making them unable to recognize new syndromes or detect new pathogens with the existing diagnostic tests. Such conditions have generated a growing interest in discovering and diagnosing these novel diseases. They have generated the rapid implementation of novel practical applications for molecular diagnostic tools, drugs, and vaccines, targeting them in a more specific way [11].

Genomic and Molecular Research Tools

Culture Methods for Virus

Although tissue culture for virus detection is sometimes considered a cumbersome, expensive, and somewhat archaic method, it continues to be used for a few reasons. First, the growth of an agent in culture provides an excellent source of enriched template for molecular characterization, while on the other hand, such cultures help to have both an inoculum for studies of animal models and cell assays for serology or a titled stock for neutralization tests in order to search for evidence of the cause of disease, in addition to in vitro evaluation of vaccines and to test drug candidates and visualization of an agent by electron microscopy, because in it, a more optimal quantity can be obtained than in a clinical sample. However, because it is not possible to cultivate these agents in tissue culture, the only system that may be useful for virus amplification is animal inoculation [11].

Imaging

The technique used as the first step toward identifying candidates for molecular assays was immunohistochemistry until the use of rapid, inexpensive unbiased high-throughput sequencing. An example of this process was applied in 1999,

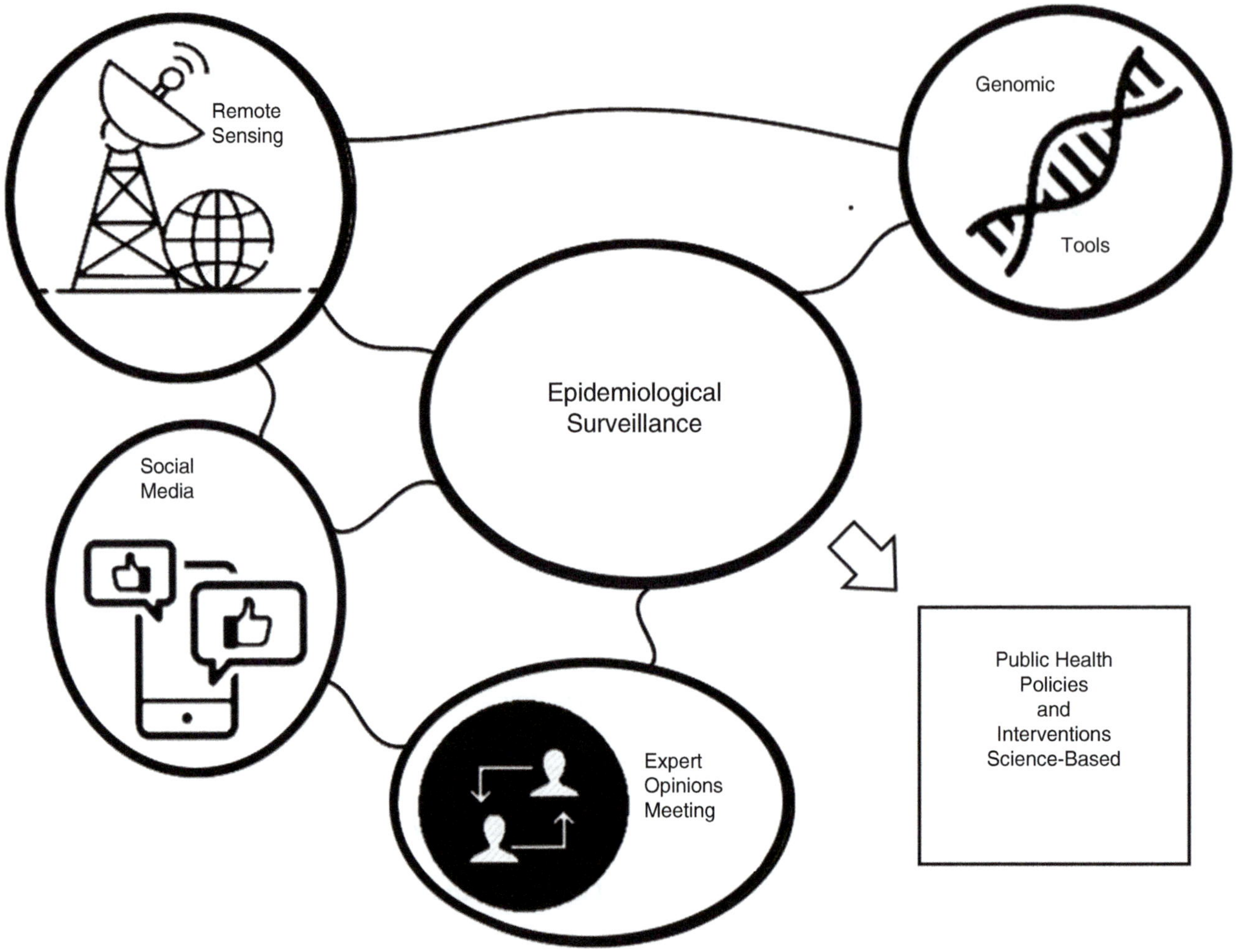

Fig. 17.1 Novel technologies for public health and epidemiological surveillance

at the Centers for Disease Control and Prevention in Atlanta, by Sherif Zaki, before conducting flavivirus consensus PCR assays, previously used immunoassays with brain extract from victims of the encephalitis outbreak in New York. Currently, microscopy images have been used mainly to examine the association between an infectious agent and a disease by testing its presence at the site of the pathology. However, although there are rapid protocols for electron microscopy to visualize viruses in a few hours, substantial operator experience and a high concentration of viral agents were required. Other faster methods can be immunohistochemical tests using serum from an infected individual, individuals with similar pathology, or antibodies generated in animal models immunized with viral proteins. However, in the case of new viral agents, this type of analysis takes time due to the need to create immunological reagents, so an alternative method is in situ hydration. In this relatively rapid method, the genetic sequence of the pathogen can be used to design specific probes to detect it in infected tissues [11] (Table 17.1).

High-Performance Sequencing

Culture-independent methods used to search for virus discovery and characterization, surveillance, and outbreak investigation were achieved by implementing fast and inexpensive platforms for DNA sequencing. In the last decade, the cost per base for sequencing has dropped 10,000 times, from $ 5000 per megabase using capillary electrophoresis to $ 0.5 per megabase using the Illumina platform. Consequently, the challenge has shifted from sequence acquisition to analysis. The complexity of the bioinformatic analysis lies in the read length, which ranges from 150 bp on the Illumina HiSeq or MiSeq instruments to 700 bp on the Roche GS FLX Titanium pyrosequencing [11]. The shorter the read length, the greater the number of calculations required to assemble continuous chains of genomic sequence.

Moreover, in recent years, single-molecule sequencers are being developed to increase fidelity and extended sequences [11]. Regardless of the platform used, the process that follows sequence acquisition is similar for pathogen

Table 17.1 Comparison and main characteristics of the different testing methods for diagnosing infectious diseases

Diagnostic	Advantages	Disadvantages	References
Molecular Biology Research Tools			
Serology	Potential for diagnosis after acute infection Inexpensive cost	May be harmful and inefficient during early infection False negatives in humoral immunodeficiencies False positives under certain conditions	[7]
Microscopy and staining (such as Gram stain, auramine–rhodamine, calcofluor-white)	Rapid Inexpensive	To have better efficiency, low sensitivity must be a significant burden of disease or sample of the etiologic agent Low specificity	[7]
Culture	Capable of holding large sample volumes Best-studied model Inexpensive	Limited sensitivity due to the use of antibiotics and antifungals Limited sensitivity for fastidious organisms Long time to result in the generation of acid-fast and fungal cultures Limited use in viral testing	[7]
Matrix-assisted laser desorption/ionization time-of-flight mass spectrometry	High specificity Rapid after culture	Requires positive-culture isolate forcibly	[7]
Genomic Research Tools			
Conventional PCR	Simple Rapid Inexpensive Potential for quantitative PCR	Depends on the hypothesis Requires primers that may not always work Limited to a tiny portion of the genome Requires specific conditions to work	[7]
Multiplex PCR	Rapid Can detect multiple organisms	Low specificity and false positives for many organisms due to difficulty in quantitation Often requires more than one amplification Requires specific conditions to work Limited to a small portion of the genome Requires primers that may not always work	[7]
Targeted universal multiplex PCR (such as 16S, ITS) for Sanger sequencing	Able to differentiate multiple species within one pathogen type	Requires primers that may not always work Limited to a tiny portion of the genome	[7]
Targeted universal multiplex PCR (such as 16S, ITS) for NGS	Can differentiate multiple species within one pathogen type Multiplexing capability Potential for quantitation	Requires primers that may not always work Expensive and time-consuming Often requires more than one amplification Limited to a tiny portion of the genome	[7]
Targeted NGS	Sensitive detection for selected organism types Potential for quantitation Potential to be combined with 16S NGS	Sequencing library preparation more complex, typically with more than one amplification Limited to a small portion of the genome Expensive and time-consuming Vulnerable to contamination with environmental species	[7]
Metagenomic NGS	Hypothesis-free, or unbiased, testing Discovery of new or unexpected organisms Potential for quantitation Able to detect any portion of the genome	Must also sequence human host background Expensive Time-consuming Not all genomes are available Susceptible to contamination with environmental species	[7]

discovery: contiguous sequences (the host sequence is subtracted) and residual sequences (examined for similarities with known microbial sequences nucleotide or amino acid). When there are no similarities, nucleotide composition or order is examined for patterns consistent with viral genera and host species. This approach can have the critical advantage of examining fecal samples in which the sequences can represent both viruses that infect the host or an animal consumed by the host. Interestingly, the methods for preparation and sequencing of samples are becoming more straightforward and more efficient. In the distant future, sequencing will be used primarily for clinical diagnosis and discovery of new etiological agents [11].

Virus Outbreaks and Genomic Surveillance

Whole-genome sequencing (WGS) is a standard tool for pathogen identification and monitoring, transmission routes, and outbreak control. Advances in this field over recent years have been quite remarkable; for instance, in late 2019, metagenomic RNA sequencing was used to identify the etiology of an unknown respiratory disease that was present in only one patient in Wuhan, China. Subsequently, the etiological agent causing this syndrome was identified as a new coronavirus by reconstructing the viral genome of the bronchoalveolar lavage fluid sample. In early January 2020, the virus genome sequence was released, facilitating rapid molecular diagnostic assays in other laboratories worldwide; this virus was later named SARS-CoV-2 [10]. In March 2020, a national genomic surveillance network was created in the United Kingdom to track viral transmission, identify viral mutations, and integrate viral data with health data. Interestingly, by June 2020, approximately 20,000 SARS-CoV-2 genomes were sequenced, and transmission lineages were defined based on phylogeny [13]. As of June 2020, about >57,000 SARS-CoV-2 genomes from about 100 countries have been uploaded to the GISAID database [10].

Between 2016 and 2018, São Paulo had its largest outbreak of yellow fever virus (YFV) in decades. To contextualize human cases, identify epizootic foci, and discover the rate and direction of the spread of YFV in São Paulo, genomic data of the virus were generated and analyzed. Thanks to current personalized genomic tools such as MinION genome sequencing, 51 new virus genomes from positive YFV cases were identified in 23 different municipalities in São Paulo, mostly sampled between October 2016 and January 2018. The findings of this study suggest that wild transmission of YFV took place in highly fragmented forested regions of the state of São Paulo, so it is essential to carry out continuous surveillance for zoonotic pathogens in sentinel species such as non-human primates (PNH) [14]. On the other hand, genomic surveillance became very important during Ebola virus outbreaks to elucidate the transmission chains and developed diagnostic tests. For instance, in 2020, Postigo-Hidalgo et al. combined 31 parallel PCR assays with Illumina/MinION-based sequencing, allowing genomic surveillance of the Ebola virus outbreak in Sudan, and this new methodology allowed rapid genomic surveillance of both such virus outbreaks, regardless of variant divergence [15]. Another compelling example of using genomic data with the help of Global Positioning System (GPS) technology has been used to monitor rabies outbreaks to understand the processes that affect viral spread, evolution, and host restriction and develop intervention strategies for rabies in the future [16].

The information generated on the sequence of influenza virus genomes helped to the development of new methods of multiplex reverse transcription PCR (RT-PCR) of the influenza A and B virus of single reaction (Flu A/B) that made it possible to amplify the most critical genomic segments, such as hemagglutinin (HA), neuraminidase (NA) and matrix (M) of seasonal influenza viruses A and B, regardless of viral type, subtype, or lineage. This will help to understand more about the mechanisms that support antigenic evolution and antiviral resistance. However, due to the sequence diversity and dynamics of influenza virus evolution, rapid and high-throughput sequencing of influenza viruses remains a challenge [17].

Another example is the Epstein–Barr virus (EBV), a tumor virus related to different malignant neoplastic processes, such as Burkitt's lymphoma and nasopharyngeal carcinomas, which is an endemic trend and has a skewed geographical distribution around the world. Recent advances in deep sequencing technology have enabled high-throughput sequencing of the EBV genome from clinical specimens. In addition, it has become possible to perform both rapid cloning and sequencing of cancer-derived EBV genomes, followed by reconstitution of the infectious virus. These advances have made it possible to find that several EBV strains are distributed differentially throughout the world. There is a different behavior of EBV strains derived from cancer concerning the prototype EBV strain of non-cancerous origin [18]. On the other side, genomic surveillance has the potential for monitoring HIV drug resistance during the continued expansion of antiretroviral therapy and the deployment of pre-exposure prophylaxis. Furthermore, single genome sequencing has helped characterize HIV integration sites and clonal expansions of infected cells [19].

Bacterial Outbreaks and Genomic Surveillance

During the mid-1990s, to carry out epidemiological surveillance on pathogenic bacteria, the so-called standardized molecular subtypes based on pulsed-field gel electrophoresis (PFGE) were used to identify and study outbreaks. Over time, this has transitioned from PFGE to WGS [20, 21]. Comparing both methodologies, WGS offers better resolution by generating sequences of three to six million base pairs, compared to a gel pattern with ten to twenty bands, which show slight changes in parts of the genome. On the other hand, the WGS data are entirely digital and standardized and do not depend on the choice of a laboratory protocol. In addition, the results of this technique show evolutionary relationships between bacterial isolates, allowing us to understand more about the transmission and the relationship between the cases [22]. Among the advantages of WGS is that it predicts phenotypic characteristics, which include virulence, serotype, and antimicrobial resistance.

Regarding costs, the WGS is priced at approximately $200–$250 for insulation, higher than the PFGE, priced at about $ 100. Nevertheless, these high costs can be partially or entirely offset since the need for traditional phenotyping assays is eliminated, and automation can further reduce costs [22]. A remarkable example of the benefits of genomic surveillance is that on common foodborne pathogens where *Listeria* sp. surveillance has been encouraging since in the first years of WGS implementation (2013–2016), 18 listeriosis outbreaks (6 per year) were resolved with a median of 4 cases per outbreak [23].

Mycobacterium Tuberculosis

An exciting example of genomic surveillance applications is *Mycobacterium tuberculosis* (MTB). Since closely related strains are identified to detect cases and their possible recent transmission, various DNA fingerprint technologies have been used for subtyping strains from this pathogen [24, 25]. As mentioned above, the WGS offers a more acceptable resolution and greater fidelity and confidence in the inferred relationships between cases. For instance, the US tuberculosis control program has expanded sequencing isolates from all culture-confirmed cases in the United States, allowing public health workers to rebut more than half of the suspected outbreaks initially identified by conventional genotyping in California, saving time and resources.

Interestingly, other countries (United Kingdom [25, 26], Canada [27, 28], and the Netherlands) have implemented effectively the use of WGS in their tuberculosis programs [29] with massive success for defining outbreaks more precisely. Providing information on the dynamics of transmission and the presence of previously unidentified cases or possible "super-spreaders" should be prioritized for isolation and treatment [22]. Another advantage of the WGS is the capability to indicate whether recurrent cases are due to reactivation or reinfection, providing helpful information to evaluate the effectiveness of the public health programs [22].

A different application of MTB sequencing in low-income countries is its diagnostics directly from sputum [30], thus avoiding the use of expensive techniques such as spreading the infectious agent in culture [22]. Such methodology allows a rapid inference of the identified strains concerning drug susceptibility rapidly available for appropriate treatment with efficient drugs; additionally, NGS will reduce the need for routine phenotypic tests and complex, time-consuming challenges to perform in laboratories with limited resources [22]. Moreover, these advances have led to laboratories in both the New York State Department of Health and Public Health England receiving regulatory approval to avoid the use of traditional drug susceptibility tests of isolates [31]. Since WGS predicts a greater efficiency of Tuberculous meningitis (TBM)'s susceptibility to the first-line drugs from between 70% and 80% of all isolates [31, 32], however, it should be noted that for a sequence-based method for inferring drug susceptibility, it is essential the continuous updating of databases with correlated genotypic and phenotypic data [31].

Clostridium Perfringens

Clostridium perfringens is another medically necessary enteric pathogen as the etiologic agent of significant gastroenteritis outbreaks in most adult humans [33]. Limited studies based on WGS have been conducted, allowing information about the virulence genes of *C. perfringens* strains. Under this approach, phylogenomic analysis of human and food isolates obtained from cases reported in Wales between 2011 and 2017 showed a remarkable discriminatory capacity of such genomic approach in the profiling of *C. perfringens* strains, compared to the current fluorescence amplified fragment length polymorphism test, which is a reference test proposed by public health authorities in England [34]. Other findings from this large-scale genomic study suggest three main genotypes encoding *cpe* (toxinotype F) implicated in the outbreaks: plasmid pCPF5603, plasmid pCPF4969, and chromosome-cpe strains [34]. Such results suggest that more studies should be carried out to probe the dissemination and regional reservoirs of this enteric pathogen in a more profound way, allowing the design of prevention strategies to reduce the burden of morbidity from food poisoning in vulnerable patients.

Mycobacterium Leprae

Leprosy, a chronic infectious disease caused by *Mycobacterium leprae (M. leprae)*, has shown great importance in public health in tropical countries. Although the burden of disease has decreased in recent decades, thanks to the implementation of WHO multi-drug therapy (MDT) in countries such as India, about 120 334 new cases of leprosy were reported during the year 2018–2019, which yields an annual new case detection rate of 9.27 per 100,000 individuals [35]. Unlike the free-living cultivable bacteria, *M. leprae* has the particularity of being an obligate pathogen (it needs a suitable host to carry out its life cycle), and this has been a limitation for its laboratory isolation and the understanding of the molecular diversity of strains and their genomics, fundamental aspects of the transmission of the disease, and its biology. However, recent advances in molecular techniques have allowed the identification of a variable number of tandem repeats (VNTRs) [36] and single nucleotide polymorphisms (SNPs) [37] in the *M. leprae* genome, which has helped to understand the diversity of strains in addition to determining the more excellent geographic distribution of these strains. Moreover, WGS is employed to characterize the genetic background and trace the origin of *M. leprae* strains circulating the world. Interestingly, this methodology has been used extensively in Madagascar and the Comoros, two islands where leprosy is considered a public health problem and is monitored as part of a drug resistance surveillance program [38].

Genomic Surveillance over Other Infectious Diseases

Parasites include several diseases such as leishmaniasis, malaria, taeniasis, or dengue, just to mention a few examples. Leishmaniasis is a neglected tropical disease spread across 98 countries from different continents, caused by several species of the parasitic intracellular protozoa *Leishmania* spp. by the sandfly bites. Leishmania genomic studies are pretty essential to evaluate its mechanism of drug resistance and variability. However, currently, genomic analysis is being performed from cultured parasites, which causes sampling biases [39]. Nevertheless, there are several efforts being performed to directly sequencing from visceral leishmaniasis [39]. Another essential disease is malaria caused by four different *Plasmodium* spp. transmitted by the infective female of the *Anopheles* mosquito. Interestingly, *Plasmodium vivax*, the second most abundant cause of malaria [40], can survive for several months of years as dormant hypnozoites in the liver where they can cause blood-stage infection later at any other time; therefore, relapses cannot be distinguished from reinfections, and genotyping does not effectively resolve the relationship of lineages. In this context, there is the need for more data from more genotypes around the world and novel techniques such as the DNA-parasite enrichment methods. To this end, several programs have started sequencing more than 4000 genotypes of *Plasmodium* spp. [41]. Cysticercosis is caused by eating undercooked beef or pork that contains tapeworm eggs (*Taenia solium*, *Taenia saginata*, and *Taenia asiatica*). The most potentially lethal ways of such infections are neuro-, ocular, and subcutaneous cysticercosis; interestingly, when *T. solium* was analyzed, they were divided into two clades, Asian and Afro/American, with a difference in pathogenicity and antigenicity for both genotypes [42]. Interestingly, several candidate nuclear genes have been suggested to differentiate genotypes, which is quite essential since several inconsistencies between genotypes have been reported. Such results suggest the importance of analyzing molecularly such parasites since hybrids of the genotypes have been reported [43], and identifying the location and geographical distribution of strains becomes crucial for the correct surveillance of such disease.

Dengue virus is transmitted to humans by the mosquito *Aedes aegypti*, which, due to the rapid urbanization and climate change, has increased its availability to transmit such disease [44]. Dengue virus generally causes a mild disease; however, there are severe forms of the infection such as hemorrhagic fever and hemorrhagic fever with shock, which is a fatal hypovolemic shock that could result in death, particularly in susceptible populations [45]. Amplification and sequencing dengue virus is becoming quite common in many laboratories analyzing genetic variations and viral evolution. In this sense, there are five genotypes through different geographical regions [46]; epidemiological data suggest that specific viral genotypes differ in their availability to cause severe forms of the disease; moreover, WGS analyses suggest that there are two different genotypes (Asian and American) with different rates of dissemination. Additionally, results suggest that selective pressure imposed by specific viral genes differ between mosquito and human host [47]; further genomic studies will help us to generate more information to correlate genotypes with virulence and have more precise control of the disease.

Chronic Diseases and Genomic Surveillance

Millions of people suffer chronic diseases, such as cancer, cardiovascular diseases, diabetes, obesity, and neurodegenerative diseases, among others [48]. These diseases are highly complex to genomic surveillance since most of their etiological background depends on environmental–behavior situations (epigenetics). They have been associated with multiple non-coding genomic regions that make it hard to distinguish common haplotypes for epidemiological surveillance [48]. In this context, genome-wide association studies (GWASs) are pretty helpful for the detection of genomic variants in the whole genome associated with a particular trait in a population [49]. Employing these studies, it is now possible to evaluate the risk an individual has for developing a specific disease [50]. In this sense, the development of the so-called *polygenic risk scores* becomes essential since such scores can describe how a person's risk compares to others with a different genetic constitution by a correlation measure. In the present section, we will cover some of the primary chronic diseases and their advances in genomics that could be used for epidemiological surveillance.

Cancer

Cancer is the second leading cause of death worldwide; therefore, it is a significant concern across healthcare providers [51]. Cancer is a given name to a collection of almost 200 diseases, characterized by the body's cells beginning to divide without control and spread to surrounding tissues. Several alterations characterize it at different molecular levels, such as DNA, RNA, proteins, or metabolite.

Cancer cells are characterized by having high mutation rates and incidence, promoting cell division and tumor growth [52]. Diverse omic approaches, such as genomics, transcriptomics, proteomics and metabolomics, enabled identifying mutations, altered pathways, and expressed molecules, and potential therapeutic targets also named biomarkers for understanding such a disease [51]. For instance, WGS allows identifying mutational signatures present in different kinds of cancer, enabling patients to predict specific types of treatments [53]. On the other hand, transcriptomic technologies, such as RNA-seq or microarrays, contribute to differential gene expression studies and non-coding RNA molecules, such as miRNAs and lncRNAs, known to have essential roles in cancer [53]. Moreover, RNA-seq enables the characterization of splice variants solely expressed in disease contexts [51]. Recently, single-cell RNA-seq (scRNA-seq) has helped portray intratumor cellular heterogeneity. Likewise, proteomic technologies have been implicated in cancer studies since they represent the functional molecules and provide information on signaling pathways [51]. Proteomic technologies, such as mass spectrometry, supply details on protein expression levels, intracellular localization, post-translational modifications, and protein–protein interaction networks [51].

Furthermore, methylation profiling permits the characterization of aberrant DNA methylation hallmarks, known to be related to cancer [51]. Similarly, ChIP-seq enables characterizing transcription factor binding sites and nucleosome occupancy, providing information on regulatory events leading to disease [53]. In this sense, bioinformatic tools such as the Cancer Genome Atlas (TCGA) project compile results from these platforms to freely access its results [53].

Genomic studies have revealed alterations in cancer patients' genomes. Acquired mutations are a common cancer cause, which may be due to different environmental conditions such as smoking, radiation, ageing, or viruses [54]. Mutations in tumor suppressor genes prevent the cell from dividing normally. They instead begin dividing uncontrollably, giving rise to the appearance of tumors [54]. Early identification of mutations in these genes may help prevent the appearance of the tumor, as well as in early diagnosis and promising prognosis [54]. On the other hand, studies performed in the transcriptome in cancer patients showed expression signatures correlated with survival and chemotherapy efficacy and drug responses [55]. Circulating tumor cells of patients with metastatic breast cancer show overexpression of TFF1. Furthermore, there have been found expression signatures of PGR, GABRP, ESR1, TBC1D9, SLC39A6, and LRBA associated with mortality and with recurrence of cancer, proving transcriptomes' usefulness in patients' prognosis. Finally, there have been similar findings for other classes of cancer, including signatures of DUP6, MMD, STAT1, ERBB3, and LCK in lung cancer and HELLS and NOLC1 in colorectal cancer, which was proven to reduce tumor growth [55].

Neurodegenerative Diseases

Adult-onset neurodegenerative diseases are characterized by neuron degeneration, causing cognitive, motor, and emotional impairment, depression, apathy, sleep alterations, and anxiety [56, 57]. Furthermore, both diseases present overlap in the activation of diverse pathways, including MAPK and GSK3B, associated with the toxicity caused by β-amyloid plaques and tau proteins in Alzheimer's disease and abnormal α-synuclein filaments in Parkinson's disease [57]. Despite current attempts at finding treatments and diagnosis and prevention tools, there are still very few available, circumventing the possibility of halting disease progression [56]. For this reason, it is necessary to find new biomarkers that may help in early disease diagnosis and disease progression tracking and those that may be used as new therapeutic targets [57].

In neurodegenerative diseases, neurons show activated pathways that may result in the finding of new valuable biomarkers. Some of these are misguided apoptosis and autophagy, mitochondrial functioning and cytoskeleton impairment, and aberrant protein expression [58]. Cells show altered cell adhesion pathways, leading to impairments at the tissue level, including faulty neurotransmission and cell proliferation [58]. Furthermore, additional neurodegeneration manifestations include exacerbated immunological and inflammatory responses in microglia and astrocytes, along with increased cytokine expression [57]. Thus, anti-inflammatory treatments may be promising for neurodegenerative diseases treatment, but none has been effective to date [57, 58].

GWASs have revealed hundreds of genetic variants linked to risk for AD and PD onset. Most of them present in non-coding regions of the genome, such as the HLA-DR locus, which has common variants in many neurodegenerative diseases. Furthermore, H3K27ac, an active enhancer histone mark, and DNA methylation analysis have revealed genomic regions contributing to disease heritability [56]. Tissue samples from neurodegenerative disorder show differential H3K27ac and DNA methylation patterns mainly in GWAS genetic variants' adjacent regions, demonstrating that these variants are often found in gene regulatory regions [56].

On the other hand, transcriptomic technologies have revealed gene expression patterns intersecting between several neurodegenerative diseases, including Alzheimer's and Parkinson's disease [58]. Many of them were highly enriched in pathways related to the cellular response to hypoxia, downregulated apoptosis, upregulated angiogenesis and cytokines, and extracellular matrix structure [58].

Post-mortem brain tissues from Parkinson's disease and healthy controls have revealed altered gene expression related to dopamine transmission and synapse, mitochondrial function, and protein degradation [59]. For example, VMAT2, a dopamine transporter, shows a significant downregulation [59]. Furthermore, mitochondrial dysfunction, caused by aberrant gene expression in genes such as the ATP synthase and the cytochrome C, results in severe outcomes in dopaminergic cells, as the whole ATP production is downregulated and neurotoxic ROSs are generated [59]. Finally, tissues exhibit changes in the ubiquitin–proteasome system, altering the unfolded protein response [59]. Similarly, blood samples from Alzheimer's disease patients and healthy subjects reveal alterations in transcripts related to inflammatory responses and fatty acid metabolism [60]. In addition, several studies performed in different human samples have revealed essential roles of miRNAs in the abnormal expression of genes contributing to the disease, instead of by being upregulated and lowering the expression of "protective" genes or by being downregulated and allowing higher expression of "harmful" genes [61]. miRNAs' altered expression is commonly associated with inflammation, cell survival, apoptosis, and neuroprotective pathways [61].

In contrast, proteomic approaches, including mass spectrometry, enable the characterization and quantification of abnormal protein expression, as well as unusual isoforms and protein localization. In PD, alterations in the unfolded protein response conduct to misguided protein expression [59]. Downregulation of the parkin protein, an E3 ligase that guides tagged proteins for degradation in the proteasome, increases the accumulation of misfolded and damaged proteins in dopaminergic cells, severely altering their correct functioning [59]. In AD, proteins related to the processing pathway of the β-amyloid peptide and neuronal cell adhesion pathways are frequently found, suggesting essential roles for these proteins in AD pathogenesis [61]. The β-amyloid peptide has been found with diverse isoforms due to different alternative splicing pathways of the amyloid precursor protein (APP) [61]. Furthermore, protein–protein interaction studies revealed essential interactions between the APP and other proteins, including the brain transglutaminase, which may contribute to the formation of peptide aggregates in the disease [61].

Finally, in the context of metabolomics, glucose shows region-dependent increases both in Alzheimer's and Parkinson's disease, including in the brain cortex, and glucose metabolism decreases [62]; this indicates that glucose could be a robust biomarker for diagnosis. However, it is also likely that glucose has etiological roles in these diseases [62]. This could explain possible links between diabetes and neurodegenerative diseases and its associated ROS increase and mitochondrial dysfunction [62]. Nevertheless, further investigation is still required [62]. On the other hand, uric acid was found to be at lower concentrations, consistent with its antioxidant roles [62].

Cardiovascular Diseases

Cardiovascular diseases are the leading causes of the dead across the world [63]. Furthermore, it is expected that by 2030, almost half of the adult population will have cardiovascular disease [64]. Cardiovascular diseases encompass a wide range of pathologies affecting the heart and the blood vessels, including heart failure, cardiomyopathies, and coronary artery diseases, among many others [65]. Several factors have been associated with these diseases; some of them are smoking, obesity, hypertension, and hypercholesterolemia, increasing the risk for their development [65]. However, it is essential to be noted that, since these diseases are complex, none of these factors is sufficient nor necessary for its appearance [65]. For this reason, it is now of great interest to identify genetic regions associated with the occurrence of cardiovascular diseases, as well as the interplay between the genome and the environment, usually manifested in the epigenome [65].

GWASs are essential because significantly few cardiovascular diseases have been associated with a single gene, including some classes of premature myocardial infarction, hypertrophic cardiomyopathy, heart failure, long QT syndrome, and aortic aneurysms, among others [66]. Although these diseases are more easily treatable, they are relatively rare in the population [66]. Therefore, GWASs have served as an essential implement in the discovery of cardiovascular diseases' genetic background. These studies have yielded thousands of related genome loci. Some examples are loci 9p21.3 and TCF21, associated with coronary artery disease [66, 67].

Transcriptomic approaches have revealed significant expression patterns in cardiovascular diseases in specific tissues and cell types, including blood, lymphoblastoid cell lines, and peripheral blood mononuclear cells [68]. Furthermore, leukocytes have been proved to show altered gene expression in response context, serving as a potential diagnose tool [68]. Transcriptome analysis in heart biopsies from heart failure patients revealed differentially expressed genes from patients and controls with functions associated with cardiac muscle contraction, oxidative phosphorylation, and cell and matrix composition and organization, as well as the inhibition of STAT4, SRF, and p53 and activation of HEY2 and KDM5A [69]. Similarly, it has been proven that HDAC9 (a histone deacetylase) is overexpressed in cardiac muscle; this exhibits a link between gene expression and function since one of these enzyme's targets, MEF2, has been proven to be implicated in hypertrophic cardiomyopathy [70]. Additionally, the presence of other classes of RNA molecules, including miRNAs and lncRNAs preferentially expressed in heart tissue and with functions in its function, could serve as a diagnostic tool, such as miR-1, miR133a, miR-208a/b, and miR-499 [64].

On the other hand, proteomic technologies are a promising area for discovering serum biomarkers [71]. For example, troponin assays have served as diagnostic tools for acute coronary diseases [71]. Myocardiopathy is the preferred cardiovascular disease for research [72]. In comparison with coronary disease patients, two-dimensional gels from biopsies of dilated cardiomyopathy patients revealed downregulation of several proteins, including desmin, ATP synthase, creatine kinase, myosin, HSP60, and HSP70, among others [72]. Blood samples have yielded relevant information as well [73]. For example, mass spectrometry and protein array assays suggested IGFBP1) and ApoA-1/HDL-C, isoforms of haptoglobin, and ApoC1, ApoC2, and ApoE as potential biomarkers for abdominal aortic aneurysm, acute myocardial infarction, and acute ST-elevation myocardial infarction, respectively [73].

A highly relevant area for cardiovascular disease research is epigenetics since these modifications have been proved to serve as the intercommunication between environment and phenotype [64]. Nonetheless, epigenetic modifications are harder to assess since approaches are diverse. It is necessary to use a combination of them to obtain a more comprehensive view of the epigenetic landscape [64]. Furthermore, a study showed a 46% reduction of cardiovascular disease onset in subjects with genetic predisposition and a healthy lifestyle compared to those with an unhealthy one (considering diet, exercise, and tobacco and alcohol consumption, among others) [64, 74]. For these reasons, the study of the exposome has gained increasing attention [64]. Examples of these measurements include devices able to monitor chemical compounds and microorganisms in the environment and early markers of disease [64]. Additionally, it is expected that in future years, artificial intelligence algorithms may help in the identification of a subject's lifestyle based on data published in social media, such as photos and quotes of their meals or habits in Facebook, Instagram, or Twitter, which may help predict disease risks [64].

It is highly likely that in the future, people may have access to this type of information, not only from genomic, transcriptomic, or proteomic studies but also from the exposome and the epigenome [64]. However, it is still compulsory that more research and investment be made in cardiovascular diseases [64]. Furthermore, the new creation of databases that encompass diverse data from omic studies, such as HeartBioPortal, is a requisite for the gathering of all of the massive data published every day on the topic [63].

Other Chronic Diseases

Other diseases have also had fundamental advances; for example, in diabetes, it is thought that SNPs in the HLA-DR3/4 and HLA-DQ2/8 alleles in chromosome 6 in CD4 cells are responsible for the creation of antibodies against pancreatic cells in type 1 diabetes [75]. Furthermore, SNPs in CDKAL1, KCNQ1, and CDKN2A/B are associated with type 2 diabetes and may serve as blood biomarkers since they are expressed in this tissue [75]. Similarly, transcriptomic approaches have found differential expression of several miRNAs, commonly found in patients' plasma [75]. Many of these miRNAs are in charge of the regulation of autoantigens in type 1 diabetes, therefore providing an immediate link between the genome and its expression [75]. Type 2 diabetes cells also exhibit miRNA aberrant expression, many of them in charge of the regulation of insulin resistance genes [75]. Finally and with great interest, metagenomic analysis has found alterations in the microbiome of diabetes patients in comparison with healthy subjects, highlighting the presence of Actinobacteria, Bacteroidetes, and Proteobacteria and *Bacteroides caccae*, *Desulfovibrio*, *Eggerthella lenta*, and *Escherichia coli* in type 1 and type 2 diabetes, respectively [75].

Another example is chronic obstructive pulmonary disease (COPD) [76]. GWASs have found FAM13A, CHRNA3/CHRNA5 IREB2, and HHIP as relevant disease-associated loci [76]. Interestingly, CHRNA3/CHRNA5 is associated with tobacco use, suggesting associations between smoking and disease COPD [76]. Furthermore, SNPs associated with lung function (measured with FEV1 and FVC) such as FAM13A, HHIP, and HTR4 have been found to increase COPD risk, which may suggest changes in lung function influenced by COPD risk [76]. Epigenomic analysis in COPD revealed hypomethylation of il13, RUNX3, and TIGIT [76]. Additionally, it is speculated that changes in the epigenome associated with COPD might be partly due to tobacco use since smoking affects DNA methylation patterns [76]. Finally, metabolomic approaches have discovered molecules that could serve as early diagnosis biomarkers [76]. Examples include higher expression of sphingolipids in COPD smokers' sputum compared with "healthy" smokers and association of glycosphingolipids with COPD in plasma [76].

Concluding Remarks

Undoubtedly, the importance of genomic tools for disease surveillance and their impact on public health has increased rapidly. Thanks to integrating all the information in different databases around the world, it has been possible to understand the dynamics of infection of different diseases, a more evident diagnosis, and the growing use of pharmacogenomics for adequate and efficient treatments. One of the significant limitations of these tools and their applications is undoubtedly access to them. Although the costs of equipment and supplies have fallen considerably in recent years, they are still difficult to access for many health institutions in developing countries. On the other hand, the participation of government agencies and access to information on disease outbreaks and prevalence is essential for proper surveillance and to establish containment strategies to prevent contagion, particularly in the case of infectious diseases with pandemic potential.

Acknowledgments This chapter is part of a registered project at the Instituto Nacional de Geriatría with the number DI-PI-003/2018.

References

1. Molster CM, Bowman FL, Bilkey GA, Cho AS, Burns BL, Nowak KJ, Dawkins HJS. The evolution of public health genomics: exploring its past, present, and future. Front Public Health. 2018;6:247.
2. Wake DT, Ilbawi N, Dunnenberger HM, Hulick PJ. Pharmacogenomics: prescribing precisely. Med Clin North Am. 2019;103:977–90.
3. Chan WS, Au CH, Chung Y, Leung HCM, Ho DN, Wong EYL, Lam TW, Chan TL, Ma ESK, Tang BSF. Rapid and economical drug resistance profiling with Nanopore MinION for clinical specimens with low bacillary burden of mycobacterium tuberculosis. BMC Res Notes. 2020;13:444.
4. NIHR Global Health Research Unit on Genomic Surveillance of AMR. Whole-genome sequencing as part of national and international surveillance programmes for antimicrobial resistance: a roadmap. BMJ Glob Health. 2020; https://doi.org/10.1136/bmjgh-2019-002244.
5. Gardy JL, Loman NJ. Towards a genomics-informed, real-time, global pathogen surveillance system. Nat Rev Genet. 2018;19:9–20.
6. Khoury MJ, Bowen MS, Clyne M, Dotson WD, Gwinn ML, Green RF, Kolor K, Rodriguez JL, Wulf A, Yu W. From public health genomics to precision public health: a 20-year journey. Genet Med. 2018;20:574–82.
7. Gu W, Miller S, Chiu CY. Clinical metagenomic next-generation sequencing for pathogen detection. Annu Rev Pathol Mech Dis. 2019;14:319–38.
8. Bakulski KM, Fallin MD. Epigenetic epidemiology: promises for public health research. Environ Mol Mutagen. 2014;55:171–83.
9. Saran S, Singh P, Kumar V, Chauhan P. Review of geospatial technology for infectious disease surveillance: use case on COVID-19. J Indian Soc Remote Sens. 2020. https://doi.org/10.1007/s12524-020-01140-5.
10. Lo SW, Jamrozy D. Genomics and epidemiological surveillance. Nat Rev Microbiol. 2020;18:478.
11. Lipkin WI, Firth C. Viral surveillance and discovery. Curr Opin Virol. 2013;3:199–204.
12. Ganesh B, Rajakumar T, Malathi M, Manikandan N, Nagaraj J, Santhakumar A, Elangovan A, Malik YS. Epidemiology and pathobiology of SARS-CoV-2 (COVID-19) in comparison with SARS, MERS: an updated overview of current knowledge and future perspectives. Clin Epidemiol Glob Health. 2021;10:100694.
13. Rambaut A, Holmes EC, O'Toole Á, Hill V, McCrone JT, Ruis C, du Plessis L, Pybus OG. A dynamic nomenclature proposal for SARS-CoV-2 lineages to assist genomic epidemiology. Nat Microbiol. 2020;5:1403–7.
14. Hill SC, de Souza R, Thézé J, et al. Genomic surveillance of yellow fever virus epizootic in São Paulo, Brazil, 2016–2018. PLoS Pathog. 2020;16:e1008699.
15. Postigo-Hidalgo I, Fischer C, Moreira-Soto A, Tscheak P, Nagel M, Eickmann M, Drexler JF. Pre-emptive genomic surveillance of emerging ebolaviruses. Euro Surveill. 2020. https://doi.org/10.2807/1560-7917.ES.2020.25.3.1900765.
16. Brunker K, Nadin-Davis S, Biek R. Genomic sequencing, evolution and molecular epidemiology of rabies virus. Rev Sci Tech. 2018;37:401–8.
17. Zhou B, Deng Y-M, Barnes JR, et al. Multiplex reverse transcription-PCR for simultaneous surveillance of influenza A and B viruses. J Clin Microbiol. 2017;55:3492–501.
18. Kanda T, Yajima M, Ikuta K. Epstein-Barr virus strain variation and cancer. Cancer Sci. 2019;110:1132–9.
19. Parikh UM, McCormick K, van Zyl G, Mellors JW. Future technologies for monitoring HIV drug resistance and cure. Curr Opin HIV AIDS. 2017;12:182–9.
20. Carleton HA, Gerner-Smidt P. Whole-genome sequencing is taking over foodborne disease surveillance. Microbe Magazine. 2016;11:311–7.
21. Lindsey RL, Pouseele H, Chen JC, Strockbine NA, Carleton HA. Implementation of whole genome sequencing (WGS) for identification and characterization of Shiga toxin-producing Escherichia coli (STEC) in the United States. Front Microbiol. 2016;7:766.
22. Armstrong GL, MacCannell DR, Taylor J, Carleton HA, Neuhaus EB, Bradbury RS, Posey JE, Gwinn M. Pathogen genomics in public health. N Engl J Med. 2019;381:2569–80.
23. Besser J, Carleton HA, Gerner-Smidt P, Lindsey RL, Trees E. Next-generation sequencing technologies and their application to the study and control of bacterial infections. Clin Microbiol Infect. 2018;24:335–41.
24. Guthrie JL, Gardy JL. A brief primer on genomic epidemiology: lessons learned from Mycobacterium tuberculosis. Ann N Y Acad Sci. 2017;1388:59–77.
25. Althomsons SP, Hill AN, Harrist AV, France AM, Powell KM, Posey JE, Cowan LS, Navin TR. Statistical method to detect tuberculosis outbreaks among endemic clusters in a low-incidence setting. Emerg Infect Dis. 2018;24:573–5.
26. Walker TM, Ip CLC, Harrell RH, et al. Whole-genome sequencing to delineate Mycobacterium tuberculosis outbreaks: a retrospective observational study. Lancet Infect Dis. 2013;13:137–46.
27. Gardy JL, Johnston JC, Ho Sui SJ, et al. Whole-genome sequencing and social-network analysis of a tuberculosis outbreak. N Engl J Med. 2011;364:730–9.
28. Guthrie JL, Delli Pizzi A, Roth D, Kong C, Jorgensen D, Rodrigues M, Tang P, Cook VJ, Johnston J, Gardy JL. Genotyping and whole-genome sequencing to identify tuberculosis transmission to pediatric patients in British Columbia, Canada, 2005–2014. J Infect Dis. 2018;218:1155–63.
29. Jajou R, de Neeling A, van Hunen R, de Vries G, Schimmel H, Mulder A, Anthony R, van der Hoek W, van Soolingen D. Correction: epidemiological links between tuberculosis cases identified twice as efficiently by whole genome sequencing than conventional molecular typing: a population-based study. PLoS One. 2018;13:e0197556.
30. Luo T, Yang C, Peng Y, et al. Whole-genome sequencing to detect recent transmission of Mycobacterium tuberculosis in settings with a high burden of tuberculosis. Tuberculosis. 2014;94:434–40.
31. CRyPTIC Consortium and the 100,000 Genomes Project, Allix-Béguec C, Arandjelovic I, et al. Prediction of susceptibility to first-line tuberculosis drugs by DNA sequencing. N Engl J Med. 2018;379:1403–15.
32. Doyle RM, Burgess C, Williams R, et al. Direct whole-genome sequencing of sputum accurately identifies drug-resistant mycobacterium tuberculosis faster than MGIT culture sequencing. J Clin Microbiol. 2018. https://doi.org/10.1128/JCM.00666-18.
33. Kiu R, Hall LJ. An update on the human and animal enteric pathogen Clostridium perfringens. Emerg Microbes Infect. 2018;7:141.

34. Kiu R, Caim S, Painset A, Pickard D, Swift C, Dougan G, Mather AE, Amar C, Hall LJ. Phylogenomic analysis of gastroenteritis-associated Clostridium perfringens in England and Wales over 7 years indicates distribution of clonal toxigenic strains in multiple outbreaks and extensive involvement of enterotoxin-encoding (CPE) plasmids. Microb Genom. 2019. https://doi.org/10.1099/mgen.0.000297.
35. Weekly Epidemiological Record, 30 August 2019, vol. 94, 35/36 (pp. 389–412). https://www.paho.org/en/node/65061. Accessed 25 Mar 2021.
36. Sakamuri RM, Kimura M, Li W, et al. Population-based molecular epidemiology of leprosy in Cebu, Philippines. J Clin Microbiol. 2009;47:2844–54.
37. Cole S, Monot M, Honoré N. On the origin of leprosy. BMC proceedings. 2008. https://doi.org/10.1186/1753-6561-2-s1-s6.
38. Avanzi C, Lécorché E, Rakotomalala FA, et al. Population genomics of Mycobacterium leprae reveals a new genotype Madagascar and the Comoros. Front Microbiol. 2020. https://doi.org/10.3389/fmicb.2020.00711.
39. Domagalska MA, Imamura H, Sanders M, et al. Genomes of Leishmania parasites directly sequenced from patients with visceral leishmaniasis in the Indian subcontinent. PLoS Negl Trop Dis. 2019;13:e0007900.
40. Wesolowski A, Taylor AR, Chang H-H, Verity R, Tessema S, Bailey JA, Alex Perkins T, Neafsey DE, Greenhouse B, Buckee CO. Mapping malaria by combining parasite genomic and epidemiologic data. BMC Med. 2018;16:190.
41. Website. Pf3K Project, https://www.malariagen.net/data/pf3k-pilot-data-release-3. Accessed 22 Apr 2021
42. Ito A, Yanagida T, Nakao M. Recent advances and perspectives in molecular epidemiology of Taenia solium cysticercosis. Infect Genet Evol. 2016;40:357–67.
43. Yanagida T, Yuzawa I, Joshi DD, Sako Y, Nakao M, Nakaya K, Kawano N, Oka H, Fujii K, Ito A. Neurocysticercosis: assessing where the infection was acquired from. J Travel Med. 2010;17:206–8.
44. Hotez PJ, Fenwick A, Savioli L, Molyneux DH. Rescuing the bottom billion through control of neglected tropical diseases. Lancet. 2009;373:1570–5.
45. Ahmad Z, Poh CL. The conserved molecular determinants of virulence in dengue virus. Int J Med Sci. 2019;16:355–65.
46. Vasilakis N, Fokam EB, Hanson CT, Weinberg E, Sall AA, Whitehead SS, Hanley KA, Weaver SC. Genetic and phenotypic characterization of sylvatic dengue virus type 2 strains. Virology. 2008;377:296–307.
47. Sim S, Hibberd ML. Genomic approaches for understanding dengue: insights from the virus, vector, and host. Genome Biol. 2016;17:38.
48. Scheuner MT, Sieverding P, Shekelle PG. Delivery of genomic medicine for common chronic adult diseases: a systematic review. JAMA. 2008;299:1320–34.
49. Visscher PM, Brown MA, McCarthy MI, Yang J. Five years of GWAS discovery. Am J Hum Genet. 2012;90:7–24.
50. Rappaport SM. Genetic factors are not the major causes of chronic diseases. PLoS One. 2016;11:e0154387.
51. Yoo BC, Kim K-H, Woo SM, Myung JK. Clinical multi-omics strategies for the effective cancer management. J Proteomics. 2018;188:97–106.
52. Vucic EA, Thu KL, Robison K, Rybaczyk LA, Chari R, Alvarez CE, Lam WL. Translating cancer "omics" to improved outcomes. Genome Res. 2012;22:188–95.
53. Chakraborty S, Hosen MI, Ahmed M, Shekhar HU. Onco-multi-OMICS approach: a new frontier in cancer research. Biomed Res Int. 2018. https://doi.org/10.1155/2018/9836256.
54. The genetics of cancer. 2012. https://www.cancer.net/navigating-cancer-care/cancer-basics/genetics/genetics-cancer. Accessed 21 Apr 2021.
55. Sager M, Yeat NC, Pajaro-Van der Stadt S, Lin C, Ren Q, Lin J. Transcriptomics in cancer diagnostics: developments in technology, clinical research and commercialization. Expert Rev Mol Diagn. 2015;15:1589–603.
56. Diaz-Ortiz ME, Chen-Plotkin AS. Omics in neurodegenerative disease: hope or hype? Trends Genet. 2020;36:152–9.
57. Krokidis MG, Exarchos TP, Vlamos P. Data-driven biomarker analysis using computational omics approaches to assess neurodegenerative disease progression. Math Biosci Eng. 2021;18:1813–32.
58. Ruffini N, Klingenberg S, Schweiger S, Gerber S. Common factors in neurodegeneration: a meta-study revealing shared patterns on a multi-omics scale. Cell. 2020. https://doi.org/10.3390/cells9122642.
59. Caudle WM, Bammler TK, Lin Y, Pan S, Zhang J. Using "omics" to define pathogenesis and biomarkers of Parkinson's disease. Expert Rev Neurother. 2010;10:925–42.
60. Xicota L, Ichou F, Lejeune F-X, et al. Multi-omics signature of brain amyloid deposition in asymptomatic individuals at-risk for Alzheimer's disease: the INSIGHT-preAD study. EBioMedicine. 2019;47:518–28.
61. Sancesario GM, Bernardini S. Alzheimer's disease in the omics era. Clin Biochem. 2018;59:9–16.
62. Scholefield M, Unwin RD, Cooper GJS. Shared perturbations in the metallome and metabolome of Alzheimer's, Parkinson's, Huntington's, and dementia with Lewy bodies: a systematic review. Ageing Res Rev. 2020;63:101152.
63. Khomtchouk B, Vand KA, Koehler WC, Tran D-T, Middlebrook K, Sudhakaran S, Gozani O, Assimes T. HeartBioPortal: an internet-of-omics for human cardiovascular disease data. https://doi.org/10.1101/487744.
64. Khomtchouk BB, Tran D-T, Vand KA, Might M, Gozani O, Assimes TL. Cardioinformatics: the nexus of bioinformatics and precision cardiology. Brief Bioinform. 2020;21:2031–51.
65. Paone C, Diofano F, Park D-D, Rottbauer W, Just S. Genetics of cardiovascular disease: fishing for causality. Front Cardiovasc Med. 2018. https://doi.org/10.3389/fcvm.2018.00060
66. O'Donnell CJ, Nabel EG. Genomics of cardiovascular disease. N Engl J Med. 2011;365:2098–109.
67. Yamada Y, Yasukochi Y. Genetics and genomics of coronary artery disease. Cardiovasc Genet Genom. 2018:661–678.
68. Pedrotty DM, Morley MP, Cappola TP. Transcriptomic biomarkers of cardiovascular disease. Prog Cardiovasc Dis. 2012;55:64–9.
69. Das S, Frisk C, Eriksson MJ, et al. Transcriptomics of cardiac biopsies reveals differences in patients with or without diagnostic parameters for heart failure with preserved ejection fraction. Sci Rep. 2019. https://doi.org/10.1038/s41598-019-39445-2.
70. Irvin MR, Zhi D, Joehanes R, et al. Epigenome-wide association study of fasting blood lipids in the Genetics of Lipid-lowering Drugs and Diet Network study. Circulation. 2014;130:565–72.
71. Lindsey ML, Mayr M, Gomes AV, et al. Transformative impact of proteomics on cardiovascular health and disease: a scientific statement from the American Heart Association. Circulation. 2015;132:852–72.
72. Vivanco F, López-Bescós L, Tuñón J, Egido J. Proteómica y enfermedad cardiovascular. Rev Esp Cardiol. 2003;56:289–302.
73. Mokou M, Lygirou V, Vlahou A, Mischak H. Proteomics in cardiovascular disease: recent progress and clinical implication and implementation. Expert Rev Proteomics. 2017;14:117–36.
74. Khera AV, Emdin CA, Drake I, et al. Genetic risk, adherence to a healthy lifestyle, and coronary disease. N Engl J Med. 2016;375:2349–58.
75. Gan WZ, Ramachandran V, Lim CSY, Koh RY. Omics-based biomarkers in the diagnosis of diabetes. J Basic Clin Physiol Pharmacol. 2019. https://doi.org/10.1515/jbcpp-2019-0120.
76. Kan M, Shumyatcher M, Himes BE. Using omics approaches to understand pulmonary diseases. Respir Res. 2017;18:149.

Index

J. C. Gomez-Verjan, N. A. Rivero-Segura (eds.), *Principles of Genetics and Molecular Epidemiology*,
https://doi.org/10.1007/978-3-030-89601-0

GPSR Compliance

The European Union's (EU) General Product Safety Regulation (GPSR) is a set of rules that requires consumer products to be safe and our obligations to ensure this.

If you have any concerns about our products, you can contact us on ProductSafety@springernature.com

In case Publisher is established outside the EU, the EU authorized representative is:

Springer Nature Customer Service Center GmbH
Europaplatz 3
69115 Heidelberg, Germany

Batch number: 10731528

Printed by Printforce, the Netherlands